CONSERVATION BIOLOGY

Conservation Biology

Edited by
Ian F. Spellerberg

Director, Centre for Resource Management, Lincoln University, New Zealand & Visiting Professor, Centre for Environmental Sciences, University of Southampton, UK

LONGMAN

Longman
Longman Group Limited
Edinburgh Gate, Harlow
Essex CM20 2JE, England
and Associated Companies throughout the world

First published 1996

British Library Cataloguing in Publication Data
A catalogue entry for this title is available from the British Library.

ISBN 0-582-22865-4

Library of Congress Cataloging-in-Publication data
A catalog entry for this title is available from the Library of Congress

Set by 22 in 9/11pt Times Roman
Printed in Singapore
Produced by Longman Singapore Publishers (Pte) Ltd.

Contents

List of contributors

G.W. Barrett, Department of Zoology, University of New England, Armidale, NSW 2351, Australia

Mark N. Boulton, International Centre for Conservation Education, Greenfield House, Guiting Power, Cheltenham, Glos. GL54 5TZ, UK

H.A. Ford, Department of Zoology, University of New England, Armidale, NSW 2351, Australia

Alan J. Gray, Institute of Terrestrial Ecology, Furzebrook Research Station, Furzebrook Road, Wareham, Dorset BH20 5AS, UK

Michael J. Hutchings, School of Biological Sciences, Biology Building, University of Sussex, Falmer, Brighton BN1 9QG, UK

Keith J. Kirby, English Nature, Northminster House, Peterborough PE1 1UA, UK

David Knight, Department of Adult Continuing Education, University of Southampton, Southampton SO17 1BJ, UK

Nigel Maxted, Plant Sciences, School of Biological Sciences, The University of Birmingham, Edgbaston, Birmingham B15 2TT, UK

Sue McIntyre, CSIRO Division of Tropical Crops and Pastures, 306 Carmody Street, St Lucia, Queensland 4067, Australia

Jeffrey A. McNeely, Chief Biodiversity Officer, IUCN, The World Conservation Union, Rue Mauverney 28, CH-1196 Gland, Switzerland

Michael O'Connell, Director, Habitat Conservation Planning, The Nature Conservancy, 2699 Lee Road, Suite 500, Winter Park, FL 32789, USA

Hugh Possingham, Centre for Population Biology, Imperial College at Silwood Park, Ascot, Berks. SL5 7PY, UK and Department of Applied Mathematics, The University of Adelaide, GPO Box 498, Adelaide, South Australia 5001, Australia

R.L. (Bob) Pressey, New South Wales National Parks and Wildlife Service, PO Box 402, Armidale, NSW 2350, Australia

P.D. Putwain, Applied Ecology Research Group, Department of Environmental and Evolutionary Biology, Liverpool University, PO Box 147, Liverpool L69 3BX, UK

D. Graham Pyatt, The Forestry Authority, Research Division, Northern Research Station, Roslin, Midlothian EH25 9SY, UK

Richard F. Pywell, Monks Wood, NERC Institute of Terrestrial Ecology, Abbots Ripton, Huntingdon PE17 2LS, UK

Ian F. Spellerberg, Centre for Resource Management, Box 56, Lincoln University, Canterbury, New Zealand

Alan J.A. Stewart, School of Biological Sciences, Biology Building, University of Sussex, Falmer, Brighton BN1 9QG, UK
David W.H. Walton, British Antarctic Survey, Madingley Road, High Cross, Cambridge CB3 0ET, UK
David Worley, 27 Malthouse Lane, Overbury, Tewkesbury, Glos. GL20 7PQ, UK

Foreword

Biodiversity is more than a haphazard collection of plants and animals. The chain that links one organism to another binds each into an interdependent community or ecosystem in which all living creatures, however small, have their place and function.

Biodiversity is part of our daily lives and livelihoods and constitutes the resource upon which families, communities, nations and future generations depend. Every gene, species or ecosystem lost represents the loss of an available option to adapt to local and global change.

Even today, despite the destruction we have clearly inflicted on our natural capital – biodiversity, its resilience is taken for granted. The question that all of us have to ask ourselves is: Has the environment reached the limits of its endurance? What can each one of us do to conserve it?

It is in this light that we have to see the value of education, of sensitizing people to the dangers of environmental destruction. An education that empowers and enables people to collectively seek ways to overcome current destructive trends is a critical component of any successful strategy for achieving a sustainable future.

Dr Ian Spellerberg has done the environmental community a great service by bringing out this book on 'Conservation Biology'.

The book presents a sample of the vast array of threads that make up the tapestry that is our contemporary view of biodiversity. Some of the articles are historical, others scientific, others legal but all are enlightening.

Whatever we may think about any particular part of this tapestry, it is important for us to stop and contemplate it. If we are to conserve our natural bounty, we will have to learn more about it and the nature of our interactions with it.

I am pleased to note that this book is targeted to an impressionable and effective segment of the society – the students as well as educators and teachers. Taken together, the articles in the book bring out the multi-dimensional challenges that the issue of biodiversity conservation poses to policy makers, scientists and the lay public alike.

As we go about implementing the provisions of the Convention on Biological Diversity and strive to behave in environmentally responsible ways as individual citizens, I am sure this book will prove to be a compendium of knowledge on biodiversity issues to all those who have a stake in conserving our natural bounty.

Elizabeth Dowdeswell
Executive Director, The United Nations Environment Programme

Preface

The term 'biological conservation' is used here to mean a branch of the science of biology dealing with the conservation of living organisms and the interactions between those organisms and the natural environment. How did the term 'conservation biology' arise and why use it when we already have the similar term 'biological conservation'? The term conservation biology was given much prominence in 1978 at the first International Conference on Conservation Biology but has been used in the biological literature for several decades. Conservation biology uses an integration of both social and science disciplines to achieve its objective of conservation of biological diversity. Conservation biology is therefore interdisciplinary, that is it integrates aspects of traditional disciplines such as economics, politics, law, education, taxonomy, genetics, ecology and biogeography. Of all these disciplines, the science of ecology provides the essential information (but not the processes) for putting conservation into practice.

In the western world, the natural historians and wildlife managers from the early part of this century have provided the foundations for the conservation biology movement. Their philosophies voiced through their own writings and societies provide us with an intriguing story concerning the changes in perceptions of interactions between humans and their environment. Only later were biology and ecology (having emerged as a science from natural history) to have an influence on the development of conservation biology as an interdisciplinary subject in its own right.

The journals and books on natural history and wildlife management published throughout this century provided the first forum for discussing practical ideas for conservation of wildlife. Later, books such as Rachel Carson's *Silent Spring* (1962) and organizations such as the World Wide Fund for Nature (previously the World Wildlife Fund), established in 1961, did much to popularize conservation issues. Scientific journals have influenced the nature of conservation biology. For example, the journal *Biological Conservation*, first published in 1968, had and continues to have an important influence not only on the kind of material being published but also on the accessibility of that material to an audience extending beyond research centres and academia. The most important recent landmark occurred when the Society for Conservation Biology (established in 1985) held its first annual meeting in the USA in 1987. The contents of the Society's journal *Conservation Biology* demonstrate the interdisciplinary nature of the subject and are indicative of the efforts to bridge the gap between research and practice.

Over the last two decades there has been a growth in conservation biology educational programmes. That growth has been prompted by growing concerns about our use of and impact on the natural environment, wildlife and biological diversity. Those concerns are apparent at many different levels from local to global. But it is not only concern about conservation for conservation's sake. Many people are increasingly recognizing that all economies and the quality of life they support are dependent upon how successfully living organisms and the environment are managed and conserved. At the same time there has been a steady growth in demand for people who have taken educational programmes in conservation biology. That in turn has information resource implications and textbooks are an important part of that information resource.

From the outset it was apparent that a book on conservation biology (as defined here) could not easily be written by one person. The nature of the discipline required the collaborative experience and expertise of many people. All contributors to this book are well qualified to write on their particular topics and have done so with enthusiasm and dedication. Unfortunately the constraints of space have demanded that the contributors be selective in their choice of material. Their choice and interpretation of that material has been influenced partly by their personal experiences, educational background and culture. Having to be selective brings with it an advantage because it has allowed us the opportunity to provide a comprehensive overview which will be useful for readers who are approaching the subject for the first time as well as for more experienced readers. The overall strategy has been to produce a style of writing and level of material which will be attractive to a wide range of people including those with no background in biology. All contributors have structured their contributions to include introductory material which then leads on to some specialist aspects. There is a glossary of technical terms at the end of the book.

The contributors describe many of the exciting developments in conservation biology and provide a source of research and further reading material both in the contents and in the references. The material comes from a large range of sources, but in most cases from the experience of the 20 contributors. There are many exciting projects and initiatives in conservation – there is a lot going on. It has been our aim to synthesize this information and make it readily accessible in an attractive manner, and to provide an enjoyable and stimulating source of material which will inspire many people from many backgrounds. The number of references at the end of each chapter is not unusual for modern textbooks, especially when there is an expectation of students to pursue various topics in a more specialized manner. Some of the references also provide that all-important need to substantiate what has been said in the text.

This book has three sections: A, Conservation Biology Issues; B, Conservation Biology in Perspective; and C, Conservation Biology in Practice. Section A explores perceptions of nature, wildlife and biological diversity, some aspects of cultural and social issues relevant to conservation practices, the role and functions of biodiversity and the need for evaluation and strategies for conservation. These first three chapters serve as an introduction to the rest of the book but also serve (with the glossary) to introduce readers to biological terms and concepts.

The second section puts conservation biology into perspective by describing the limitations and the constraints. These are the political and economic aspects, the legislation, policies and the role of organizations, education, the physical environment and finally taxonomy and systematics. The last section deals with conservation in practice at different levels and on different scales, from the genetic level to the ecosystem level and from dealing with immediate problems to modelling for the future.

Ian Spellerberg
Lincoln University,
New Zealand
1995

Acknowledgements

Nineteen of my colleagues made this book possible. My gratitude goes to all my colleagues who have so kindly and so tolerantly responded to my many requests. I am grateful to all of them for firstly agreeing to contribute to this volume. That was the easy part. Obtaining the completed manuscript in a form which is acceptable to the rationale of the book was not always easy but negotiations were always in good spirit. I thank them all.

Many other people have made this project possible. John Barkham, Kathleen Henderson, Ken Hughey and John Sawyer kindly read parts of the manuscript and suggested ways of improving the text. I am grateful to all my colleagues at CRM who have been very patient while I spent so much time finishing a project which commenced in England.

The idea for this book was prompted by Alexandra Seabrook at Longmans and I feel very proud to have been able to steer this project in a direction which I felt was most relevant to conservation biology. Many thanks to her and her colleagues at Longman House who willingly gave encouragement throughout the project.

My special thanks to Elizabeth Dowdeswell, Executive Director, UNEP for so kindly writing the Foreword. It is important to me that this book contributes to conservation biology at a world scale as well as at a local level.

Finally, I owe so much to the patience, company and loyalty of Tess and Megan. Without them, this project would have taken much longer to complete.

We are grateful to the following for permission to reproduce copyright material:

Cambridge University Press for box 2.1; the Ministry of the Environment, New Zealand for box 3.1; the Ministry of Agriculture, Fisheries and Food, UK for box 6.1; the Ecological Society of America for figure 6.2; the Royal Botanic Gardens, Kew for figure 15.2; the Chicago Zoological Society for box 15.1; the Zoological Society, London for box 15.2; Chapman and Hall for plate 1; the Ordnance Survey for plate 6 (© Crown copyright).

Whilst every effort has been made to trace the owners of copyright material, in a few cases this has proved impossible, and we take this opportunity to offer our apologies to any copyright holders whose rights we may have unwittingly infringed.

SECTION A Conservation biology issues

There is the conservation movement and the environmental movement, and there are conservationists and environmentalists. There is the ecology movement and there is the science of ecology. Movements and people in certain disciplines tend to be labelled the same thing and there has also been a tendency to use some terms in a generic sense, thus blurring distinctions between ecology and conservation. This is not to say that ecology, conservation and biological diversity have not found a place on the political platform. The debate about conservation of biological diversity has reached the political platform and there are signs of an increasing controversy about the significance and implications of all the efforts to conserve biological diversity. Perhaps some people are all too easily drawn into movements and support debates without question. There are many reports drawing attention to the extent and rate of losses in biological diversity. Some of those reports are emotive and provide little or no justification for the claims being made. Other reports attempt to be objective and provide a rationale for the argument being presented. Those who oppose or question the efforts being made to conserve biological diversity can also be emotive. Are most environmental groups really defending economic interests and not nature (report of an interview by John Maier in *Time*, 16 September 1991, pp. 54–5)? Can biological diversity really be dismissed as a luxury (C. Krauthammer, *Time*, 17 June 1991, p. 82)?

Differences in perceptions and differences in understanding of terms and concepts may be contributing to these arguments. The lesson to be learnt here, therefore, is to be careful to explain what you mean and explain why you have adopted a certain approach and belief. Therefore, the basis of this section is three questions:

1 What is biological diversity? (The main focus for conservation biology)
2 What is happening to biological diversity?
3 What is being done to conserve biological diversity?

By addressing these three main questions, Section A sets the scene for the following two sections and via the references acts as a source of further research and reading material.

CHAPTER 1

Themes, terms and concepts

IAN F. SPELLERBERG

Introduction

On the island of Rodrigues there is a single example of the tree known as the Café Marron tree. Its scientific name is *Ramosmania heterophylla*. That sole remaining example of that tree is sterile and so also are the cuttings taken from it. For the Café Marron tree, extinction may be inevitable. For many other kinds of trees, other plants, other kinds of organisms and for many biological communities, extinction is also inevitable; they will have gone from this earth for ever. Does it matter that humans are modifying and destroying elements of nature? If it does matter then how can living organisms and the natural environment best be conserved? How can economic development continue with consideration of human impact on nature? Conservation biology provides the opportunity to address these questions.

The aim of conservation in a biological sense is to ensure the continuing existence of species, habitats and biological communities and the interactions between species and with ecosystems. It does not necessarily mean the same as preservation in the sense of preventing any kind of human impact. Conservation, in the sense of conserving nature and the natural environment (as opposed to conservation of art and cultural heritage), is not new and has been practised widely throughout the world for a long time. Although the conservation of old buildings, archive material, landscapes, plants and animals and their habitats has become widely publicized, the term 'conservation' does not easily translate into non-European languages. Furthermore, whereas the environment is commonly perceived as something separate from ourselves, for many indigenous peoples the people and the environment are one. Acknowledging that the concept of 'conservation' is not new and has been practised widely throughout the world for centuries helps to put the approach and contents of this book into perspective.

If conservation biology were to have a mission it would probably be the conservation of biological diversity, that is conservation of the diversity of life. But what do all these terms such as 'ecology' and 'biological diversity' mean and do they have the same meaning for everyone? The aim of this chapter is to introduce themes, terms and concepts which are used throughout the book and which also provide a basis for the chapters in Section B (Conservation Biology in Perspective) and Section C (Conservation Biology in Practice). There are four main recurring themes introduced in this chapter: 1, values; 2, changes in perceptions and attitudes towards conservation; 3, differences in interpretation of events and processes; and 4, the interdisciplinary character of conservation biology. The terms and concepts which are introduced in this chapter are presented in a manner which assumes little or no knowledge of biology. These terms and concepts include environment, ecology, wildlife, species, and biological diversity (the glossary at the end of the book provides definitions for other technical terms used here and elsewhere in this book).

Themes

Values

Conservation in its broadest sense can apply to more than just nature; it can apply to books, paintings, buildings, etc. There sometimes seems to be more concern about conservation of art than conservation of nature. Huge prices are paid for works of art depicting nature: nearly £25 million for one painting by Van Gogh (Spellerberg, 1992). Who would be prepared to spend so much on conserving one species? There are sometimes restrictions on export of art but not so for 'works of nature'. For example, Brown (1991) drew attention to the fact that whereas licences for artworks (considered part of the national heritage) can be delayed to allow people to raise money to keep the work in the country, no such controls could be granted for natural objects. The natural object in question at that time was a 338 million year old reptile fossil, the oldest reptile fossil in the world.

In the south of England, conservation values came under considerable question in the early 1990s when a road was finally allowed to be built through a chalk hill (Twyford Down) on which there were designated protected areas, in this case Sites of Special Scientific Interest (SSSIs, as defined in legislation, The Wildlife and Countryside Act 1981). A tunnel through the hill would have saved these designated sites and not destroyed the protected areas nor damaged the landscape (Plate 2). In England, at about the same time (1992) when efforts to conserve Twyford Down were being lost to the decision of the bureaucrats, Windsor Castle was ravaged by fire, a fire so severe it caused damage valued at £35 million and destroyed many works of art. Some people wept at the damage to the castle and loss of the works of art, but the question might be asked 'who would weep for Twyford Down?' Can we put a price on the loss of sites of scientific importance and a price on damage to landscapes? Is there more concern about the loss of human artifacts than the loss of nature?

Britain is a small country with a high population density and there are more and more frequent conflicts about the future of protected areas. Proposals for new railway lines and motorways are just two causes of these conflicts which have raised questions about what is important in terms of both land use and types of habitats being conserved. The choice of a route for a new rail link between London and the Channel Tunnel not only caused conflicts about the best route but also raised questions about how to weigh up benefits of different kinds of wildlife habitats. One question asked in the national press was 'how do you measure an ancient woodland against an internationally important wetland?' What criteria could be used? Perhaps it is impossible to compare the conservation importance of different habitats and therefore the question may not be the right one to ask.

So, can we put a value on nature? The values of some species (commercial stocks of fish) and of some ecosystems (forests as sources of timber, wetlands as sources of fish and as a buffer against flooding) seem easy to quantify and indeed there has been a growing interest in this area (see for example Barbier *et al.*, 1994). Assigning economic values to all species, habitats and ecosystems is not straightforward but it may be essential to help promote conservation of biological diversity (see Chapter 4).

Attempts to conserve species and natural areas will always be questioned, especially when there seems to be more concern for wildlife and natural habitats than for jobs, the local economy and people. There will always be debates about conservation of nature and the priorities of conservation of nature in relation to other forms of conservation. This is because we all have different values.

Changes in perceptions and attitudes

Conservation biology has emerged as a popular subject in the western world where a predominantly white, middle class society has expressed concern about the degradation of wildlife and wilderness areas. This is certainly not to say that conservation does not have its origins in the far distant past and amongst many indigenous peoples. The contemporary western conservation movement has now reached many countries, many peoples and many cultures. The need for conservation has become widely accepted but nevertheless the concept of conservation means many things and is practised in many ways. There is a wide spectrum of views and attitudes towards conservation which embraces politics, culture, spirituality and religion. Conservation and biological diversity are now in the political arena. Conservationists may seem by some to be extremists and by others to be 'managers of the earth'. Some of these conflicts over conservation and environmental issues have led to bitter and sometimes tragic ends. For example, the 1985 sinking of Greenpeace's flagship *Rainbow Warrior* in Auckland Harbour (New Zealand) is one legacy of an ongoing, bitter debate. The people who have died for conservation of wildlife are all too many (see examples in Day, 1989). As Clark (1993) suggests, environmental rhetoric can be dangerous and any merely sectarian, abusive response to environmental issues may get us nowhere.

The term 'biological conservation' is sometimes equated with ecology and environmentalism and colloquial terms associated with biological conservation and conservationists include 'green', 'earth muffins' and 'bunnyhuggers' (terms from newspapers in 1994). Biological conservation is not the same as ecology, and ecology is not the same as environmentalism. There is the science of ecology, there is an ecological movement and there is the conservation movement; they all have different origins and different objectives. Use of the prefix 'eco' (ecotourism, ecofriendly) and use of colloquial terms tend to undermine the importance of conservation and the integrity of ecology as a science. Conservation education therefore has an important role to play (see examples in Chapter 7).

Differences in interpretation

Disagreement amongst scientists has sometimes been used to discredit scientific claims and is also occasionally used as a humorous characterization of scientists' behaviour. In all areas of science there are differences in interpretation. For example there are different interpretations about climate change and about the rate at which the earth's resources are being used. There are differences of opinion about the number of species on earth and the rates of extinction. There are different views about land use and resource use. The resources we use include light energy and heat energy, wind, minerals, soils, water, air and living organisms. Some of these resources are finite or not renewable and it is the use of those resources in particular which has been at the centre of much concern. That concern has grown steadily and what were often local concerns have now become global concerns. World publications such as *The World Conservation Strategy* (IUCN *et al.*, 1980) and its sequel *Caring for the Earth. A strategy for sustainable living* (IUCN *et al.*, 1991) and the products of the 1992 United Nations Earth Summit (see details in Chapter 3, Box 3.1) are testament to the global extent of the discussion and concern about resources, sustainability and conservation issues. Not surprisingly, for such a large and complex topic, there are differences of opinion about rates of resource use and the implications for future generations. For example, in 1984 Simon and Kahn published their book *The Resourceful Earth*. The executive summary of that book included the following:

> If present trends continue, the world in 2000 will be less crowded (though more populated), less polluted, more stable ecologically, and less vulnerable to resource-supply disruption than the world we live in now. Stresses involving population, resources, and the environment will be less in the future than now.... The world's people will be richer in most ways than they are today ... the outlook for food and for other necessities of life will be better ... life for most people on earth will be less precarious economically than it is now.

That was written in response to two paragraphs of the *Global 2000 Report* (Barney, 1980–81). The first of those two paragraphs was as follows:

> If present trends continue, the world in 2000 will be more crowded, more polluted, less stable ecologically, and more vulnerable to disruption than the world we live in now. Serious stresses involving population, resources, and environment are clearly visible ahead. Despite greater material output, the world's people will be poorer in many ways than they are today.

These two views were both written after considerable research, consultation and discussion yet they seem to be contrasting views. It is important to acknowledge that there are these differences and to try and understand why. The complexities of the natural environment and the small amount of information from long-term studies may contribute to these differences of opinion. Attempts to fill some of the gaps in this knowledge are sometimes made by interest groups and it is not unusual for unsubstantiated claims to arise (Chapter 4). There are many other conservation and environmental issues open to different interpretations and one aim of this book is to ensure that some of these differences are discussed. Meanwhile, however, we might note the salutary comments made in 1981 by Frankel and Soulé in their book *Conservation and Evolution*:

> Conservationists cannot afford the luxury and excitement of adversary science. The weakness of this parochial style of intellectual progress is that years or decades may pass before a clear resolution is reached and before timid technocrats or politicians decide that action will not bring a storm of criticism.

Interdisciplinarity

Conservation biology is an interdisciplinary subject embracing aspects of taxonomy, genetics, ecology and biogeography but within the context or framework of some aspects of economics, politics, law, and education. Conservation of biological diversity in practice depends not just on science, education and biology alone. A simplistic view might be that basic ecological research leads directly to conservation. There are social, economic and political aspects; there are many groups who should and in some cases have to be consulted (see Chapters 6, 12 and 13 for more discussion about the process of achieving conservation and the importance of involving local communities).

The interfaces between disciplines play an equally important role as does the interdisciplinary focus. For example there is a continuing need to bridge science and policy, and the World Resources Institute (established in 1982) is one independent research and policy institute which undertakes and publishes policy research (for example the Global Biodiversity Strategy mentioned later in this chapter in Table 1.1). There have been many efforts to bridge the gap between ecology and economics and only more recently has the joint discipline ecological economics come of age. Institutes such as the Beijer International Institute of Ecological Economics (established in 1991 in Sweden) have made important contributions to better use of biological diversity through their research and publications (see, for example, *Paradise Lost? The ecological economics of biodiversity* by Barbier *et al.*, 1994).

Terms and concepts

Some terms are defined in other parts of this book. There is also a glossary of technical terms at the end of this book. The aim here is to explore terms and concepts which have come into popular use and which are perceived in different ways. It would seem important to acknowledge and respect these differences in perception and understand why there are differences. For some, terms such as 'ecosystem' have been viewed with fear and even apprehension (Barker, 1993). When putting conservation into practice it is important to know what different people mean by different terms. It would also seem important that definitions of terms as given in policy documents, conventions and legislation be acknowledged and be used as a basis for wider understanding of these terms.

Ecology

Like so many other technical terms, 'ecology' has become popularized, so much so that there is a wide variation in perceptions of what ecology is, who uses it and what ecologists do. A simple definition of ecology is the science of the study of interrelationships between living organisms and their environment. Ecologists study nature's patterns and processes, collecting data to try and explain why certain species are found where they are, assembling data which may explain changes that occur in biological communities over time, and looking at cycles and flows of energy and elements through natural systems. Ecology is a broad science and consequently ecologists tend to specialize in one or just a few areas; for example there are population ecologists, community ecologists, plant ecologists, wildlife ecologists, ecological geneticists and wetland ecologists.

The term ecology has become popularized and so also has the expression ecological value. In ecological terms, an ecological value may be considered to be a measurable attribute such as species richness (the number of species in an area) or biomass (the total weight or total volume of living material in a certain area). In another sense, ecological value or values has come to mean the value an individual may give to a place or to a species. People appreciate and value the environment, natural places and wildlife for different reasons. This value of nature has been extended to what is seen or perceived as ecology. In Chapter 3 there is a discussion about valuation and evaluation of nature.

Environment

The term 'environment' has a scientific meaning and a popular meaning. In a scientific sense, the environment consists of all the surroundings, that is everything around us. The environment includes us, our products and constructions, the physical components (energy, air, water, soil, minerals) and the biotic components (plants, animals and other living organisms). In a popular sense there are different kinds of environments such as our home environment, work environment, urban environment and outdoor environment. For some, the environment is something detached from humans and not part of our activities. For others, environment is perceived as being natural (as opposed to being a product of human activity). Very often the term environment is equated with other terms such as wild areas and the wilderness (in the sense that the environment is somewhere you can go to but you leave it behind when you return home). In some cultures, humans are believed to be an integral and inseparable part of the environment. These different perceptions of the environment are determined in part by cultural, religious and educational differences and such different perceptions influence the way humans use their environment.

Ecosystems (natural and modified)

The terms 'biological community' and 'ecosystem' are becoming more and more widely used. A biological community is an assemblage of species living in a certain area and interacting with each other. The difference between a community and an ecosystem is that an ecosystem encompasses the living organisms *and* the physical environment (soil, water, climate, atmosphere). In that sense there is really no distinction between the two terms because a biological community cannot exist without interacting with the physical environment. The limits of a biological community may be defined in terms of the dominant vegetation or plants of that community, thus a grassland community, a woodland community. Alternatively the extent of a biological community may be qualified by the physical features, thus a wetland community, a desert community, an alpine community. On a map, a line can be drawn around a biological community such as a swamp, a lake or a woodland but it is not possible to draw a line around an ecosystem. This is because the concept of ecosystem includes the populations and species, the interactions between these and interactions between the living communities and the physical environment. There are inputs and outputs from the swamp, lake or woodland. Water, nutrients, energy and even some organisms flow in and out of the biological community. It is therefore difficult to define exactly where the boundaries are for an ecosystem, whereas the boundaries for a biological community can be defined by using the identifiable periphery to vegetation or physical features. The use of the term ecosystem serves to emphasize that the characteristics and components of biological communities are present and maintained by ecological processes such as the flow of energy, minerals and nutrients. (See also Chapters 13 and 16 for more detailed accounts of ecosystems, communities and types of communities.)

We all live in ecosystems, whose components are forever changing in time and space: there are processes taking place, such as flows of energy, minerals, gases, water and nutrients; there are changes taking place in the species composition; and there are successional changes in different communities (the plant and animal species change and the structure of the vegetation changes). For example, as trees grow old, then collapse, the rotting wood provides another essential source of resources (shelter and nutrients) for a variety of woodland organisms. There are natural disturbances which affect ecosystems. In a woodland, for example, natural events such as wind and lightning may cause clearings which provide opportunities for new plants and animals to colonize (as part of the process of ecological succession).

For centuries, humans have impacted on and modified ecosystems. It is because of the many interconnections within ecosystems that our impacts may affect more than one species. For example, pollutants released into ecosystems may be transferred from species to species within those ecosystems and accumulate in some species. The transference of DDT in ecosystems from insects through other organisms to birds of prey was one of the incentives for Rachel Carson to publish her book *Silent Spring* in 1962. Human management of ecosystems has been directed largely at simplifying them and reducing the variety of components and number of (trophic) levels. For example agricultural ecosystems lack variety of species and may have a few or just one main plant species (monoculture). Agricultural or forest plantation monocultures cannot function without human management and intervention. We control the pest populations and diseases and we supply water, nutrients and energy (for example energy to produce and distribute fertilizers and energy required for machinery and transportation). On both a world basis and a regional basis, the justification for the amount of energy and the amount of non-renewable resources such as oil that are required to sustain some forms of intensive agricultural practices has come into question (see Barbier *et al.*, 1994).

Wildlife

Wildlife is a term commonly used to refer to non-domesticated animals. However, wildlife in a biological sense refers to all the kinds of non-domesticated living organisms on the earth and encompasses plants, animals (including mammals, birds, reptiles, amphibians, fish and invertebrates) and other organisms such as fungi, bacteria and viruses. The more well known groups of animals, plants and fungi are listed in Chapter 9 (Table 9.1).

Species

The disciplines of systematics and taxonomy deal with the classification and naming of organisms (see Chapter 9 for details). They are the scientific disciplines which are the foundation for conservation biology. A species is a group of individuals of a certain kind that do not normally interbreed with individuals of another kind (Chapter 9). Each species is given a Latin binomial name consisting of two parts (a generic name and a specific name) according to the system of binomial nomenclature; for example the common bean, the french bean and the pole bean are all varieties of the species *Phaseolus vulgaris*. There are at least 30 species of bean in the genus *Phaseolus*, most of which have not been studied and consequently we know little about their potential as a resource.

In biology, as elsewhere, attempts to establish clear and distinct categories do not always work. Variations and exceptions in nature do not allow that. Consequently there are some controversies and disagreements. There are some species which interbreed under certain circumstances and there are some products of interbreeding between species which are difficult to classify as a 'true' species. There are occasional differences of opinion about whether or not some species are distinct from each other or should be grouped into one species. These are exceptions and the disagreements between taxonomists do not undermine the importance, integrity and value of systematics and taxonomy and the people who practise these disciplines.

Fig. 1.1 Edward Wilson, Author of books on biological diversity including *The Diversity of Life. (Photograph kindly supplied by Professor Edward O. Wilson, Museum of Comparative Zoology, Harvard University.)*

Biological diversity: biological view

Had this book been written about 30 years ago, it is likely that the emphasis would have been on the conservation of species of wildlife, especially plants and animals such as birds and some large mammals. Written ten years ago the emphasis would have been on conservation of habitats rather than species. Written five years ago the emphasis might have been on conservation of entire ecosystems. Now, the emphasis on conservation efforts is on biological diversity (embracing species, habitats and ecosystems). Many questions might be asked: where has the term 'biological diversity' (abbreviated to biodiversity) come from, what does it mean to a conservation biologist, how is the term perceived in the wider community, why has the term been promoted, what is important about biological diversity?

The term 'biological diversity' as used here has been in existence since at least 1980. In 1986 Norse *et al.* referred to biological diversity at various levels (genetic, species and ecological communities). The proceedings of a National Forum (held in Washington in 1986) edited by Wilson and Peter (1988), and indeed Edward Wilson (Figure 1.1) in his own writings (e.g. Wilson, 1992), have done much to promote the concept of biodiversity (the diversity of life). In his book *The Diversity of Life* (1992) he defines biodiversity as 'The variety of organisms considered at all levels, from genetic variants belonging to the same species through arrays of species to arrays of genera, families, and still higher taxonomic levels; including the variety of ecosystems, which comprise both communities of organisms within particular habitats and the physical conditions under which they live'. It was the 1992 Convention on Biological Diversity (see Chapter 5) and subsequent documents which have promoted the importance of biological diversity.

Perhaps Edward Wilson's definition could be expanded to include other biological levels of organization such as molecular, cultivar and breed, population, habitat, community and possibly biogeographical units such as biomes (see Chapter 2,

p. 14). The definition could also include species interactions and ecosystem processes. Conservation of biological diversity is therefore not just about conserving certain popular animal species such as the panda, colourful plants such as orchids, examples of some few remaining ecosystems such as rainforests or wetlands, or some 'green areas' such as urban parks. Conservation of biological diversity means conserving variety, interactions between species and processes in ecosystems: conserving genetic variety; conserving variety of rare breeds and cultivars; conserving variety of populations of single species; conserving variety of life which occurs in the habitat of a single species; conserving forest and wetland ecosystem processes in bio-geographical regions. At the risk of quoting from Edward Wilson out of context, 'Biological diversity – "biodiversity" in the new parlance – is the key to the maintenance of the world as we know it'.

Reference to biological diversity should, where appropriate, be qualified by one of the levels of biological organization. The most common and easily studied level of biological diversity is the variety of species. It is for this reason that biological diversity is sometimes incorrectly thought to be a measure only of the number of species (in an area), or species richness. For example, in 1991 a leader article in the journal *Economist* noted:

> SPECIES GALORE
> Avoiding extinctions should not be an overriding goal for environmentalists.
> What is the measure of life? Increasingly, conservationists are saying that it is biodiversity, a measure of the number of species that inhabit the planet . . . .

Of the many definitions of biological diversity amongst biologists there have been at least two which have equated biological diversity with species richness, e.g.

1 'Species richness within areas 10 000–40 000 ha in area'.
2 'The variety of living beings in an area or on the planet'.

In addition to equating the concept with species richness, some biologists have equated biological diversity solely with genetic diversity and some solely with species diversity; yet others have suggested it would be useful to equate it with wildlife. Brussard *et al.* (1992), for example, have argued that 'wildlife' should be used in the public domain rather than 'biodiversity' and that it would be a good thing to make biodiversity and wildlife synonymous. They suggested that wildlife is meaningful to almost everyone and that most people have no idea of what biodiversity means and are unlikely to learn in the near future.

The biological meaning of biological diversity needs to be made clearer to a wider audience. It is that variety at different biological levels which is so important. Conservation of biological diversity is about ensuring that our life support ecosystems are sustained, that we use genetic variation to develop new strains or varieties of food crops but conserve the wild species, that we use living resources in a sustainable manner. Conservation of biological diversity is about ensuring that future generations will enjoy the same levels of diversity of life or better than present generations.

Biological diversity: popular view

The term biological diversity, especially the abbreviated version biodiversity, has found its way into popular language and is often equated with nature, wildlife or the natural environment. Conservation of biological diversity is widely understood as meaning conservation of rare or endangered species or rare and endangered communities such as wetlands. Not surprisingly, different people think of diversity in nature or variety of life in different ways. Different people have different perceptions of what is meant by diversity, whether it be diversity of species or diversity within a landscape. We all relate to our environment and to landscapes in different ways (see Chapter 12 about the ways the landscape is perceived).

Examples of a high degree of variety of life or diversity in nature could be perceived as being any of the following:

1 Variety of size, shape and texture of different kinds of leaves on different kinds of plants
2 Variety of modes of reproduction
3 Variety in soil types
4 Variety of colour amongst the flowers in a garden
5 Variety of features in a landscape
6 Variety of kinds of corals in a reef
7 Variety of kinds of trees in a park

The locations of some of the highest levels of diversity are not so easily visible as in the examples above. The diversity of organisms in some soils can be very high and it is that diversity upon which the soil structure, its properties and nutrients depends.

Fig. 1.2 The Brandenburg Coppice, Lincoln, New Zealand. This collection of plants consists of species from around the world. The coppice is diverse in terms of species, and in terms of the vegetation structure, but the coppice is not a natural plant community, or natural assemblage of plant species. (Photograph: Ian Spellerberg.)

By way of contrast, examples of low diversity or low variety of life could be perceived as any of the following:

1 A forest plantation
2 A field of wheat
3 A well-kept lawn
4 Land which has been affected by drought
5 Polluted water
6 Inner city areas

All of these examples of low diversity would have been brought about either directly or indirectly by human actions. In many parts of the world, intensive agriculture has destroyed the diversity of life in soils and eventually that can lead to loss of fertile ground for growing crops. Amongst the examples of high diversity there are some which have been highly modified by human action. The variety of trees in a park, for example, is the result of human design and is therefore not a natural biological community (natural in the sense of the species that would occur together in undisturbed areas). Areas set aside for restoration of the vegetation range from collections of plants from different regions of the world to the re-creation of vegetation communities which previously were found in the area. The simplest and easiest method of vegetation restoration is to establish plants from different regions (see for example the Brandenburg Coppice shown in Figure 1.2). Another simple option is to establish collections of native species of vegetation. But which species and why? The most complex is to try and establish trees and shrubs which in terms of species abundance and age classes resemble the vegetation community which previously existed in the area. That task is made difficult by the amount of ecological information which is required to achieve that objective and by the fact that a decision has to be taken about what to aim for, in this case the historical condition or something approximating that.

To achieve conservation of biological diversity it may be necessary to establish protected areas, to reintroduce some species, to restore some ecosystems and to manage or eradicate previously introduced plants and animals. It may be considered necessary to adopt a principle of planting of indigenous ('native') plants in favour of species not native to the area. It may be considered important to try and re-establish whole communities and not just random collections of indigenous species.

The 1992 Convention on Biological Diversity

The products of the United Nations 1992 Earth Summit are described in Chapter 3 on p. 26. The concept of biological diversity is discussed in more than one product of that Earth Summit; in Agenda 21, for example, Chapter 15 deals with 'Conservation of Biological Diversity'. But it was the Convention which was the focal point for discussion (Chapter 4, Box 4.1). Not surprisingly the definition used in the 1992 Convention on Biological Diversity (Box 1.1) has become widely adopted in post-Earth Summit discussions and it also seems logical that this definition should become widely used. The Convention was signed by 152 nations. The USA was notable by its refusal to sign, but later a new administration decided to ratify it. That Convention came into force in December 1993 as a result of ratification by more than 30 states. The Convention has not escaped criticism and, while some of that criticism is based around the apparent conflict between conservation of of biological diversity and sustainable use of biological diversity as a resource, most of the concern has been about genetic resources. For example, Kothari (1994) in a series of essays expresses concern about inequity in access to genetic resources and the products coming from the applications of biotechnology to genetic resources and also the limited extent to which the Convention addresses conservation for domesticated varieties. The fact is that we now have this Convention and countries throughout the world are now considering how it might be implemented. A comprehensive and valuable guide to the Convention has been prepared by the IUCN (Glowka *et al.*, 1994).

Many international meetings, initiatives and projects on biological diversity have taken place, as is indicated by the growing number of published reports about biological diversity (Table 1.1). These events are not taking place in isolation. The concept of the 'global biodiversity forum', for example, was initiated during the preparation of the Global Biodiversity Strategy and was greatly influenced by the discussions on the Convention on Biological Diversity and preparation of the sections in Agenda 21 on biological diversity (Grubb *et al.*, 1993; Johnson, 1993).

Box 1.1 Diversity in nature

Diversity or variety in nature can mean many different things to many people. It is important, however, to adopt a precise meaning of biological diversity in order to provide a basis for policy formulation and decision making at both national and international levels.

Biological diversity was defined by the 1992 Convention on Biological Diversity as:

> the variability among living organisms from all sources including, *inter alia*, terrestrial, marine and other aquatic ecosystems and the ecological complexes of which they are part; this includes diversity within species, between species and of ecosystems.

In ecology, the following terms are used as measures of 'variety':

- Species richness: the number of species in an area.
- Species composition: the species assemblage or a list of the species present.
- Species diversity: the relative abundance of species. For example in a particular habitat there may be ten species. The number of individuals of each species could vary from the same number for each species to one species being represented by large numbers of individuals and the rest by only a few individuals. The former is a high species diversity and the latter is a low species diversity.

Indices of species diversity have been devised so as to provide data in a simple and collective manner. These indices are useful for monitoring ecological change but all have their limitations as well as advantages (Spellerberg, 1991). One index of diversity is calculated from the expression Σp_i^2 and this has also been used to calculate genetic diversity (Chapter 10).

In community ecology use is made of the terms alpha diversity, beta diversity and gamma diversity:

- Alpha diversity is the diversity of species in a habitat.
- Beta diversity is a measure of the rate and extent of change in species along a gradient from one habitat to another.
- Gamma diversity is the diversity of species within a geographical area.

The range of organizations involved in these publications is notable. What is not obvious is that for some of the publications a large number of people have been involved in a prolonged process. For

Table 1.1 Some of the many international publications which have used the term 'Biological Diversity' in their title. This is an indication of the growing interest in the concept

1988. *Biodiversity. The key role of plants.* By the joint IUCN–WWF Plants Conservation Programme.

1989. *The Importance of Biological Diversity.* Edited by P.S. Wachtel, L. Alyanak, S.-K. Chang and J. Hug. WWF, Gland, Switzerland.

1989. *Loss of Biological Diversity: a global crisis requiring international solutions.* A report to the National Science Board. Committee on International Science's Task Force on Global Diversity, report NSB-89-171, National Science Foundation, Washington, DC.

1990. *Conserving the World's Biological Diversity.* By J.A. McNeely *et al.* IUCN, WRI, CI, WWF-US and the World Bank, Gland, Switzerland and Washington, DC.

1991. *Biological Diversity and Developing Countries. Issues and options.* ODA, London.

1991. *Biodiversity. Social and ecological perspectives.* By V. Shiva, P. Anderson, H. Schucking, A. Gray, L. Lohmann and D. Cooper. World Rainforest Movement, Penang, Malaysia.

1991. Biodiversity. Scientific issues and collaborative research proposals. *Mab Digest* **9**, 77. UNESCO, Paris.

1992. *Global Biodiversity and Global Change.* International Union of Biological Sciences.

1992. *Putting Biodiversity on the Map.* ICBP, Cambridge.

1992. *Tropical Forests and Biological Diversity.* USAID, Washington, DC.

1992. *Global Biodiversity Strategy. Guidelines for action to save, study, and use Earth's biotic wealth sustainably and equitably.* WRI, IUCN, UNEP in consultation with FAO and UNESCO.

1992. *Global Biodiversity. Status of the Earth's living resources.* Edited by Brian Groombridge. World Conservation Monitoring Centre. Chapman & Hall, London.

1992. *The Convention on Biological Diversity.* Ratified in December 1993.

1993. *Biological Diversity Conservation and the Law. Legal mechanisms for conserving species and ecosystems.* By C. de Klemm, in collaboration with Clare Shine. IUCN Environmental Policy and Law Paper no. 29, IUCN, Gland, Switzerland.

1993. *Global Marine Biological Diversity.* By E.A. Norse (ed.) Island Press, Washington, DC.

1994. *A Guide to the Convention on Biological Diversity.* By L. Glowka, F. Burhenne-Guilmin and H. Synge, in collaboration with J.A. McNeely and L. Gundling. IUCN Environmental Policy and Law Paper no. 30, IUCN, Gland, Switzerland.

1994. *Report of the Global Biodiversity Forum.* IUCN, Gland, Switzerland.

1995. In 1993 UNEP approved the Global Biodiversity Assessment project with the expectation that the book will be published in 1995.

example the Global Biodiversity Strategy was developed through a process of research and consultation beginning in 1989 and involved six consultations, six workshops and more than 500 people. The Convention on Biological Diversity had an even longer period of consultation and involved many more people. Initial drafts of a treaty were prepared in 1987 by the IUCN in response to a resolution adopted at its 16th General Assembly in 1984. The UNCED process was also a drawn-out process (Chapter 3, Box 3.1).

At an international scale the UNEP has an ongoing Global Biodiversity Assessment Project which it is hoped will provide an essential foundation for decision making to help meet the objectives of the Convention. The Global Biodiversity Assessment Project is comprehensive as is indicated by the twelve sections:

- Characterization of biodiversity
- Generation, maintenance and loss of Biodiversity
- Magnitude and distribution of biodiversity
- Inventory and monitoring
- Biodiversity and ecosystem function: basic principles
- Biodiversity and ecosystem function: ecosystem analyses
- Economic values of biodiversity
- Multiple values of biodiversity
- Human influences on biodiversity
- Measures for conservation of biodiversity and sustainable use of its components
- Biotechnology
- Data and information management and communication

The emergence of the concept of biological diversity and its popularization by world meetings has meant that biological diversity has found a place in politics. Indeed biological diversity now has a political dimension. The challenge for this new global paradigm is acceptance and support at a local level. Conservation biologists will have roles to play in the implementation, monitoring and sustainable use of biodiversity.

Conclusions

It is important to acknowledge that we all have different values when it comes to conservation of art, artifacts and nature. The popularization of nature conservation has sometimes led to a lack of distinction between the meanings of conservation, ecology and environmentalism. The focus of conservation has changed from species to habitats, from habitats to ecosystems and then to variety within nature. The importance of variety at different levels is why the term biological diversity has been introduced.

The significance of change in emphasis from conservation of wildlife or nature to conservation of biological diversity has not been well explained, yet signatory nations to the 1992 Convention on Biological Diversity have launched many 'biodiversity' campaigns. Whereas many people with no background in biology understand and may relate to a 'Save the Tiger Campaign' or a 'Save the Wetland Campaign', because they know what a tiger or a wetland looks like, they may not so easily understand the biological diversity of which the tiger or the wetland is one part. It is not suggested that we should spend a lot of time debating the different perceptions of biological diversity. Rather, it is important to appreciate that variety of life or diversity within nature means many things to many people. People's understanding of diversity is surely influenced by their education, culture, experiences, religion and the way biological diversity is used by different societies. It seems important therefore that as well as having a strict definition of biological diversity for policy formulation, the wider acceptance of biological conservation can only be achieved if the significance of introducing the concept of biological diversity is widely understood. Therein lies a challenge for conservation biology and conservation biologists.

References

Barbier, E.B., J.C. Burgess and C. Folke (1994) *Paradise lost? The ecological economics of biodiversity.* London: Earthscan.

Barker, R. (1993) *Saving all the parts. Reconciling economics and the Endangered Species Act*, Washington: Island Press.

Barney, G.O. (1980–81) *Global 2000 Report to the President* (3 vols). Washington: Government Printing Office.

Brown, W. (1991) Export controls on 'works of nature' rejected. *New Scientist*, **131**(1782), 7.

Brussard, P.F., D.D. Murphy and R.F. Noss (1992) Strategy and tactics for conserving biological diversity in the United States. *Conservation Biology*, **6**, 157–9.

Carson, R. (1962) *Silent spring.* Boston: Houghton Mifflin.

Clark, S.R.L. (1993) *How to think about the earth. Philosophical and theological models for ecology.* London: Mowbray.

Day, D. (1989) *The Eco wars. A layman's guide to the ecology movement.* London: Harrap.

Frankel, O.H. and M.E. Soulé (1981) *Conservation and evolution.* Cambridge: Cambridge University Press.

Glowka, L., F. Burhenne-Guilmin, H. Synge, J.A. McNeely and L. Gundling (1994) *A guide to the convention on biological diversity.* Gland: IUCN.

Grubb, M., M. Koch, A. Munson, F. Sullivan and K. Thomson (1993) *The Earth Summit agreements. A guide and assessment.* London: Earthscan.

IUCN, UNEP, WWF (1980) *The World Conservation Strategy. Living resource conservation for sustainable development.* Gland: IUCN.

IUCN, UNEP, WWF (1991) *Caring for the earth. A strategy for sustainable living.* Gland: IUCN.

Johnson, S.P. (1993) *The Earth Summit: The United Nations Conference on Environment and Development (UNCED).* London: Graham & Trotman/Martinus Nijhoff.

Kothari, A.S. (1994) *Conserving life: implications of the Biodiversity Convention for India.* New Delhi: C17/A Munirka.

Norse, E.A., K.L. Rosenbaum, D.S. Wilcove, B.A. Wilcox, W.H. Romme, D.W. Johnston and M.L. Stout 1986 *Conserving biological diversity in our national forests.* Washington: The Wilderness Society.

Simon, J.L. and H. Kahn (1984) *The resourceful earth. A response to Global 2000.* Oxford: Basil Blackwell.

Spellerberg, I.F. (1991) *Monitoring ecological change.* Cambridge: Cambridge University Press.

Spellerberg, I.F. (1992) *Evaluation and assessment for conservation.* London: Chapman & Hall.

Wilson, E.O. and F.M. Peter (1988) *Biodiversity*, Washington: National Academy Press.

Wilson, E.O. (1992) *The Diversity of Life.* Harvard: Belknap Press of Harvard University Press.

CHAPTER 2

Changes in biological diversity

IAN F. SPELLERBERG

Introduction

Much is known about the variety of species throughout many parts of the world. However, there are areas of land, inland waters and particularly the sea which have not been subjected to systematic species inventories. New species are occasionally found in countries well known for their long history of studies in natural history and ecology. Incomplete species inventories exist partly because no one has looked in the right places and partly because there are not enough people trained to undertake species inventories. Some taxonomic groups such as birds, some groups of fish, butterflies and some families of flowering plants are well studied but there are many groups, especially amongst the invertebrates, which are poorly known. As well as gaps in our knowledge about the variety of species, there are gaps in our knowledge about interactions between species, changes in biological communities and rates of loss of biological diversity at different levels. Consequently there are differences of opinion and in interpretation about those losses. For example, claims and statements about the rates of losses of habitats and communities (most often about tropical forests) and rates of extinction are found in a wide range of literature, from popular magazines to scientific journals. Different estimates are given and all too often claims are made with little or no substantiation (a lesson for conservation education? – see Chapter 7). The different estimates and sometimes wild claims have not gone unnoticed. For example, in 1989 Robert Whelan made the following delightful comment in his book *Mounting Greenery*:

> ... if Lewis Carroll were alive today he would clearly be an ideal candidate for presidency of the WWF, with his unique talent for mixing science with fantasy.

Lewis Carroll was the pen name of Charles Dodgson who wrote children's books (for example *Alice's Adventures in Wonderland*, first published in 1865) which appealed to adults not only for their humour but also for their logic and innovative absurdity.

Whelan's comments seem to have been prompted by the many and varied estimates (some absurd) of the number of species which are alive today and of the rates of extinction. The number of named species can be given with some confidence (see Chapter 9, Table 9.1) despite some uncertainties about the appropriate classification of some groups. There are many species as yet unnamed and as yet undiscovered; how many is difficult to estimate. Even more difficult to estimate are extinction rates but there have been some objective and rational attempts (see for example Lawton and May, 1995).

In this chapter, some levels of biological diversity are discussed, some examples of changes (losses and gains) in biological diversity are described and finally the implications of losses in biological diversity are considered.

Different levels of biological diversity

There seems no disagreement that there are different levels of biological diversity and a commonly quoted range is from the genetic diversity within populations of a species to communities and ecosystems. There has been some discussion about the extent of this range and how best to compartmentalize the different levels. Harper and Hawksworth (1994) have suggested the use of the adjectives 'genetic', 'organismal' (for species and higher taxonomic groups) and 'ecological'. They also question reference to biodiversity of an ecosystem because it seems to devalue two useful concepts in the same phrase. Whatever the feeling about compartmentalization of biological diversity, it is important, when biological diversity is used in a technical sense, that the level be specified.

Communities and ecosystems

There is no one standard or unique classification of ecosystems at the global level and therefore it has been difficult to know how best to quantify diversity at this level. However, we can make use of certain categories such as deserts, tropical moist forests and wetlands. The extent of diversity within large units at a global scale could also be assessed in terms of the biogeographical units, sub-components and processes; that is biomes, sub-biogeographical regions, habitats, number of taxonomic groups, number of age classes within species and number of trophic levels in a given area.

Worldwide patterns of ecosystem and community diversity have been described in many ways and have been a main element of study by biogeographers (see Chapter 12 for descriptions of surveys at smaller scales, such as habitats). For many years a popular classification has been based on the term 'biome'. Terrestrial biomes are areas defined by the dominant vegetation and the climatic region (see Chapter 8). Tundra, temperate deciduous forest and desert are examples of three biomes. The IUCN and UNESCO have made use of Udvardy's (1975) terrestrial biogeographical regions of the world as a guideline for conservation planning purposes (Plate 1).

Species

The most commonly quoted level of biological diversity is the number of species in an area (in ecological terms, the species richness). This is used because it is relatively easy and practical to count species and because conservation efforts in the past have concentrated on species protection. Quantifying biological diversity at the species level is done simply by counting the species. However, there is more to species richness than just the number of species. There are species interactions; some species exist only because they have specialist interactions with other species (see page 31). Different groups of species are found together in different places: a species list needs to be accompanied by the names of the species. This would be particularly relevant when comparing the biological diversity at species level of different sites. There could be two sites with the same number of species, but at one site all the species could be closely related whereas the other site could include species ranging from bacteria to mammals. The number of species, the species composition and the taxonomic levels combined seem to be a useful way of quantifying and monitoring change in biological diversity at the species level.

There are some species which are represented by only a few individuals but, generally speaking, species are made up of different populations and each population is made up of different individuals. Apart from those individuals produced by parthenogenesis and identical twins, all individuals of a population are different. Diversity measurements within a species could therefore include an analysis of populations and individuals within populations.

The world's species richness

Since the mid-nineteenth century many people have pondered on the question 'how many species are there?' Today, much has been written about the question but it remains largely unanswered. It seems surprising that there is no definitive list, no complete inventory of all the species living on earth. There have been some attempts to devise methods for estimating the total number of species on earth and some people such as Erwin (1982) have put forward an interesting and testable hypothesis (Box 2.1). The fact is that so few studies have attempted to investigate species richness amongst even the more well-known taxonomic groups. From recent studies of insects in the tropics, for example, it has been suggested that there are probably many more species of insects than were previously thought. But it is not just insects which seem high in species richness. It has been suggested by Hawksworth (1991) that there may be as many as 1.6 million species of fungi rather than 40 000. This estimate is based on surveys of the ratio between numbers of vascular plants and species of fungi in well-studied areas. The known number of marine fungi has risen from 209 species in 1979 to 321 in 1990 (Lasserre, 1994).

At country level, there are many instances of well-researched inventories of some groups of plants and animals. There are few countries which have attempted more comprehensive inventories. However, a National Biological Survey has been proposed for the United States (National Research Council, 1993). On a world scale it may be impractical or impossible to establish a complete inventory of all species on land, in the inland waters and in the sea,

Box 2.1 Methods of estimating number of species on earth

Ratio of temperate and tropical species

The number of species in temperate regions is fairly well documented for most groups, whereas in tropical regions only the birds and mammals have been studied in detail. It has been found that there are 2–3 times as many tropical birds and mammals as temperate ones. It is therefore assumed that all other tropical groups are 2–3 times as numerous. An overall estimate based on this ratio puts the figure between 3 and 5 million.

Host-specific species

This estimate is based on a study of insects by Erwin in 1982. He collected more than 1200 species of beetles from the canopy of one species of tree in Panama and made the assumption that 13.5% were host-specific (only found on the tree species under study). Since beetles represent about 40% of all arthropods there must be about 400 host-specific species of arthropods in the canopy, and since canopy arthropods are twice as numerous as forest floor ones the total number of host-specific species of arthropods for the study tree is 600. Using an estimate of 50 000 tree species he concluded that there were approximately 30 million tropical arthropod species alone.

Length and species number

For species over 10 mm there is a linear relationship between log (length) and log (number of species) (see figure). Below 10 mm this relationship breaks down. It is argued, though, that our knowledge of the smaller creatures is far from complete. Studies on parasites, for example, have been limited to humans and economically important host species. If every species was host to at least one specific parasite then the relationship might also apply below 10 mm. Extrapolation of the graph to 5 mm places the total species figure at over 10 million (see figure).

Predictions based on the above methods vary widely. Each has used fairly crude assumptions which may not be true. Erwin's estimate is dependent on the assumption that 13.5% of arthropods are host-specific. If he had chosen a value of 5% a revised estimate would be about 11 million. The last method is also highly speculative and highlights the problems of extrapolation, especially with a log scale. If the line was continued to include all sizes down to 1 mm, this is log (length) = 0, the total species number would be 100 million. Despite these difficulties it would be fair to conclude that there are significantly more species alive than have yet been recorded, particularly in the smaller groups.

Source: Spellerberg and Hardes (1992), with kind permission of Cambridge University Press.

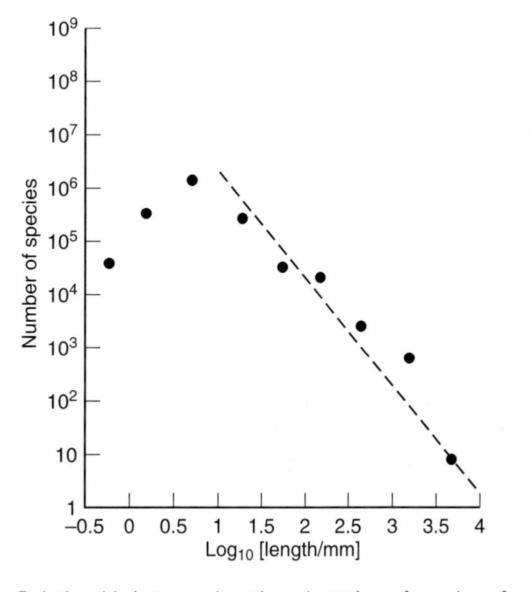

Relationship between length and number of species of terrestrial animals (from R.M. May (1988), How many species are there on earth? *Science*, **241**, 1441–9).

because there are not enough taxonomists to undertake the task, few of those taxonomists are familiar with some taxonomic groups, and there are so many species yet to be identified and described and many more yet unknown to science. The amount of knowledge, expertise and resources required to establish a complete inventory would be very large indeed and furthermore the storage of the information in a practical and usable manner might require further advances in computer technology and information handling. The financial costs would be immense, relative to biological research projects, though possibly not relative to large science and technology projects.

Despite the enormity of such a task and indeed seemingly insurmountable difficulties, there are some initiatives being taken to attempt a complete inventory, for example Systematics Agenda 2000: charting the biosphere. Systematics Agenda 2000 is a consortium of the American Society of Plant Taxonomists, the Society of Systematic Biologists and the Willi Hennig Society in cooperation with the Association of Systematics Collections with financial support from the US National Science Foundation. This is a global, 25-year initiative to discover, describe and classify the world's millions of species, including microorganisms. There are three missions:

- to inventory species diversity by surveying the major marine, terrestrial and freshwater ecosystems
- to analyse and synthesize the information
- to organize the information so that it can be accessed efficiently to meet the needs of science and society

But will species inventories have a useful purpose? Renner and Ricklefs (1994) question the practical value of inventories and have suggested that the demands of inventory work will undermine, rather than support, development of the discipline of systematics which is the foundation of all inventories. The passing of this biodiversity inventory fad, they suggest, will then enable conservationists to concentrate on the practical issues of conservation and rational development of natural systems. However, if we have no inventory then how will we know what is happening and how fast it is happening?

Number of described species

There are between 1.6 and 1.7 million named species (Chapter 9). However, it should be borne in mind that the ratio of named to unnamed species is unequal between taxonomic groups. There are some (taxonomic) groups of plants and animals which have been well studied and for which it is relatively easy to provide information on numbers of species. Birds, for example, are a well studied and easily observable group for which there is little dissension about the total number of named species. There are, however, some minor groups (minor in the sense of having relatively few species) and some poorly studied groups for which there continue to be discussions about the number of species.

Patterns of species richness

The types of species distribution and patterns of species richness are tremendously varied. For example, the European Starling (*Sturnus vulgaris*) has a widespread distribution, having been introduced from Europe and Asia to North America and elsewhere. At the opposite extreme, some species are restricted to small areas, for example the African Violet (*Saintpaulia ionantha*), a popular house-plant, is almost extinct in its wild habitat in the mountains of Tanzania. These extremes in patterns of distribution have been brought about partly by natural evolutionary and ecological processes but mainly by human activities. In some regions of the world there are few species and in other regions there are large numbers of species. Species which are native to an area are known as indigenous species; those found in only one area and nowhere else are known as endemic species.

On land, patterns of species richness vary with area, latitude and altitude. Some of the greatest concentrations of species occur as soil flora and invertebrate fauna, an observation often clouded by the popular image of many species in rainforests. In the sea, concentrations of species occur on coral reefs, around deep hydrothermal vents and where cool and warm currents meet and cause upwellings of bottom material. For island fauna and flora there tends to be an increase in number of species (of certain groups) with increasing area. Similarly on land there is generally a decrease in species richness with increasing altitude and latitude. All these patterns are easily observed but not easily explained.

Centres of endemism, that is areas where there are high numbers of endemic species, have attracted many studies. For example Myers (1988) has identified ten areas of tropical forests that have exceptional concentrations of species and high levels of endemism. Sadly, these areas are experiencing unusually rapid rates of depletion. The areas are Madagascar, the Atlantic coast of Brazil, western Ecuador, the Colombian Choco, the uplands of western Amazonia, the eastern Himalayas, the Malaysian peninsula, northern Borneo, the Philippines and New Caledonia.

Centres of endemism seem appropriate places to focus conservation efforts and have been of interest to conservation organizations. For example the Biodiversity Project of Birdlife International (previously the International Council for Bird Preservation) has identified areas with concentrations of bird species with restricted ranges; areas with two or more bird species entirely restricted to them are considered of primary importance and have been called Endemic Bird Areas (EBAs). In 1992 a total of 221 EBAs had been identified throughout the world and over 95% of restricted range birds occur in those areas.

Genetic diversity

Genetic diversity is of central importance to all levels of biological diversity and has been the basis of evolutionary processes. Genetic variation within species populations can be measured in many ways (see Chapter 10 for examples) and such measurements do provide possibilities for quantifying biological diversity. Such measurements may only be open to the skills of geneticists and may require a certain expertise, but nevertheless it should not be forgotten that genetic diversity is the prerequisite for all other kinds of biological diversity. Variation in external appearance within a species is sometimes easy to observe. The natural variety within one species can sometimes easily be detected in the size, shape, morphology and colour of individuals. It is the genetic variation within a species that determines the variation that we can observe and which we have used to produce artificial variation. The many varieties of cultivated fruits, garden flowers and domestic animals have all been made possible only by selecting different genetic characteristics.

The selective breeding of plants and animals has produced a large number of cultivars (plant varieties in cultivation) and breeds of animals. Changes in preferences by consumers and economic forces have resulted in some varieties becoming very rare and almost lost. Not surprisingly there are now societies and organizations which work towards the conservation of rare animal breeds and rare plant cultivars; this is just one more example of conservation of a particular level of biological diversity.

Human impact on biological diversity

Many kinds of human activities result in disturbance to biotic communities, such as deliberate extermination of species, hunting, tourism and recreation, commercial exploitation of species, effects of pollution on species and ecosystems, and effects of changes in land use on species, biological communities and ecosystems (see detailed examples in Chapter 13, Table 13.2). The continued damage, loss and fragmentation of natural and semi-natural areas (see example below in Box 2.2) is a major contributor to losses in biological diversity at all levels and has dominated thinking in conservation biology (see Chapter 13 for classification of habitat fragmentation). The biological consequences of habitat fragmentation are many in both the short term and the long term (Saunders *et al.*, 1991; see also Chapter 13 for examples). Intensive agricultural and forestry practices have contributed much to the changes in land use resulting in modified and simplified ecosystems. Establishment of monocultures causes losses in biological diversity both directly and indirectly. There is the direct effect brought about by replacing a diverse ecosystem with a monoculture (excluding some species from inhabiting the area) and there is the deliberate focus on the use of only a few species for crop plants.

Not all species which are perceived as being attractive or which contribute to the attractiveness of a country's wildlife and countryside are native species (some are introduced). Nevertheless a product of human intervention with nature and one which in some countries is a major threat to biological diversity is the introduction of alien or exotic species (for example *Rhododendron ponticum* in the UK, *Robinias* in Central Europe, mammals on the islands of New Zealand). Some alien species are accidental introductions but many are the result of deliberate programmes of acclimatizing species to new countries (there are many texts on this subject: see for example Mooney and Drake, 1986; Macdonald, 1994). More recently, the practice of dumping ship's ballast water near coasts is posing a threat to coastal and marine wildlife. The damage caused by alien species has been addressed by many countries for many years, at a huge cost.

Changes in biological diversity

Communities and ecosystems

There are many examples of biological communities and ecosystems which have been greatly reduced in area. For example, natural grasslands throughout the world have been modified or completely destroyed. In the USA, for example, virtually 100% of natural grasslands have been lost since 1492. One of the most exploited terrestrial ecosystems has been wetlands. There are many kinds or types of wetlands and they are typically rich in species. Traditionally, human settlements have occurred around wetlands, estuaries and rivers and consequently wetlands have been drained for agriculture, peat extraction, human settlements, etc. For New Zealand Dugan (1990) has estimated that over 90% of natural wetlands have been lost since European settlement. That loss, as elsewhere, still continues despite the considerable international efforts prompted by the Ramsar Convention (see Chapter 5 for more details of the Ramsar Convention).

Table 2.1 Preliminary FAO (1991) estimates of forest area and deforestation for 87 countries. (From *Global Biodiversity, status of the earth's living resources* by WCMC after FAO. With kind permission of WCMC and Chapman & Hall)

Continent	No. of countries studied	Total land area (km^2)	Forest area 1980 (km^2)	Forest area 1990 (km^2)	Annual deforest. 1981–90 (km^2)	Deforest. rate 1981–90 (%)
Latin America	32	16 756 000	9 229 000	8 399 000	84 000	0.9
Central America and Mexico	7	2 453 000	770 000	635 000	14 000	1.8
Caribbean Sub-region	18	695 000	488 000	471 000	2 000	0.4
Tropical South America	7	13 608 000	7 971 000	7 293 000	68 000	0.8
Asia	15	8 966 000	3 108 000	2 748 000	35 000	1.2
South Asia	6	4 456 000	706 000	662 000	4 000	0.6
Continental SE Asia	5	1 929 000	832 000	697 000	13 000	1.6
Insular SE Asia	4	2 581 000	1 570 000	1 389 000	18 000	1.2
Africa	40	22 433 000	6 504 000	6 001 000	51 000	0.8
West Sahelian Africa	8	5 280 000	419 000	380 000	4 000	0.9
East Sahelian Africa	6	4 896 000	923 000	853 000	7 000	0.8
West Africa	8	2 032 000	552 000	434 000	12 000	2.1
Central Africa	7	4 064 000	2 301 000	2 154 000	15 000	0.6
Tropical Southern Africa	10	5 579 000	2 177 000	2 063 000	11 000	0.5
Insular Africa	1	582 000	132 000	117 000	2 000	1.2
Total	87	48 155 000	18 841 000	17 148 000	170 000	0.9

Source: FAO (1991), *Second Interim Report on the State of Tropical Forests by Forest Resources Assessment 1990 Project.* Tenth World Forestry Congress, September 1991, Paris.

Throughout history and throughout the world, forests have been exploited and continue to be exploited, largely for their timber products. In some areas the forests have been replaced with monoculture plantations but in general the land has been converted to other agricultural use. In either case there has been loss of biological diversity. Of the many kinds of natural forests, tropical forests have received most publicity in the last decade and indeed 'saving the tropical forests' has become a common catch-phrase for environmental and conservation campaigns. Temperate native forests have largely been destroyed in Europe and continue to be destroyed in Australia and New Zealand.

But what of the estimates of forest losses: how are they derived and are they to be believed? There are many and varied estimates of current rates of losses of forests. Very often these estimates (expressed in terms of loss of forest equivalent to *x* hectares per day or the equivalent of the area of country *y* per year) are made without substantiation and a description of the method used for estimation. Even amongst those estimates which do provide an account of the methods used, the figures vary. Those differences can be attributed largely to a lack of defining what kind of forest is being examined and what is meant by deforestation. In the latter case deforestation can mean either total removal of forest or degradation of forests by logging.

Today, satellites and recent advances in computer technology provide us with the tools to monitor the loss of forests and other biological communities. For example in 1990 Fearnside concluded that 'forest loss in 1988 was proceeding at 20×10^3 km^2 per year'. The work of UNEP and the FAO has provided the most comprehensive reports of rates of deforestation (Table 2.1). These estimates come from very detailed and comprehensive studies based on careful classification of forests and careful appraisal of the methods of data collection.

The detailed extent of this work by FAO and UNEP was demonstrated when it was shown that comparisons between 1980 and 1990 estimates of forest loss from 76 countries showed that annual rates had increased from 113 000 km^2 to 169 000 km^2. The FAO is considering the following reasons:

- an actual increase in the rate of deforestation
- an underestimate of the rate of deforestation in the first assessment
- an overestimate of the rate of deforestation in the 1990 assessment

Even with satellite technology it can be difficult to quantify the extent and rate of losses in the world's communities, ecosystems and habitats and even more difficult to quantify the extent of fragmentation. We know from historical records that losses have been taking place for centuries and we know that this seems to be increasing. There are, however, some communities which have been well studied and which clearly show the extent and rate of loss, fragmentation and isolation. One well-documented community is the lowland heathlands of England (see Chapter 16). These heathlands, made popular by the English author Thomas Hardy (1840–1928), have been the subject of detailed ecological study (Box 2.2).

This pattern of fragmentation at possibly similar rates is taking place for many other communities throughout the world.

Rates of species extinctions

Extinction of species is a natural process and an integral part of evolution. Extinctions have taken place since life began and throughout millions of years of evolutionary history. Species continue to become extinct but the rate is now far higher than has ever occurred before and the cause is almost always human activity (exceptions are those caused by natural perturbations such as volcanic eruptions). With species extinctions go species associations and species interactions; it is not just species being lost but the integral components of ecosystems.

What is the rate of extinction and does it vary in different regions of the land and sea? These are difficult questions to answer because of a lack of comprehensive studies and there has been a tendency to popularize the estimates. Perhaps not surprisingly this has led to some interesting discussions about not only the rates of extinctions but also the integrity of conservation biologists and ecologists. The *New York Times* for 13 May 1993 carried an article by Julian Simon (a professor of business) and Aaron Wildavsky (a professor of political science). In that article they questioned the estimates of rates of extinction made by various people such as Paul Ehrlich, Thomas Lovejoy and Edward Wilson and suggested that the standard source for all estimates was Myers' book *The Sinking Ark* published in 1979.

There are numerous, accurate records of actual extinctions among many taxonomic groups and much work has been undertaken on the implications for extinctions as populations of species become smaller and isolated in habitats which are themselves becoming smaller and fragmented. Perhaps there has been too much emphasis placed on the answer to the simple question 'just how many species are becoming extinct?' It is more important to consider that enhanced rates of extinction are a symptom of continued human impact on the natural environment, and perhaps it is more important to consider the implications of losses of species. That is a pragmatic response to the continued debate about rates of extinction and we should not forget that there is also an ethical side to this. Do we have a responsibility to ensure that species do not become extinct as a result of human activities?

Populations and genetics

The amount of genetic diversity within populations of a species is constantly changing. However, losses in genetic diversity can occur as populations diminish and/or become fragmented, and for some species there is justifiably much concern expressed about the low levels of genetic diversity within the remaining individuals (see Chapters 10 and 11). Species most at risk are likely to be those with small geographical distributions and fragmented populations. As well as being concerned about threats of extinction to many species, we should perhaps have the same concern about losses in genetic diversity. It is one thing to conserve the last remaining individuals of a species but it is quite another to conserve genetic diversity, whether it be by *in situ* or *ex situ* means (see Chapters 10, 11 and 15).

Functions of biological diversity

The environment and life

All life depends on the natural environment because the environment and its ecosystems provide a life-supporting atmosphere and life-supporting sources of energy and materials. The ecosystems are those functional aspects of the environment which help to maintain the balance between carbon dioxide (CO_2) and oxygen (O_2) in the atmosphere, act as a source of minerals and nutrients and function as a sink for our waste products. These last two sentences are neatly summarized in Leopold's comment made in 1939, 'So complex, so conditioned by interwoven cooperations and competitions that no man can say where utility begins or ends'.

Box 2.2 Fragmentation of a biological community

The maps show the successive fragmentation of lowland heathland in the Poole Basin, Dorset, southern England (see inset for location), a plagioclimax community contained within the zone of Köppen's Cfb climate, moist temperate with mild winters (see Chapter 8). See also Plate 6.

Rate of loss:

1811–1896 0.3% per year over 85 years
1896–1934 0.5% per year over 38 years
1934–1960 1.7% per year over 26 years
1960–1974 3.0% per year over 14 years
1974–1978 0.9% per year over 4 years
1978–1987 <0.5% per year over 9 years

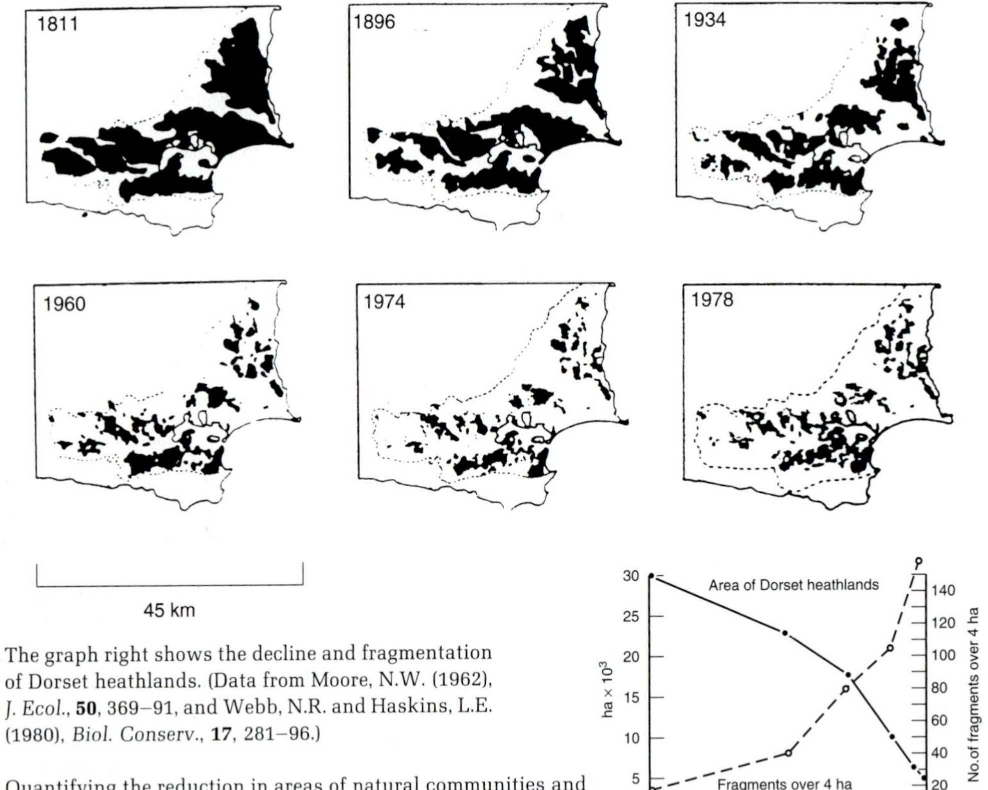

The graph right shows the decline and fragmentation of Dorset heathlands. (Data from Moore, N.W. (1962), *J. Ecol.*, **50**, 369–91, and Webb, N.R. and Haskins, L.E. (1980), *Biol. Conserv.*, **17**, 281–96.)

Quantifying the reduction in areas of natural communities and other changes in our landscape has long been the objective of many census and monitoring programmes. Although sophisticated remote sensing techniques provide a basis for good broad-scale monitoring, the effects of fragmentation and isolation (of woodlands, heathlands and other habitats) on the fauna and flora need to be monitored because 'insularization' is perhaps the greatest threat to our living environment. As yet we know very little about the effects of 'insularization' despite the continuous process of reduction, fragmentation and isolation of biotic communities.

Source: Maps of 1811–1974 from Webb, N. (1986) *Heathlands*. London: Collins; 1987 map from Webb, N.R. (1990) *Biological Conservation*, **51**, 273–86. Material supplied, redrawn and reproduced by kind permission of Nigel Webb, Institute of Terrestrial Ecology, UK.

We depend on and interact with wildlife no matter where we are or what we are doing. It is always part of our surroundings. We use wildlife both directly and indirectly. We may dislike some aspects of it, we may have a close affinity and liking for some forms, we may exploit it, manage it and destroy it. We use wildlife or parts of wildlife in both raw and extracted forms for many reasons: as a source of enjoyment, education, food, fuel, energy, construction materials and pharmaceutical products.

Ecosystem and community diversity

Why is it important to conserve a diversity of ecosystems and communities? One answer is because many kinds of ecosystems play a role in determining the speed and direction at which natural processes take place. For example, the straightening of rivers, the removal of forests and the draining of swamps all affect the rate at which water moves across or through the land. Throughout the world, diminished levels of ecosystem diversity are increasingly being suggested as contributing to floods and erosion. Natural ecosystems can provide sources of materials and can provide services such as removing pollutants from water. For example, some natural or restored wetland plant communities could help remove low levels of industrial pollutants from watercourses (possibly more cheaply than any chemical process). The capacity to function in that way depends on the diversity of species within the wetland plant community. There is of course a limit to the amount of pollution that can be absorbed by natural communities, and the most common example of abuse of natural systems as a sink is that of coastal waters for receiving sewage.

Why be concerned about some few remaining examples of some biological communities? There are many types of biological communities and within each type there are many different kinds of communities: different types of woodland, forest, wetland, estuary, river and lake community. There is not just one kind of wetland, there are many kinds with their own particular assemblage of species and unique processes. The same applies to other communities such as woodlands. In Europe, for example, there are many and varied kinds of woodlands and only recently have they been classified on the basis of the species and species assemblages that occur within them (see Chapter 11 with reference to Britain's National Vegetation Classification).

Species variety

Conservation of variety of species could be defended on moral grounds. Should we not ensure that future generations will be able to enjoy the same variety of species that exists today? There are other, pragmatic reasons. For example, few species, even of plants, have been studied for their potential use benefits. Conservation of as many species as possible ensures that we retain those potentially useful species for future generations.

The variety of species is the building material for the biological community in an ecosystem, and in any one ecosystem there are many interconnections between the communities and the species. The importance of diversity within a system is beautifully summed up by Edward Wilson in his book *The Diversity of Life*: 'Life in a local site struck down by a passing storm springs back quickly because enough diversity still exists. Opportunistic species evolved for just such an occasion rush in to fill the spaces. They entrain the succession that circles back to something resembling the original state of the environment.' For some time now there has been an interest in how diversity of species may help to ensure the survival and stability of an ecosystem. More recent empirical studies have suggested that diversity of species in biological communities may indeed have important stabilizing effects on the way communities and ecosystems function (Collins, 1995). If that is true then there are important implications for the use of the variety of species in monitoring and ensuring the sustainable use of biological communities and ecosystems.

Variety of species, coupled with measures of relative abundance of each species (sometimes expressed as a species diversity index: Box 1.1), is often used as a way of measuring 'environmental health'. For example in some river systems, a high level of variety of insect species is an indicator of low levels of pollution. As pollution levels increase then the variety declines until there are only a few hardy species left. During that process the relative abundance of different species changes. Ecological monitoring of the environment is both an art and a science which has an increasing role to play in ensuring sustainable use of ecosystems (examples in Spellerberg, 1991).

In a few biological communities, there is a natural dominance of one plant species, for example bracken (*Pteridium aquilinum*) stands in Europe and southern beech (*Nothofagus*) forest stands in New Zealand.

There may be dominance of one species but there are other naturally occurring plant species in those communities. Monocultures (also consisting mainly of one plant species) are managed so as to ensure the lack of other species and are highly modified communities. We manage agriculture and forestry monocultures to obtain the greatest possible crop yields and uniformity of produce quality. Monocultures have to be managed because the dominance of a single species attracts pests and diseases and there is a drain on the soil nutrients. Management of agricultural monocultures requires large energy consumption by way of manufacture of chemicals and nutrients, machinery and transportation. There is a high nutrient input and a high chemical input. Being simplified components of ecosystems, their existence is dependent on artificial inputs such as nutrients, fungicides and insecticides. While intensive agriculture has provided increased yield, there have been long-term implications in terms of increased levels of pest resistance to chemicals and in terms of excess nutrients entering waterways and aquifers. Whereas there has in the past been much emphasis on either chemical or biological control of pests, the integration of these methods has become an area of considerable research efforts. Ways of attracting a variety of beneficial organisms (such as pollinators or predators of pests) have become the objectives of many research programmes. The simple concept of increasing species variety within monocultures such as agricultural ecosystems has, in some instances, helped to maintain populations of beneficial organisms (see for example Bunce and Howard, 1990; Dent, 1995). In other words, introducing increased levels of species variety could contribute towards integrated pest management and help to lessen our dependence on chemicals and so alleviate problems of chemical residues in ecosystems.

Population diversity

In 1936 Leopold commented 'that there are grizzlies in Alaska is no excuse for letting the species disappear from New Mexico'. That moral stance is as true today as it was in the 1930s. There are also important yet simple pragmatic reasons for conserving different populations of the same species. Environmental impacts may increase the vulnerability of some populations; fire, for example, may destroy the habitat of one population and the species' survival is therefore dependent on there being more than one population. From ecological studies of populations scattered throughout the landscape, the theory of metapopulations has emerged (populations of populations linked in a functional way and sustained by individuals dispersing between the sub-populations; see Chapter 11). It seems that some species have evolved to survive as many populations scattered throughout patches of habitats. It would be wrong to assume that small populations are dispensable, because they may have an important role in maintaining larger populations.

Genetic diversity

Within populations of a species there is variety in shape, colour and behaviour as well as in levels of resistance to disease and tolerance to adverse conditions such as drought. Throughout the millions of years of evolution, this variety has been a prerequisite for natural selection and survival of the species. Even within shorter time scales, genetic diversity is essential for populations if they are to survive changes in the environment, pathogens and parasites.

Advances in our understanding of population genetics have highlighted the importance of variety within a species at the population level (see Chapters 10, 11 and 15). Genetic considerations are a prerequisite for the management of populations both in captivity and in the wild because the loss of genetic diversity in small populations is likely to be a major cause of species vulnerability to extinction. A knowledge of genetic diversity within a population may be an important part of species recovery programmes involving reintroductions and controlled breeding.

For thousands of years, humans have selected for certain biological characteristics of utilitarian species. Industries such as horticulture, agriculture and forestry depend on the development and selection of certain characteristics within a species. There has long been the use of varieties (cultivars and breeds) which allows for the selection of drought-resistant forms, disease-resistant forms and of course high-yielding forms. That process of selecting new forms continues. In apples, for example, new varieties might be selected on the basis of the tree growth, shape and structure (providing a basis for different methods of harvesting), different-sized fruit, fruiting period, taste, colour and storage. It is possible to select for these characters only because of the use of genetic diversity, and that new or novel genetic diversity is to be found in the wild forms (though for some species there are few wild forms remaining).

Ownership and knowledge

The importance of biological diversity in its many forms has been highlighted by the 1992 Convention on Biological Diversity and in so doing there are now questions about ownership of biological diversity.

There are difficult areas ahead with questions being asked about who owns biodiversity and whether it can be patented. For example, if a rare and protected species is found on private land, does someone have ownership of the representatives of that species? If some new form of biological diversity in a developing country is found by someone from an industrialized country to have material benefits, who gains from this discovery? Who owns the material, the intellectual property rights, and should the benefits of developing the newly discovered useful biodiversity go to the country of its origin and the people living there? Indigenous peoples have a vast wealth of knowledge of their environment; whereas they have not opposed the use of their knowledge, it is the way the information has been taken from them that is of great concern (Gray, 1991). A particular concern is that patenting could result in greater commercial secrecy as certain material becomes more and more valuable. A contrasting argument is that if material belongs to anyone it should belong to the country in which the species is indigenous (Given and Harris, 1994).

Conclusions

The extent of variety in nature presents us with a great challenge in terms of quantifying and qualifying it. There has perhaps been much wasted effort on simple aspects such as efforts to determine how many species there are. Species variety is just one aspect of biological diversity.

The human impact on biological diversity is sometimes difficult to quantify, but there are clear examples which show that both the rate and the extent of such impacts are rapidly growing. These impacts on biological diversity have implications for sustainability and for quality of life, for both current and future generations.

Conserving biological diversity is not just about conserving wildlife. It is also about safeguarding genetic variation upon which we depend for improvements to plant and animal varieties in horticulture and agriculture; it is about providing a richer source for deriving the world's food, materials and medicines; it is about ensuring sustainable use of the soil; it is about maintaining water quality; and it is about ensuring that natural ecosystems continue to buffer us from floods and storms and provide us with our life-support systems.

The conservation of living organisms, natural environments and biological diversity is important for us all. There is now a global commitment to safeguarding biological diversity and there is a commitment to ensure that the use of biological diversity is sustainable and that rewards are shared with equity and with future generations in mind.

References

Bunce, R.G.H. and D.C. Howard (eds) (1990) *Species dispersal in agricultural habitats*. London: Belhaven Press.

Collins, S.L. (1995) The measurement of stability in grassland. *Trends in Ecology and Evolution*, **10**, 95–6.

Dent, D. (1995) *Integrated pest management*. London: Chapman & Hall.

Dugan, P.J. (ed.) (1990) *Wetland conservation: a review of current issues and required action*. Gland: IUCN.

Erwin, T.L. (1982) Tropical forests: their richness in Coleoptera and other arthropod species. *Coleopteran Bulletin*, **36**, 74–82.

Fearnside, P.M. (1990) The rate and extent of deforestation in Brazilian Amazonia. *Environmental Conservation*, **17**, 213–26.

Given, D.R. and W. Harris (1994) *Techniques and methods of ethnobotany as an aid to the study, evaluation, conservation and sustainable use of biodiversity*. London: The Commonwealth Secretariat.

Gray, A. (1991) The impact of biodiversity conservation on indigenous peoples. In V. Shiva *et al.* (eds) *Biodiversity. Social and ecological perspectives*. Penang, Malaysia: World Rainforest Movement, pp. 59–76.

Harper, J.L. and D.L. Hawksworth (1994) Biodiversity: measurement and estimation. *Philosophical Transactions of the Royal Society, London, B* **345**, 5–12.

Hawksworth, D.L. (1991) The fungal dimensions of biodiversity: magnitude, significance and conservation. *Mycology Research*, **95**, 641–55.

Lasserre, P. (1994) The role of biodiversity in marine ecosystems. In O.T. Solbrig, H.M. van Emden and P.G.W.J. van Oordt (eds) *Biodiversity and global change*. Wallingford: CAB International in association with IUBS, pp. 108–32.

Lawton, J.H. and R.M. May (1995) *Extinction rates*. Oxford: Oxford University Press.

Leopold, A. (1936) Threatened species: a proposal to the Wildlife Conference for an inventory of the needs of near-extinct birds and animals. *American Forests*, **42**, 116–19.

Leopold, A. (1939) A biotic view of the land. *Journal of Forestry*, **37**, 727–30.

Macdonald, I.A.W. (1994) Global change and alien invasions: implications for biodiversity and protected area management. In O.T. Solbrig, H.M. van Emden and P.G.W.J. van Oordt (eds) *Biodiversity and global change*. Wallingford: CAB International in association with IUBS, pp. 199–209.

Mooney, H.A. and J.A. Drake (eds) (1986) *Ecology of biological invasions of North America and Hawaii*. New York: Springer-Verlag.

Myers, N. (1979) *The sinking ark: a new look at the problem of disappearing species*. Oxford: Pergamon Press.

Myers, N. (1988) Threatened biotas: 'hot spots' in tropical forests. *The Environmentalist*, **8**, 187–208.

National Research Council (1993) *A biological survey for the nation*. Washington: National Academy Press.

Renner, S.S. and R.E. Ricklefs (1994) *Trends in Ecology and Evolution*, **9**, 78.

Saunders, D.A., R.J. Hobbs and C.R. Margules (1991) Biological consequences of ecosystem fragmentation: a review. *Conservation Biology*, **5**, 18–32.

Spellerberg, I.F. (1991) *Monitoring ecological change*, Cambridge: Cambridge University Press.

Spellerberg, I.F. and S.R. Hardes (1992) *Biological conservation*. Cambridge: Cambridge University Press.

Udvardy, M.D.F. (1975) *A classification of the biogeographical provinces of the world*. IUCN Occasional paper No. 18. Gland: IUCN.

Whelan, R. (1989) *Mounting greenery. A short view of the green phenomenon*. London: Institute of Economic Affairs.

Wilson, E.O. (1992) *The diversity of life*. Harvard: Belknap Press of Harvard University Press.

CHAPTER 3 Conserving biological diversity

IAN F. SPELLERBERG

Introduction

Texts on biological conservation tend to emphasize losses in biological diversity. We cannot escape the fact that the losses taking place now are greater than have ever taken place before and that the implications for ourselves and future generations are serious. There are indeed losses but there are also many attempts to try and reduce the rate of those losses.

Putting conservation into practice requires resources and costs money to implement. There are direct monetary costs (staff, research, maintenance of protected areas, etc.) and indirect costs. Acquisition or lease of sites for conservation costs money, and in some countries there are payments made as compensation for not developing areas which contain sites of conservation importance. Furthermore, the resources available for conservation, relative to the demands on those resources, are becoming more and more limited. Consequently there are choices to be made, strategies to be devised and priorities to be established.

What methods are there to help us make these choices and strategies and to determine such priorities? Evaluation for conservation is about deciding which plant and animal populations and which species to conserve, which habitats to manage, and in which areas of land, water and sea and which regions of the earth to concentrate most resources for conservation. Ecological evaluation for conservation refers to the use of ecological criteria for evaluation. The term assessment (for conservation) is used here in a different sense and refers to the acquisition of information which could later be used in developing management programmes for conservation (an ecological assessment would be undertaken using ecological criteria). Evaluations and assessments have different methodologies, and different rationales are used for the different methods of conservation.

In this chapter we look at different examples of conservation efforts; institutional efforts are mainly considered here because the scientific efforts are described in Section C. We then consider some aspects of the costs of conservation and go on to look at the criteria used for evaluating species and areas for conservation.

Conservation efforts

We have already outlined some of the many losses in biological diversity taking place throughout the world. Efforts to slow those losses include the introduction of legislation, the establishment of conservation policies, designation of protected areas, introduction of controls over 'ecotourism', conservation education programmes, taking agricultural land out of production, and ecological restoration and conservation gain projects. For example, there have been ecological restoration projects on heathlands in Britain (see Chapter 16) and on grasslands in North America. Grassland plantings provide some of the best developed examples of ecological restoration in North America (Howell and Jordan, 1991; see also Chapter 16).

There is much research on various levels of biological diversity. For example, there is research on methods of conserving genetic variety in crop plants. The number of different varieties of crop plants currently in use is in marked contrast to the number of species and cultivars which could be used as crop plants. Indeed we have been highly selective and it has been estimated by Prescott-Allen and Prescott-Allen (1991) that only 103 species of plants are used to feed the world. One organization, the International Board for Plant Genetic Resources (IBPGR), was established (within FAO) in 1974 as a response to the loss of substantial diversity within species produced by domestication.

There are many threats to biological diversity and various organizations are now undertaking programmes of research to address those threats. For example, the growing threat of introduced species has prompted the World Conservation Union (IUCN, International Union for the Conservation of Nature) to establish an appropriate new specialist group, the Invasive Species Specialist Group (ISSG). The mission of this specialist group is to reduce the threats caused by invasive species through increased awareness of invasive species and how to deal with them. This will be done through a network of experts who specialize in the impacts of invasive species on local fauna and flora and methods of controlling them. The ISSG has a newsletter called *Aliens*.

Over the last 90 years there has been a steady growth in conservation effort. The number of national organizations involved in conservation has increased, while the number of voluntary conservation organizations and the size of the membership of some of those organizations have increased in an exponential manner. There are more laws and more conventions relevant to conservation than ever before. The growth in the number of protected areas (nature reserves, wildlife sanctuaries, etc.) and the total extent of all protected areas has increased exponentially over the last 20 years. That is not to say that there has been an equivalent increase in conservation, because protected areas, their management and design have limitations (see Chapter 14).

Perhaps the clearest evidence of increasing international conservation effort comes in the form of world environmental initiatives which have taken place since the early 1970s. For example the United Nations Environment Programme (UNEP) was established soon after the 1972 United Nations Conference on the Human Environment held at Stockholm in Sweden. Amongst many other things, UNEP today continues to help integrate biological diversity studies into many environmental programmes and major issues including climate change and desertification (the spread of desert-like conditions). Two other major initiatives were: 1, the establishment in 1948 of the International Union for the Conservation of Nature and Natural Resources (IUCN), now known as the World Conservation Union; 2, the establishment of the World Wildlife Fund (now the World Wide Fund for Nature) 13 years later. Both were very important developments for giving wider publicity to conservation (see Chapter 6 for more information about IUCN). In 1995, IUCN and UNEP signed a partnership agreement intended to further strengthen the worldwide cooperation that both organizations have undertaken for the last 22 years in the areas of resource conservation and sustainable development. There is a lot going on and there are many people involved at many different levels of organization. Perhaps the most recent and important culmination of all these efforts was the UN Conference on Environment and Development (UNCED), or the Earth Summit, held in Rio de Janeiro in June 1992 (Box 3.1).

Five documents emerged from the Earth Summit (two major international conventions, an agenda on worldwide sustainable development, and two statements of principles):

1 The Framework Convention on Climate Change
2 The Convention on Biological Diversity
3 Agenda 21
4 The Rio Declaration
5 The Forest Principles

The Framework Convention on Climate Change lacks binding policy commitments but it was agreed that 'developed' countries needed to stabilize greenhouse gas emissions at 1990 levels by 2000 (the 'Climate Convention' is not about halting climate change but about restricting its magnitude of change with respect to global warming, floods, droughts and extreme climatic events). The Convention on Biological Diversity requires that signatory countries adopt ways and means of conserving biological diversity and ensure that there is equity of benefits from biological diversity (see Chapter 5 for more details). Agenda 21 is a blueprint to encourage sustainable development socially, economically and environmentally into the twenty-first century. The 21 principles of the Rio Declaration define the rights and responsibilities of nations as they pursue human development and well-being. The Forest Principles are to guide the management, conservation and sustainable development of all types of forests.

Inevitably there is discussion as to whether or not all of this effort is enough, whether or not it is being directed at the most appropriate aspects of biological diversity, and whether or not it has come soon enough to ensure sustainable use of biological resources. Perhaps it is more important to ask if these efforts are being used in the most effective way. How many

Box 3.1 The UNCED process

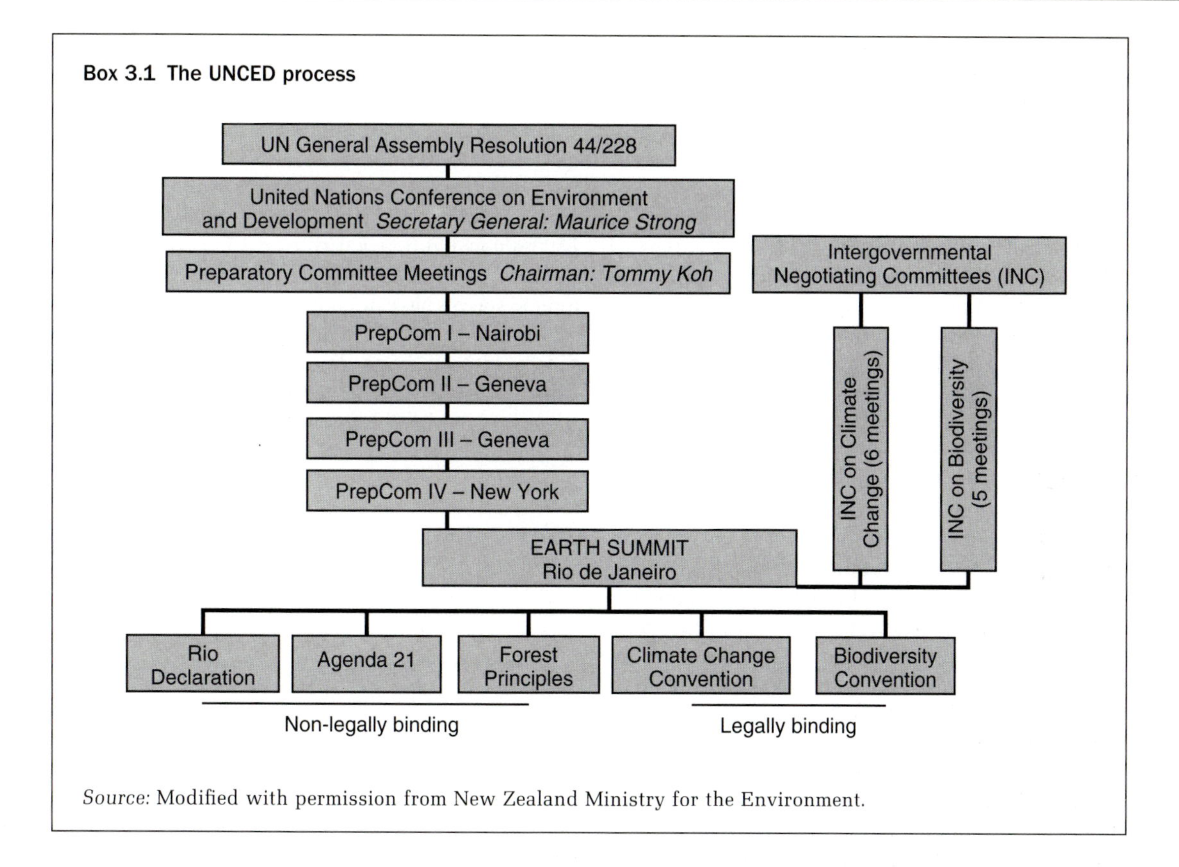

Source: Modified with permission from New Zealand Ministry for the Environment.

conservation agencies take time to consider, let alone quantify, their effectiveness? How many conservation agencies analyse the likely outcome of a conservation strategy and consider the risks and uncertainties (see Chapter 17)?

There are bound to be deficiencies in this conservation effort and there are gaps in the types of groups of people contributing to decisions about conservation. Section Three of Agenda 21 (a product of the 1992 Earth Summit) deals with 'Strengthening the role of major groups'. Two groups included were 'women in sustainable development' and 'indigenous people and their community'. There appear to be relatively fewer women than men who are well known for their work in conservation of wildlife or in environmental management. There are certainly fewer women than men who write about conservation and environmental issues. But this is an interpretation based largely on western practices. Women in the 'developing countries' have always played a very large part in conservation and have been very much at the centre of an extensive knowledge about indigenous wild species.

Plants, animals, other organisms, biological communities, ecosystems and humans are all part of the environment and we all interact with the environment. Some modern environmentalists may even suggest that ecology is not *ours* but that we are *its* (Clark, 1993). Ecology is central to conservation but there are also cultural and spiritual issues and matters concerning property rights (Shiva *et al.*, 1991). The many threats to indigenous peoples and their cultures have long been recognized (see Chapter 7) and it is this recognition which has implications for conservation biology (Oldfield and Alcorn, 1991). Whether or not indigenous peoples are viewed as conservationists or as living in harmony with nature is not so important as is perhaps the knowledge these people have. The ease by which that knowledge has been exploited by foreign interests is now beginning to be addressed at a national level. For example, Article 15 of the 1992 Convention on Biological Diversity refers to sovereign rights of states over their natural resources and authority to determine access to genetic resources. That Convention as a whole and in particular the intellectual property rights and indigenous knowledge are addressed in the book *Biodiversity. Social and ecological perspectives* by Shiva *et al.* (1991). They warn of biotechnologies which may erode biological diversity by increasing uniformity in ecological production and by imposing intellectual property rights to turn life forms into private property.

In the 1980s and more so in the 1990s there has been a small number but a steady flow of initiatives being undertaken for the rights of indigenous peoples (Kemf, 1993). For example a 1988 statement of the coordinating body for the Indigenous People's Organizations of the Amazon Basin (COICA) was later reaffirmed in February 1992, when the International Alliance of the Indigenous-Tribal Peoples of the Tropical Forests issued the Forest People's Charter. This Charter sets out a conservation policy which is based on the rights of these peoples to conserve and manage their forests.

This increasing conservation effort brings with it an increase in the amount of information being handled. That has implications for data assembly, data handling and communication of the information. How can information about conservation of biological diversity be best communicated and shared? The continuing advances in computer technology have been a major contributing factor to the more efficient and more widespread transfer of information. For example, with support from UNEP and the Brazilian government, workshops have taken place to discuss the needs and specifications for a Biodiversity Information Network (BIN). A biodiversity network is a mechanism for linking information relevant to biological diversity and making it widely available by electronic means. At the second workshop, held in Brazil in February 1994, a proposal was presented for implementing BIN21. The concept of BIN21 arose out of the Earth Summit and Agenda 21, thus the use of the expression BIN21. It was agreed that the best way for BIN21 to function was as a Special Interest Network, a new paradigm for cooperation around the world emerging on international computer networks (Canhos *et al.*, 1994). Electronic mail addresses for some biological diversity information networks may soon become widely available.

Conservation at what cost?

It costs money to design and implement conservation projects. It also costs money to protect species and recently these costs have started to be put into perspective as a basis for devising conservation strategies. In Britain, for example, the Nature Conservancy Council (as it was) commissioned a report (Whitten, 1990) on the status of all species listed on Schedules 5 and 8 of the UK Wildlife and Countryside Act 1981 together with a recovery plan with required scientific knowledge and approximate costs which would be required. Were all the recovery plans to be adopted, an annual budget of about £800 000 would be required over 15 years.

These kinds of costs for conservation of plant and animal species have not gone unnoticed and are being questioned more and more. For example, Ray and Guzzo in their book *Environmental Overkill*, published in 1993, drew attention to the amounts of money which have been spent by the US Fish and Wildlife Service (Table 3.1). They went on to say the following:

> At present, the protection of each listed 'species' costs $2.6 million per year. If more Americans knew that the Puerto Rican Cave Cockroach and the American Burying Beetle (also known as the Dung Beetle) are among those being 'protected' at this cost, would they approve? Are there perhaps more important problems on which to spend such sums?

The costs of conservation by establishing protected areas will continue to be questioned, and the costs of not putting protected areas to some other use such as agriculture or housing will also be questioned. That is, it might be argued that the costs to a country arising from designation of nature reserves could be calculated in terms of the value of that land had it been used for something else. But there is another side to this argument. Whereas the costs of a new road can be calculated in monetary terms, in terms of economic benefits to industry, other costs might be considered in terms of loss of a recreational area, of an archaeological site or of habitats for an endangered species (see page 33). In 1991 a report from the Environmental Group 2000 in Britain said that much was wrong with the Government's current cost–benefit analysis (COBA) system used to justify public spending and asked the question whether we should be putting a price on Norman churches, rare orchids or human lives when assessing the costs and benefits of road schemes. There has been much research on the methods which could be used for taking into consideration environmental values when undertaking cost–benefit analysis of road schemes. Nevertheless, and despite this good research, politics will often decide the outcome.

Some conservation activities can be used to generate money directly, and it has not gone

Table 3.1 The costs of species protection: some examples of expenditure by the US Fish and Wildlife Service to conserve and recover endangered and threatened species

Species	Cost
• Northern spotted owl	$9.7 million
• Least Bell's vireo (bird)	9.2 million
• Grizzly bear	5.9 million
• Red-cockaded woodpecker	5.2 million
• Florida panther	4.1 million
• Mojave desert tortoise	4.1 million
• Bald eagle	3.5 million
• Ocelot	3.0 million
• Jaguarundi (type of wildcat)	2.9 million
• American peregrine falcon	2.9 million
• Highest-cost insect (valley elderberry longhorn beetle)	$925 000
• Highest-cost plant (northern wild monkshood)	$226 000

Source: Ray and Guzzo (1993).

unnoticed that conservation can be a secondary benefit from other land uses which are themselves economically productive. For example, it has occasionally been suggested that nature reserves should be in the private sector and privately managed. There are of course private wildlife reserves and sanctuaries of many kinds. Is there therefore a potential customer demand which could help to sustain wildlife sanctuaries? In Australia, John Wamsley (founder of the company Earth Sanctuaries of Adelaide) has turned wildlife sanctuaries into a business. There, in Australia, as in nearby New Zealand, threats to native wildlife from introduced species have long been a problem. Earth Sanctuaries has purchased large areas of land and developed them into zones free of introduced species and protected by electric fences. Native species are introduced. The first sanctuary, Warrawong in the Adelaide Hills, was opened in 1982 and ten years later was earning $400 000 a year from visitors (Anderson, 1993).

Another method is to manage a species or habitat as a resource. Game birds and mammals are managed for sport, and through such management the habitat of the game species is managed and conserved. It has been suggested by some that rather than protecting certain species it is better to harvest the species and in so doing to conserve. This kind of approach has sometimes been advocated for some large African mammals; banning trade in ivory and in rhinoceros horns has resulted in the establishment of widespread poaching and black markets. Therefore it has been claimed that if it were legal to trade in ivory and rhinoceros horns, the species would be conserved because those people trading in these animal products would ensure that the source was protected. However, the many cases of overexploitation of commercially important fish and many other animal and plant species around the world does not provide much evidence that harvesting a species would ensure its conservation.

Not surprisingly, the need for strategies and priorities arises from the fact that the costs of conservation make it impractical to conserve all species and all communities to the same extent. If we accept the argument that it is impractical to direct equal conservation effort to all species, habitats and ecosystems and if we accept that there are likely to be some species, habitats and ecosystems in more urgent need of conservation than others, then perhaps we should consider the different methods for conservation and the methods used for identifying the greatest conservation needs. We should also be considering and calculating the likely outcomes of conservation strategies (see Chapter 17).

Rationale and criteria for making choices

In a very general sense what kinds of species, what levels of biological diversity and what circumstances would we consider when attempting to identify priorities? The rationale and criteria for establishing conservation priorities and for making decisions depends first on two important steps. First we need to know what is there in terms of the different levels of biological diversity. This implies that we must survey, map, assemble and classify information in a way which will help to determine conservation objectives (see, for example, classifications of climatic regions and of vegetation; Chapter 8, Box 8.1, Table 8.3, the National Vegetation Classification). Secondly, goals or objectives must be set. There could be one or more of many conservation objectives; is the objective to conserve a species, a particular biological community at a particular ecological successional stage, a collection of representative biological communities or life-sustaining ecosystems?

If the objective is conservation of species or of a biological community then an immediate response is to say 'endangered species' or rare species (Table 3.2). But those terms need quantifying otherwise they are subjective. There may be other criteria on which to base a decision as to whether or not to conserve a species or biological community. Other criteria could be utilitarian benefits, ethical reasons

Table 3.2 Criteria for selection of species for conservation

1 Utilitarian
- Species currently used in food, agriculture, medicine, etc.
- Species on which the above depend (in an ecological sense)
- Potential species for the above
- Species which play a role in maintenance of environmental health and quality

2 Biological
- Endemic species
- Taxonomically unique species
- Keystone species
- Species whose distribution has become fragmented
- Species whose numbers have declined
- Species of those taxonomic groups which have until present not been the focus of conservation

3 Ethical and cultural
- Flagship species
- Species which we can easily relate to
- Culturally or spiritually important species

or spiritual importance. The practice of protecting and respecting species has probably existed throughout human evolution and now occurs in many forms for many reasons. In some cultures the protected species and reasons for protection have been handed down from generation to generation by way of stories and pictures, In some countries, conflicts have arisen about conservation of certain species because of differences in cultural approaches to conservation. Customary uses of certain species which have been practised by generations and continuation of cultural or customary use may need to be considered when establishing criteria for conservation of species.

From a biological point of view, there has been support for special efforts to conserve taxonomically unique species, that is those species determined on the basis of phylogenetic relationships. For example, if a species has no near relatives and diverged from other extant species many millions of years ago, then that species sits at the end of a long branch of the phylogenetic tree (for example the Tuatara or *Sphenodon*, a lizard-like reptile found on some islands off New Zealand). It has been suggested therefore that there should be special emphasis on the protection of long-branched species because they would contain the largest amount of evolutionary novelty (see also Chapter 9 on 'target taxon summation').

In practice, species of utilitarian importance have previously not been subject to conservation (they may have been managed) unless there were some biological or ethical reasons for doing so. The exploitation of whales, the establishment of the International Whaling Commission (IWC) and the eventual conservation of whales is an example. The IWC was established originally to provide a forum for discussions about harvesting whale populations.

Broadly speaking and on a world basis, what criteria could be used to decide where there should be protected areas for conservation of biological diversity? Examples of criteria include the following (see also Chapter 14):

- Biotic communities with levels of high species richness (e.g. coral reefs).
- Unique communities not found elsewhere in the world
- Centres of endemism, that is areas where there are many species native only to that area
- Areas where there is large-scale damage and destruction to natural communities

The most commonly used initial criteria have been levels of endemism and levels of species richness. An objective and stepwise process which has become recognized as helping to ensure efficient representation of all the biodiversity was developed by Ackery and Vane-Wright (1984). This is the concept of complementarity which is used in what is called critical fauna analysis. In this approach the aim is to ensure the greatest proportion of the complement, that is a country's entire set of populations of an endemic species. The authors were assessing 158 species of milkweed butterflies (Danainae) and the highest number of endemic species were found on the island of Sulawesi (Indonesia). Conservation areas on that island could aim to conserve 21% of the species, while 29% could be conserved with protected areas on both Sulawesi and a small island of New Guinea. In their critical fauna analysis it was determined that 24 areas would be essential for the conservation of the most narrow endemic species of butterflies but that 31 would be needed to protect all species.

It is relatively simple to identify what the criteria might be. However, in practice, it is impossible to undertake these evaluations without information on the various species and their status. That is, in practice, there is a need for data collection because it is impossible to use these criteria if we do not know what is there. It has been largely the work of international organizations such as WWF, IUCN and ICBP which have assembled large data sets that has enabled these criteria to be put into practice.

Evaluation for conservation

Many methods have been used for centuries for conservation of wildlife and biological diversity. Throughout history, individual species have been protected for cultural, spiritual or religious reasons. Zoological and botanical gardens provide some opportunities to help conservation. There are also regulations, laws, conventions, codes of practice and lists of endangered species. There are many kinds of protected areas, all of which vary in their effectiveness. There are many programmes of ecological restoration. Here we discuss the criteria used for determining which species and which areas should be conserved.

Species

Ex situ conservation, that is conservation of species removed from their natural habitat, occurs throughout the world and is not a new idea (see Chapter 15). Historically, the reason why certain species were chosen for zoos or botanic gardens would have had little relevance to conservation of the species concerned (it seems strange that such collections have always consisted largely of fauna and flora selected from other countries rather than concentrating on species endangered in their own country). In recent years there has been increasing discussion about the ethics of retaining animals in captivity, as well as the costs of captive programmes, the conditions in which they are kept, and the need for international standards and protocols.

There are now zoological and botanic gardens which have dedicated programmes for conservation and which operate largely to reintroduce plants and animals to the wild. But which species are suited for *ex situ* conservation programmes? Choices have to be made and there has been increased use of refined rationale and criteria for selection of populations of animals and plants, subspecies and species. However, existing classifications at the species level or below may not reflect the natural partitioning of genetic diversity, and consequently the concept of evolutionarily significant units (ESUs) has become important in helping to identify which phylogenetic level (species, subspecies, race or single population) would contribute most to conservation in captive propagation programmes. Generally speaking, ESUs are defined on the basis of agreement between several sources of diagnostic information including genetics, biogeography, morphology and biochemistry. An alternative was to call ESUs 'Evolutionarily Significant Populations' (Ryder, 1986).

Ex situ conservation can do little to conserve some key species, that is those species on which many others depend. For example, the Brazil nut (*Bertholletia excelsa*) does not survive well by itself (see Chapter 15). The Brazil nut is pollinated by several species of large bees; they have to be large enough to lift part of the flower before gaining access to the nectar. The males of the bees are in turn dependent on certain orchids where they gather and mate. The fruit of the Brazil nut has a tough exterior and only the agouti (*Dasyprocta*) has teeth strong enough to remove the nut. The Brazil nut is dependent for its survival on a bee, an orchid and a rodent (Prance, 1991). There can, however, be justification of *ex situ* conservation and, where justified, it should be seen as supporting *in situ* efforts (see Chapter 15).

For many years now, countries throughout the world have introduced legislation to protect and conserve either particular species or groups of species (Chapter 5). A basic fault with the protection of some species, particularly wide-ranging species, by the laws of a particular country is that neighbouring countries may give no protection. This has been a problem for the conservation of migratory species of birds in Europe where they may, for example, be protected in Britain but be considered game birds in other countries. National agreements therefore play an important role in conservation of wide-ranging species.

It has not been unusual to find that the species and groups of species chosen for protection by law were not chosen on the basis of biological or ecological criteria. They were chosen largely for commercial or political reasons or because they were attractive to a certain body of people. Species considered to be 'attractive' such as birds and wild flowers rather than those considered to be 'unattractive' such as spiders or moulds have enjoyed far greater conservation efforts. Royal societies now support conservation of birds but who would suggest a royal society for the protection of spiders?

It would seem logical that species chosen for protection should be chosen on the basis of biological criteria. In some countries this is beginning to take place but before it could happen there was a need for an appropriate source of information. What are now called 'Red Books' or more appropriately 'Red Data Books' have provided that information.

Red Books, Red Data Books or just plain Red Lists (in connection with conservation of nature) have a long history and involvement with the IUCN. For more than 30 years now, the world's conservation organizations have made use of the concept of lists (Red Lists) of threatened species. The process involved the retrieval and assembly of data collection from around the world. Thereafter there was a selection process in which species could be allocated to various categories indicating their threatened status. The main rationale behind the establishment of Red Lists of species was to provide a means of assimilating and communicating information on the world's most threatened species. The IUCN Red Data Books today are but a shop window behind which there is a great deal of effort in collecting, processing and storing of data about many species. It is now the role of the World Conservation Monitoring Centre to collect, assemble and communicate information in Red Lists.

The early Red Data Book classifications were fairly arbitrary and subjective. If the status of a species could justify its inclusion in a Red Data Book then the species would fall into one of five categories. These categories seemed appropriate up to the mid-1980s but more recently there has been a reassessment of the categories (see Chapter 15). In 1984 a number of alternative options were considered by the IUCN Species Survival Commission Symposium '*The Road to Extinction*' (Fitter and Fitter, 1987). That preceded a long process of deliberation and consultation and later proposals for a revised system were published (Mace and Lande, 1991; Mace, 1995). The revision had several aims:

- to provide an explicit system that could be applied consistently by different people;
- to improve objectivity by providing those using the criteria guidance on how to evaluate different factors which affect risk of extinction;
- to provide a system which would facilitate comparisons across widely different taxa;
- to give people using threatened species lists a better understanding of how individual species were classified.

The concept has also had much support as is evident from the number of national Red Data Books: there are now more than 100 such books (Spellerberg, 1992). One milestone in national Red Data Books was the first British Red Data Book (*Vascular Plants*) published in 1977 (Perring and Farrell, 1977). The essential information for this Red Data Book came from previously published distribution maps of flowering plants based on recorded presence or absence of each species in squares of the British National Grid measuring 10 km by 10 km. If a species occurred in 15 or fewer of these squares from 1930 onwards then it was included in the Red Data Book. All these species in this Red Data Book have been assigned a threat number (from 1 to 15); the higher the number the greater the conservation needs of the species. The threat number is calculated using information on changes in distribution, a subjective assessment of the attractiveness of the species, the extent of conservation already undertaken, and the remoteness and accessibility of the locations for the species. This Red Data Book and a similar one for Ireland (Curtis and McGough, 1988) were later to have some influence on the selection of plant species included in national legislation.

Red Data Books and Lists have helped to promote data collection and data assembly as a basis for making conservation decisions. They have also helped to draw attention to the needs for conservation of many species. However, that list of species is ever growing larger and it has only been with advance in computer databases that data assemblage can be handled effectively.

While receiving much support, Red Data Books may have their limitations. That Red Data Books are species-based has been a source of criticism and there are those who suggest that we should have Red Data Books for habitats and ecosystems rather than for species. The number of species whose recovery has been brought about through Red Data Books has brought into question the whole concept. Finally, Red Data Books can all too easily be interpreted in politics as meaning that if species are not on such a red list then they do not need protection. Perhaps Red Data Books are no longer a suitable concept; perhaps it should now be Red Databases. Red is used in the context of the species being in danger. If the aim is species recovery then perhaps we should have Blue Data Books or Green Data Books for those species now out of danger; and why not Black Data Books for pest species?

Protected areas

Conservation by way of establishing protected areas (used here in a generic sense to mean all such areas from national parks to nature reserves) is nothing new and in an interesting way it is a practice which may affect people's perceptions of nature. That is, the designation of protected areas may tend to make us think that we are separate from nature and that protected areas are wildernesses which we may visit or which can be used only in certain ways. It is interesting therefore to consider that in many cultures, humans have very close cultural and spiritual connections with the environment and with nature.

Many indigenous peoples have long practised systems of law and management of nature and only in recent times have new methods of conservation been imposed. There are many instances where ancient land rights, cultural harvesting and conservation practices have been taken away from indigenous peoples. It is not before time that there are now some efforts to designate protected areas with indigenous peoples in mind. For example, the 4th World Congress on National Parks and Protected Areas, held in Caracas, Venezuela in 1992 was entitled 'The Law of the Mother: Protecting Indigenous Peoples in Protected Areas'.

In the context of western nature conservation practices, protected areas have been established for reasons of concern about protection of landscapes, wilderness areas, nature and natural resources for hundreds of years. Different sets of criteria have been developed for the selection of many kinds of protected areas. However, more recently the effectiveness of individual and isolated protected areas as a means of conserving biological diversity has been questioned (see Chapter 14).

National nature reserves

A strategy and a set of criteria for selection of nature reserves in Britain had its beginnings in the early part of the twentieth century as a result of the emerging science of ecology and the systematic work of botanists (see Chapter 6 for more details about conservation policy). By the early 1940s the key site principle for nature conservation was made clear in Cmd. 7122 (Ministry of Town and Country Planning, 1947) in a discussion of the main purposes of representative sites:

> to preserve and maintain as part of the nation's heritage places which can be regarded as reservoirs for the main types of community and kinds of wild plants and animals representative in this country, both common and rare, typical and unusual, as well as places which contain physical features of special or outstanding interest.

In the following years there was considerable discussion as to what criteria could be used to select national nature reserves. In 1977 the landmark publication *A Nature Conservation Review* (Ratcliffe, 1977) presented a detailed account of nature reserve selection based on ten criteria. These were area (size or extent), diversity (variety), naturalness, rarity, fragility, typicalness, recorded history, position in an ecological/geographical unit, potential value, and intrinsic appeal. Those criteria have been much debated, discussed and reappraised but they have become widely accepted (Spellerberg, 1992).

Ecological evaluation and ecological assessment

Evaluation for conservation is concerned with the processes, methods and criteria used to determine priorities for conservation of species and habitats. The criteria may be many and include social, cultural and scientific criteria. Ecological evaluation for conservation is the use of ecological parameters or criteria for determining the conservation needs of a species or the conservation importance of an area. Ecological assessments of species are undertaken to provide ecological information about that species which may later be used in management for conservation. An ecological assessment of a biotic community is undertaken to obtain ecological information about that community which could then be used to assess the likely effects of disturbance and pollution of the ecology of the area.

Before any ecological evaluation can be undertaken, an objective needs to be established and there has to be a survey of the area that can be undertaken at different spatial scales and directed at different groups of species (see for example Chapter 12 for details of survey methods at the habitat level).

Ecological evaluation

On page 3 reference was made to the choices for conservation being brought about by the proposals for a new rail link between London and the Channel Tunnel. One of the questions was 'how do you measure an ancient woodland against an internationally important wetland?' In this context, where the evaluation could lead to a decision about which biological community should be lost to the rail development, ecological evaluation would seem inappropriate because they are different biological communities with different species assemblages and different ecological processes. It is, however, possible to use ecological evaluation firstly to identify areas of greatest conservation value, and secondly to identify the best examples of the same kinds of biological communities. The criteria which have been used for these two purposes include:

- the species richness of various taxonomic groups
- the species diversity within certain taxonomic groups
- the species composition
- the age class distribution
- the rarity at different scales (local, regional, etc.)

Characteristics such as fragility, representativeness, naturalness and typicalness have also sometimes been used, although they are more difficult to quantify (Spellerberg, 1992). The results of the ecological evaluation may either be expressed in terms of each of the above criteria or may be combined into some form of index which may also include a weighting, where each criterion is felt to be of greater or lesser importance than the other criteria.

Ecological evaluation methods for identifying areas of greatest conservation value have been used successfully in conjunction with planning, notably in areas of Europe where there are high human population densities. In The Netherlands, for example, there has been good use made of ecological evaluation and assessment in conjunction with planning (examples in Spellerberg, 1992). These methods of evaluation are largely based on simple classifications of land use, the extent to which each category has been modified and the species richness of certain taxonomic groups such as birds and flowering plants.

Similar ecological evaluation methods have been developed to determine the conservation value of certain types of communities such as wetlands and woodlands. Many others have been developed to evaluate certain kinds of habitats such as bird habitats, and there have been many critiques of these methods. Gotmark *et al.* (1986), for example, usefully reviewed five indices for use in connection with the evaluation of avian habitats.

The validity and usefulness of an ecological evaluation is dependent not only on the ecological methods used but also on the experience of the people involved. Assembling species lists (as one measure of one level of biological diversity) requires taxonomic expertise and the value of such data depends on the experience of the people involved (see Chapter 12).

Box 3.2 Operational framework for an ecological assessment

1. Establish the reasons for and the objectives of the assessment
2. Consider different sources of information (and costs). Data and information could come from at least three sources: collected from the field, obtained from the literature or obtained from database sources.
3. Assess the type of data collection and when collected. Biological data varies according to the time of day, time of year and conditions (for example different weather conditions) under which it was collected. All sampling methods have a bias and have limitations.
4. Define the boundary of the area where the assessment is to take place and justify the positioning of the boundary. Are the effects to be considered only local or are they countrywide?
5. Define and classify the biological communities based on a standardized classification.
6. Identify the existing protected areas: type of protected area and type of legal protection.
7. Identify sensitive areas based on indicator species.
8. Identify which taxonomic groups are to be examined in more detail. These could include:
   - Protected species
   - Commercially important species
   - Species considered locally to be special
   - Indicator species
   - Red Data species
9. Identify ecological and other processes (for example hydrological) which require further examination. These could include, for example:
   - Trophic level: oligotrophic to eutrophic
   - Productivity

Ecological assessment

Ecological assessments have been used for identifying sensitive areas (sensitive to disturbance and pollution) and have also been undertaken as part of Environmental Impact Assessments where the likely impact of a proposed development is assessed in terms of the biology (what is there) and the ecology (processes) of an area. In ecological assessments, there have to be choices because it would be impractical to look at every aspect of the biology and ecology of an area. Which areas, which taxonomic groups and which ecological processes should be assessed? An operational framework for an ecological assessment is suggested in Box 3.2.

Conclusions

The growth in conservation effort continues throughout the world at many different scales from local initiatives to world projects. There are many ways of conserving species and habitats and some methods have been practised for thousands of years. Species have been protected for cultural and religious reasons, some species are conserved in captivity and the concept of protected areas has been a major conservation method. No matter which method of conservation, the increasing pressure on the environment and the costs of conservation have had implications in terms of which species and which areas are conserved.

Quite rightly there are those who ask whether conservation is being advocated at the expense of other uses of the environment. Is the cost of conservation too high, are there too many uncertainties? These questions could be asked about any form of land use but the questions have to be put into context. Perhaps it is wrong to make generalizations but nevertheless it is true to say that the livelihood and quality of life of all humans depend on the extent of conservation of biological diversity.

The many methods used for conservation rely on good decision making: the choices, the priorities, the strategies, and the assessments of risks. These in turn have been based on certain criteria and rationales, which may be biological and ecological. There are now well-researched biological and ecological criteria which can be used for ecological evaluation and assessment. The validity and usefulness of ecological evaluations depend on the experience and expertise of the people involved.

In practice, development of priorities and strategies for conservation requires more than just a knowledge of biology and ecology. There are social, cultural, educational, legal and institutional issues which need to be addressed. There are many questions to be asked; where does the information come from, who is consulted, who makes the decisions and what are the processes involved in reaching these decisions?

References

Ackery, P.R. and R.I. Vane-Wright (1984) *Milkweed butterflies*. London: British Museum (Natural History).

Anderson, I. (1993) Save a species, make a profit. *New Scientist*, **140**(1902), 7.

Canhos, D.A.L., V. Canhos and B. Kirsop (1994) *Linking mechanisms for biodiversity information*. Proceedings of the International Workshop held at the Tropical Database, Campinas, São Paulo, Brazil, 23–25 February 1994. UNEP, MMA (Ministerio do Meio Ambiente e Amazonia Legal) and RHAE/CNPq (Conselho Nacional de Desenvolvimento Cientifico e Tecnologico), São Paula.

Clark, S.R.L. (1993) *How to think about the earth. Philosophical and theological models for ecology*. London: Mowbray.

Curtis, T.G.F. and H.N. McGough (1988) *The Irish red data book*. Dublin: The Stationery Service.

Fitter, R. and M. Fitter (eds) (1987) *The road to extinction*. Gland: IUCN.

Gotmark, F., M. Ahlund and M.O.G. Eriksson 1986 Are indices reliable for assessing conservation value of natural areas? An avian case study. *Biological Conservation*, **38**, 55–73.

Howell, E.A. and W.J. Jordan III. (1991) Tall-grass prairie restoration in the North American Midwest. In I.F. Spellerberg, F.B. Goldsmith and M.G. Morris (eds) *The scientific management of temperate communities for conservation*. Oxford: Blackwell Scientific, pp. 395–414.

Kemf, E. (1993) *Indigenous peoples and protected areas: the law of Mother Earth*. London: Earthscan.

Mace, G.M. (1995) Classification of threatened species and its role in conservation planning. In J.H. Lawton and R.M. May (eds) *Extinction rates*. Oxford: Oxford University Press, pp. 197–213.

Mace, G.M. and R. Lande (1991) Assessing extinction threats: towards a reevaluation of IUCN threatened species categories. *Conservation Biology*. **5**, 148–57.

Oldfield, M.L. and J.B. Alcorn (1991) *Biodiversity: culture, conservation, and ecodevelopment*. Boulder, Colorado: Westview Press.

Perring, F.H. and L. Farrell (1977) *British red data books: 1. Vascular plants*. Nettleham: RSPNC.

Prance, G.T. (1991) Rates of loss of biological diversity. In I.F. Spellerberg, F.B. Goldsmith and M.G. Morris (eds) *The scientific management of temperate communities for conservation*. Oxford: Blackwell Scientific, pp. 27–44.

Prescott-Allen, R. and C. Prescott-Allen (1991) How many plants feed the world? *Conservation Biology*. **4**, 365–74.

Ratcliffe, D.A. (1977) *A nature conservation review* (2 vols). Cambridge: Cambridge University Press.

Ray, D.L. and L. Guzzo (1993) *Environmental overkill*, Washington: Regnery Gateway.

Ryder, O.A. (1986) Species conservation and systematics: the dilemma of subspecies. *Trends in Ecology and Evolution*, **1**(1), 9–10.

Shiva, V., P Anderson, H. Schucking, A. Gray, L. Lohmann and D. Cooper (1991) *Biodiversity. Social and ecological perspectives*. Penang, Malaysia: World Rainforest Movement.

Spellerberg, I.F. (1992) *Evaluation and assessment for conservation*. London: Chapman & Hall.

Whitten, A.J. (1990) *Recovery: A proposed programme for Britain's protected species*. Nature Conservancy Council, CSD Report No. 1089, Peterborough.

SECTION B Conservation biology in perspective

The term 'conservation biology' (as opposed to the movement) has emerged from a long evolution of natural history and wildlife management. The movement has grown out of a long history of increasing concern about the extent and growing rate of human impacts (direct and indirect) on species, habitats, biological communities and ecosystems. More and more effort is being directed towards the conservation of biological diversity but the success of that effort depends in part on how much we know about what we are attempting to conserve (what is it, where is it and how much is there?). However, conservation cannot be achieved on the basis of biological information alone. Resource management issues, debates about sustainability and the concerns about equity of resource use, including biological resources, have helped to give conservation biology an integrated and interdisciplinary perspective.

There are political, economic, legislative, policy and educational dimensions to conservation biology. Jeffrey McNeely from IUCN writes with a wealth of experience and argues that conservation biologists need to become more politically and economically astute. Michael O'Connell from the US Nature Conservancy also writes with considerable experience and suggests that conservation biologists must also be familiar with law and legislation because these will continue to be an essential ingredient of conservation. Communication and education are also important ingredients which put conservation biology into perspective. David Walton from the British Antarctic Survey has had a long association with the community relations side of the British Ecological Society and is well qualified to comment on conservation strategies. It becomes clear from his chapter that there needs to be better communication between policy makers and those who put conservation into practice. All too often poor communication and lack of consultation have been the cause of controversy in connection with conservation activities. Mark Boulton (International Centre for Conservation Education) and David Knight (University of Southampton) between them have a great breadth of educational experience and they demonstrate the need for continued support for education and training at all levels if conservation biology is to be effective. The development of new methods and materials for conservation education continues to offer exciting possibilities for both teacher and student.

The interdisciplinary nature of conservation biology has both social and science components. Abiotic variables and taxonomic information help to put conservation biology into a scientific perspective. To be able to put conservation into practice, conservation biologists first need to know what the environmental constraints are. Graham Pyatt's research while with the UK Forestry Authority has demonstrated very clearly that both physical and climatic factors must be determined before conservation strategies can be implemented. The species level of biological diversity is widely used as a focal point and therefore it is important that conservation biologists know not only what they are dealing with but how the species is determined. Nigel Maxted, a taxonomist at the University of Birmingham, provides an essential introduction to taxonomy and systematics, the fundamental biological basis of conservation.

CHAPTER 4 Politics and economics

JEFFREY A. McNEELY

Introduction

The earlier chapters in this book have presented the biological reasons why society is reaching a crisis point in its relationship with living resources and why conserving biodiversity is so important. This chapter will show that governments are responding by expanding their conservation interests beyond species and protected areas, using the term 'biodiversity' to address some of the most important political and economic issues of our time. An examination of these issues can help conservation biologists understand the political and economic context within which they must work, and suggest how they can contribute most productively to the difficult choices that must be made as society adapts to changing conditions.

Political issues in conservation biology

Conservationists – frustrated by the lack of political support – have been campaigning for many years to have their issues put on the political agenda. Under the banner of 'conservation of biological diversity', a measure of success may finally be at hand. A new Convention on Biological Diversity (Box 4.1) was signed in 1992 by political leaders from 157 nations. The Convention commits governments to conserve biological diversity, use biological resources sustainably, and promote equitable sharing of the benefits arising from the use of such resources (Glowka *et al.*, 1994).

However, action to convert this political commitment to abstract concepts into significantly changed behaviour will need to overcome a number of formidable political obstacles at both national and international levels, in both developed and developing countries (Mathews, 1991) (Box 4.2). When conservation was seen as a narrow issue of endangered species or national parks with few interests at stake, it was usually left to biologists and specialized agencies. But biologists are now sharing the conservation stage with agricultural scientists, anthropologists, ethnobiologists, pharmaceutical firms, farmers, foresters, tourism agencies, industrialists, indigenous and traditional peoples, and many others. Further, biodiversity has become an important issue in the international arena, as it is now perceived to have significant links to development. This has polarized both the developing countries, who see themselves as providers of biodiversity, and the developed countries, who depend on biological resources for their agriculture and industry. As the stage becomes more crowded with competing interest groups and significant resources are involved in the decisions to be made, politicians are put into the position of mediating among these groups to seek the course of development which best serves the larger public. Thus issues such as equitable sharing of benefits, intellectual property rights, sustainable development and national sovereignty are now at the very centre of the modern conservation scene, demanding that conservation biologists, too, become more aware of these broader political interests, seek new partnerships, and be more willing to consider new ways to achieve their objectives.

Many of the most important decisions affecting biodiversity – especially on issues of budgets, priorities, information, and resource management policies – are taken by politicians and 'non-conservation' sectors of government (ministries of finance, trade, defence, etc.); and perhaps even more important decisions are made by the private sector and rural people. Conservation biologists may see

Box 4.1 Key elements in the 1992 Convention on Biological Diversity

Major principles

- Biodiversity has intrinsic value and is a common concern of humanity.
- Governments have sovereignty over their biodiversity.
- States are responsible for conserving their biodiversity and using their biological resources in a sustainable manner.
- Causes of significant reduction of biodiversity should be attacked at their source.
- The fundamental requirement for the conservation of biodiversity is the *in situ* conservation of natural habitats and the maintenance of viable populations of species in their natural surroundings. *Ex situ* measures, preferably in the country of origin, also have an important role to play.
- Many indigenous and local communities with traditional lifestyles have a close and traditional dependence on biological resources and need to share equitably in the benefits arising from biodiversity.
- International cooperation is an important part of implementing the Convention.

Major measures

Contracting parties agree to:

- develop national biodiversity strategies, plans and programmes;
- identify and monitor important components of biodiversity;
- establish systems of protected areas, manage biological resources, rehabilitate degraded ecosystems, regulate risks of living modified organisms, control alien species, protect threatened species;
- establish facilities for *ex situ* conservation of plants, animals and microorganisms and adopt measures for the recovery, rehabilitation and reintroduction of threatened species;
- implement measures for sustainable use, including use of economic and social incentives;
- establish programmes for training, education and research, and promote access to relevant technology;
- facilitate access to genetic resources, on mutually agreed terms and under prior informed consent of the party providing such resources;
- promote access to relevant technology;
- promote technical and scientific cooperation, including exchange of information relating to biodiversity;
- provide funds to developing countries to help implement these measures.

Box 4.2 The international politics of biodiversity

The politics of biodiversity at the international level can be seen as a public struggle over who gets to use the living resources of the biosphere under what property rules, with what allocation of the costs and benefits from use. During the negotiation of the Convention on Biological Diversity, two alternative views to conserving biodiversity emerged. The major concerns of developing countries included access to technology and genetic and financial resources, economic development, consumption patterns of industrialized countries, and compensation for the opportunity costs of not using resources. The developed countries tended to focus more on protecting their commercial advantages in biotechnology and their access to genetic resources. Therefore, controversies over issues such as intellectual property rights, biosafety, access to genetic resources, technology transfer and the financial mechanism are dominating Convention discussions.

themselves as the gatekeepers to observable evidence, but when they try to apply their science to policy, governments and industries tend to treat them as just another interest group. As a result, conservation biologists are in danger of being little more than concerned bystanders when policies are formulated to address the problems of conserving biodiversity (Tobin, 1990).

This is perhaps understandable, as most problems affecting biodiversity reflect a conflict of interests between alternative ways resources should be used. For example, biologists have found that the annual runs of adult salmon in the Columbia River Basin in the USA have declined by 75 to 85%. Groups with an interest in the policy response to this observed decline include electric utilities, environmental advocates, the barge industry, recreational boaters, agricultural irrigators, logging and mining companies, the aluminium industry, government agencies, conservation biologists, and commercial, sports, and tribal fishing groups (Hyman and Wernstedt, 1991). A similar observation could be made about tropical forests or other biological resources.

Conservation biologists may contend that reliable information is the basis for sound decision-making about managing the salmon, or tropical forests. But most will admit that they have inadequate knowledge about lifecycles, natural fluctuations in populations, relations with other variables in the ecosystem, and impacts of various management regimes – indeed, many of these are surrounded by active scientific debate on which no consensus is yet available. Thus for the salmon and the forests, as for many other harvested resources in both temperate and tropical environments, an increasingly knowledgeable and sceptical public is asking questions that cannot be answered by conservation biologists with a suitable degree of certainty. Therefore, competing interest groups tend to fill the information vacuum with self-serving interpretations of 'truth', often based on incomplete knowledge or misinterpretations of the research data collected by conservation biologists. The policy environment has become highly volatile, because many mutually exclusive choices are possible, the long-term implications of management alternatives are difficult to predict, and different groups have different access to the political process. Political forces will decide what to do about the salmon and the tropical forests, but conservation biologists are most likely to have their point of view heard when they are able to couch their arguments in terms that bureaucrats and politicians find convincing, and feel that their constituents will support.

Keeping biodiversity on the public agenda is not easy. First, current practices which are depleting biodiversity often are extremely popular. The fact that the desire for consumption is far more powerful than the conservation-oriented advice of many scientists should come as no particular surprise, as incentives to consume far outweigh incentives to conserve. For example, the United Nations Commission on Sustainable Development found that worldwide the amount of money governments spent in 1992 to support environmentally destructive behaviour amounted to about US$1 trillion. Another indicator is the amount of money spent on advertising, basically encouraging people to consume more than they might otherwise consume; globally, advertising budgets in 1994 amounted to US$337 billion (more than the annual Gross National Product of Australia or The Netherlands).

Second, no easily identified opponent is available against which conservation forces can be rallied; unlike such headline-makers as Bhopal, the *Exxon Valdez* and Chernobyl, no newsworthy disasters have yet linked human welfare with the loss of biodiversity (Tobin, 1990). On the contrary, many people are making substantial profits from overexploiting biological resources, and those with the highest political profiles tend to be among those making the largest profits through overexploitation.

Third, the loss of biodiversity has no immediately observable impact on lifestyles, especially those of people living in cities far removed from the biological resources which support their consumption. If dozens or hundreds of species are being lost per day, as many experts assert (e.g. Myers, 1993; Wilson and Peter, 1988), then people are already living with the consequences of extinction without any discernible effects on their daily lives. And when conservation biologists argue that efforts to conserve endangered species deserve especially high priority, they have difficulty in linking this argument directly to the resource development issues of greater interest to politicians because these species have already been reduced to such low population levels that they often can be utilized as only symbols.

To the extent that politicians reflect public opinion, an important problem may be the perception of biological reality by the general public. While most people are at least somewhat concerned about the loss of species and disruption of ecosystems, few see any particular relationship between their own lifestyle and the loss of biodiversity. The politically influential people living in cities often give a sentimental or aesthetic value to nature, while rural people tend to see biodiversity in strictly practical terms of resources, health and welfare.

This leads to a disparity between the principle of conserving biodiversity and the practice of managing individual species or systems of resources. In many cases, preserving species such as rhinos, wolves, tigers or cranes provides primarily abstract benefits to individual members of the general – largely urban – public, while the people who are expected to make economic sacrifices by restricting their activities in the habitat of these species tend to be large-scale farmers and ranchers, forestry interests, or developers who are very effective in conveying their concerns to politicians; or small farmers or pastoralists over whom conservation agencies have little influence. So while the general public may agree with biodiversity conservation in the abstract, the support by rural people and commercial interests for specific action on threatened species tends to be much weaker because they pay more of the costs and perceive fewer of the benefits. Thus many government conservation programmes, especially if they are designed in the capital cities in response to the more powerful urban interests, face difficulties when they need to be implemented in the countryside, sometimes even causing a backlash as rural people protest about the loss of their historical responsibility as resource managers and about having to pay the opportunity costs for conserving what the world regards as its global heritage (Dang, 1991). On the other hand, conservation programmes designed by local or indigenous people or in close collaboration with them and which are designed to meet their concerns for sustainable use of biological resources can earn strong support.

Systemic solutions or small victories?

Here is the major dilemma faced by politicians, rural people, and conservation biologists alike: given increasing demands on limited resources, how can the costs and benefits of resource management be distributed most appropriately? This is not an easy question to answer. Experience suggests that the most popular public policies are those calling for modest changes in current practices to address immediate, proximate causes rather than imposing comprehensive changes in deeply embedded social behaviour (Tobin, 1990). Popular policies coincide with prevailing public opinion and do not require law-abiding people to change their lifestyles or cause them great inconvenience; they distribute material benefits to a majority or to a politically significant and effectively organized minority; they provide more benefits than costs, thereby favouring policies with easily monetized values, such as goods traded in the marketplace or development that provides jobs; and they generate concentrated immediate benefits while deferring and diffusing costs (the popularity of this approach is indicated by the budget deficits of many governments, as voters discount future benefits to enjoy present benefits).

Policies to conserve biodiversity tend to have the opposite characteristics. They call for fundamental changes in the way people relate to the environment; for example, IUCN, UNEP and WWF (1991) have recently called for limits on rates of resource use, following up the calls of Schumacher (1974) and Daly and Cobb (1989) for a minimal frugal steady state as the appropriate form of a post-industrial society. Prescriptions for a sustainable future based on principles of conservation biology tend to call for restricting access to resources, expect people to forgo material benefits, assign values to resources that are elusive or difficult to measure, and require payment today for abstract future benefits (WRI *et al.*, 1992).

Such a politically unattractive course is necessary, for any adequate action for correcting the ills of modern society must be systemic. As Rappaport (1993) says, 'we are facing such a multiplicity of quandaries, dilemmas, crises, inequities, iniquities, dangers, and stresses ranging from substance abuse, homelessness, teenage pregnancy, and prevalence of stress disease among minorities to global warming to ozone depletion that they cannot all be named, much less studied'. The traditional approach of seeking to understand systems by reducing them to components and analysing the interactions between them might facilitate 'problem-solving', but cannot provide an adequate understanding of complex systems. Conserving biodiversity requires moving towards a comprehensive view which synthesizes contributions from many sectors, but, as suggested above, putting such revolutionary changes into practice is anathema to politicians, who are serving the interests of an essentially conservative public.

Politicians – and the public they serve – are more attracted to defining manageable portions of the problem to be tackled. For most people, it is simply too overwhelming to think concurrently of whole litanies of problems; the response is to sink into passive despair. Instead, building a series of 'small wins' creates a sense of control, reduces frustration and anxiety, and fosters continued enthusiasm on the part of the public, conservation biologists and politicians (Heinen and Low, 1991). But if these 'small wins' are to be real victories, they must contribute to an overall strategy for conserving biodiversity. This requires a politically sophisticated approach involving economic concerns such as those discussed below.

Economics and conservation biology

Basic principles of economics applied to conservation biology

Because economic arguments dominate political debates, conservation biologists are well advised to consider how to use the idiom of economics, while being aware of the limitations and pitfalls of this inexact science. Modern economics views the world as consisting of autonomous self-interested individuals exchanging goods and services based on rational calculations of gain and loss in the context of an impersonal market. The market determines the prices that assign monetary value to the utility of the good or service exchanged, providing a means for comparing them with each other. The price paid for a good or service depends on the balance of supply and demand; scarce resources in high demand have higher prices than common resources with low demand. 'Market failure' occurs when prices do not completely reflect true social costs or benefits, leading to an inefficient or socially undesirable allocation of resources.

Economic theory suggests that as species such as rhinos, elephants or pandas become more rare (reduced supply), they tend to become more valuable (assuming that demand increases or remains constant). Conservation biologists may be able to mobilize support from economists in arguing that this changing balance of supply and demand may increase the value of such species alive more than their value dead. In economic terms, demands for increasing levels of conservation implicitly recognize that current levels of consuming the rhinos, elephants or pandas need to be reduced because the units saved will provide more benefit when used in the future, deferring short-term profit for long-term gain.

But economists also point out that many people apply a 'discount rate' to their decisions, preferring present returns to future returns. Discounting implies that the best long-term harvesting strategy for species with relatively low growth rates – such as trees or whales – is to liquidate the stock (Clark, 1990). The discount rate is often taken to be around 5–7%, roughly equal to or greater than the real interest rate (monetary interest rate minus inflation). When the intrinsic growth rate of a population is less than the discount rate, it makes economic sense to harvest the full stock and invest the capital earned in activities that yield a higher return. Conservation efforts become exceedingly difficult when economic motives seem to give such powerful support to overexploitation. Lande *et al.* (1994) used mathematical models to demonstrate that even when the intrinsic growth rate of a population is greater than the discount rate, various random or stochastic effects can lead to rapid extinction of the population. 'Even a small discount rate drastically diminishes the maximum expected present value of a cumulative harvest before extinction', they point out, 'and little profit can be gained in comparison with immediate harvesting to extinction; this narrow profit margin will often be eroded by fixed costs of maintaining harvesting operations'. They conclude that economic discounting should be avoided in the development of optimal strategies for sustainable use of biological resources.

Economists commonly use cost–benefit analyses in their identification and evaluation of problems and solutions, arguing that the production of a good is by definition economic only when the total benefits (and their distribution) exceed the total costs (and their distribution) (Pearce, 1986). In principle, this must include the costs of dealing with loss of biodiversity, pollution, atmospheric warming, and so forth; but in practice such factors are commonly ignored, or 'externalized', by economists; this can lead to market failure which encourages overconsumption and the loss of biodiversity. Ignoring environmental costs which are generated but not paid by the producer may in fact be the usual situation, and one of the main reasons biodiversity is being lost; when full costs are not paid, the price of a good is lower than its real value and more of it will be consumed.

Externalization of costs helps to explain why the production of a good – such as tropical timber – may well be profitable in a commercial sense even if it is not economic. For example, the way many forests are being harvested today is not economic, since the costs – including the long-term cost to local communities, disruption of watersheds, loss of genetic resources, and so forth – far exceed the benefits (Repetto and Gillis, 1988). Harvesting those forests is certainly profitable for the individuals who earn considerable commercial benefits through the process; but the market fails to force those individuals or corporations to pay the full cost of their exploitation because of an inadequate policy, regulatory or inspection system overseeing production (Daly and Cobb, 1989).

Economists therefore draw a distinction between *economic analysis*, which uses social-welfare values of an activity and eliminates distortions in prices by means of shadow prices and other devices, and *financial analysis*, which is based on the perspective of a private individual or firm, uses market prices and market interest rates, includes transfer payments such as taxes and subsidies, and externalizes many costs (Dixon and Sherman, 1990). They also recognize that some individuals or firms receive benefits which they value but do not pay for; such 'free riders' include, for example, a pharmaceutical firm that develops new drugs on the basis of medicinal plants identified by a shaman in Brazil but provides no compensation to the shaman (nor indeed to the government of Brazil).

Critical to issues of distribution of costs and benefits is consideration of property rights: who has rights over the disposition of biological resources and how are the costs and benefits to be distributed? Modern forms of development often have removed the responsibility for managing biological resources from the people who live closest to them, and instead have transferred this responsibility to government agencies located in capitals which are distant from the resources and remote from the realities of rural life. Many common property resources, such as forests and wildlife, have become the responsibility of governments rather than the grass-roots communities which historically had managed them reasonably well (Berkes, 1989); and even though the Convention on Biological Diversity recognizes the sovereignty of national governments over the biological resources found within their borders, few governments have yet been able to replace the traditional forms of management with more sustainable modern approaches. Too often, government control means that the resources become 'public goods', open to all but controlled by none – a sure recipe for overexploitation.

Given the power of the global trading system, and its accountability to commercial interests and higher levels of government rather than to local institutions (so decisions are taken on the basis of financial rather than economic analysis), it is not surprising that forests, grasslands, marine habitats, and the species they support have been grossly overexploited. The government agencies have seldom developed the resource-management capacity and political clout which is sufficient to counteract the newfound technological capacity to exploit, especially in times of shrinking government expenditures for resource management and calls for expanding consumption of resources.

Economics and loss of biodiversity

Economics should also be of great interest to conservation biologists because economic factors are driving the loss of biodiversity (see, for example, Daly and Cobb, 1989; Barbier *et al.*, 1994; Costanza, 1991; Pearce and Moran, 1994). Three such factors are particularly worth mentioning:

- *Subsidies to overexploitation.* Many governments subsidize over-exploitation of biological resources. For example, in 1991 the industrialized nations spent a total of US$322 billion on agricultural support and subsidies; much of this has had significant negative impacts on biodiversity, both at home and abroad. And in the US, national forests are so heavily subsidized that some 67% of them consistently record below-cost sales of timber (Repetto and Gillis, 1988). Vast amounts of energy are being provided to development in the form of petroleum products, pesticides, transportation, and so forth. Most of this energy is a one-time-only and non-renewable windfall, depending on the productivity of species long dead and now converted to coal or petroleum; the exploitation of these energy sources is highly subsidized. These energy subsidies enable the true cost of exploiting the resources to be hidden, because the prices paid for timber, fish or crops (for example) do not include the cost of replacing the oil consumed in harvesting, processing and marketing.
- *Focus on single products.* Where habitats such as forests, coral reefs and grasslands were once used to provide a whole range of benefits, including wildlife, medicinal plants, watershed protection, construction materials, firewood, spiritual values, and so forth, today they are increasingly being used to provide a single product, such as logs, fish or beef. This single-use focus tends to promote large areas of genetically similar species, often exotic, because they are more commercially attractive (though not necessarily more economic).
- *Response to global economic forces.* Much of the loss of native habitats has been due to changing the land to uses devoted to providing commodities for export, responding to a global, rather than a local, system of supply and demand. The scale of the economic forces driving the loss of biodiversity is no longer local, where local people will benefit or lose depending on the appropriateness of their practices. What were once locally self-sufficient and sustainable human systems have become part of much larger national and global systems whose higher productivity is both welcome and undeniable, but whose long-term sustainability is far from proven (Daly and Cobb, 1989). Further, the increased consumption facilitated by this higher productivity is also encouraging land-use practices which are unsustainable, especially deforestation and use of land for agriculture that would be more suitable for forests or other uses. When countries are all part of one system connected by powerful economic forces, it becomes very easy to overexploit any part of the global system because it is assumed that other parts will soon compensate for such overexploitation. The damage may not

even be noticed by the international community until it is too late to do anything to avoid permanent degradation. Perhaps worse, global interdependence enhances the domination of the economically powerful, yet requires a support structure which in itself can be very fragile – the global impacts of changes in oil prices, the Mexican bond market, exchange rates, interest rates, natural disasters and local wars demonstrate the point.

Valuation and conservation biology

The species contained within an ecosystem, their mass, their arrangement and the information they contain are the standing stock of an ecosystem – what might be considered nature's free goods. The functions of an ecosystem – maintaining clean air and pure water, cycling nutrients and supporting a balance of creatures that support human welfare – are nature's free services (Westman, 1977). Such free goods and services are very difficult to put into monetary terms, but assigning monetary values to biodiversity enables decision-making to be put on a more economically defensible footing. This may simplify decision-making for politicians, but it also implies that biodiversity can be reduced to the simple metric of money. Giving a cash value to biodiversity or its components 'forces the great range of unique and distinct materials and processes that together sustain or even constitute life into an arbitrary and specious equivalence' (Rappaport, 1993).

Further, not all problems affecting biodiversity can be adequately characterized or described in quantitative, let alone monetary, terms. As Morowitz (1991) put it, 'we are often left trying to balance the "good" of ethics with the "goods" of economics'. Assigning economic values to biodiversity is difficult because – at least with present knowledge – species extinction cannot be reversed no matter how much money is spent. The preferences of future generations are impossible to predict, present benefits are difficult to balance against future costs, and commodity value and moral value can be totally different. And values are not absolute, as different interests assign different values to the same resource; a forester may value a forest primarily for its timber, a farmer may value it for the clean water its streams provide, a tourism department may value it as a national park, and conservationists may value it for intrinsic reasons. Nor will values always be simply in terms of goods. For example, bird hunters in 1987 spent US$1.1 billion to harvest 5 million game birds, approximately US$216 per bird felled. Clearly, it is not the value of the meat that is motivating this effort, but rather the value of the experience of hunting.

On the other hand, finding ways to give economic value to biodiversity may be essential for conservation. Some economists contend that biodiversity is decreasing at least partly because so few genetic traits, species or ecosystems have market prices, the feedback signals which equilibrate market economies. Many economists maintain that prices which accurately reflect environmental costs would reduce consumption and keep the use of biological resources in a closer balance with their sustainable availability. If the true value of biodiversity could be included in the market system, they contend, markets could help conservation.

Today, most countries are finding it difficult to justify current, much less increased, expenditures on conserving biodiversity, especially when these costs are accompanied by local and regional opportunity costs which entail politically costly land-use conflicts with local people (Wells, 1992). In such conditions, it becomes imperative to seek economic support for conservation activities; this requires giving values to at least some of nature's goods and services.

Since policy-makers need to quantify costs and benefits, a large literature has developed around the problems of assigning economic value to biological resources, summarized in Pearce and Moran (1994) and Swanson and Barbier (1992). In addition to the relatively straightforward measurement of harvested goods which are sold in a marketplace, such as timber, fish, medicinal plants, and agricultural commodities, a number of more subtle measures are available. 'Contingent valuation methods' (CVM) use questionnaires and other interview techniques to measure the 'willingness to pay' (WTP) for a resource or 'willingness to accept compensation' (WTAC) for forgoing the use of a resource. Indirect approaches derive preferences from observed market-based information; for example, the 'travel cost approach' can measure all travel-related expenditures to protected areas, to estimate the benefit arising from that experience. The 'replacement cost technique' measures the cost of replacing or restoring a damaged biological resource to its original state. Numerous other techniques are also available.

Using such techniques, a number of studies have generated economic information on the values of biological resources. These include measures of direct use, basically through consumption; of indirect use, for example through carbon storage or watershed protection; and of non-use or option values based on an individual's willingness to pay to safeguard the possibility of using a resource at a future date, or for simply knowing that a resource exists or can be passed on to future generations (so-called existence or bequest values). The total economic value, therefore, includes the sum of direct, indirect and option values.

Several studies of direct use values indicate that using tropical forests for their non-timber values is more economic than logging. For example, Peters *et al.* (1989) estimated that sustainable harvesting in the Peruvian Amazon would yield a sustainable benefit of $1987 per hectare, while clear-felling would bring in a one-time net revenue of only $1000 per hectare. Sustainable harvesting of medicinal plants in Belize would yield a net present value of $3327 per hectare, while plantation forestry with rotation felling yields only $3184. Travel cost evaluation of tourist trips to Costa Rica's protected areas for foreign visitors amounted to US$12.5 million per year, giving the protected areas a value per hectare which was over 12 times the market price of local non-protected area land.

Such methods have been applied to many other habitats as well. Direct uses of the Hadejiia-Jama'are floodplain in Nigeria yield a net present value per hectare of $1990. Based on a combination of market values and willingness to pay for storm protection, Louisiana wetlands are valued at $1219 per hectare, while the present values per hectare of mangrove systems, based on direct use from fisheries, forestry and recreation, are estimated at $6000 in Trinidad, $4400 in Fiji and $5200 in Puerto Rico.

While many of these direct values are substantial, indirect uses often yield even greater values. Schneider (1992) gives a carbon storage value of tropical forests as $1300–5700 per hectare per year, while the total carbon storage value of the Brazilian Amazon has been calculated as $46 billion; and Western (1984) determined that each lion in Kenya's Amboseli National Park is worth US$27 000 per year in visitor attraction (the same lion would have a direct value of about $1000 as a skin).

Economic parameters can also be used to support the management of protected areas. In 1987, Canadians spent C$5.1 billion on wildlife-related activities, leading to gross business production of $10.7 billion, gross domestic product of $6.5 billion, government revenue from taxes of $2.5 billion, personal income of $3.7 billion, and 159 000 jobs; government expenditures in this field amounted to less than $1 billion. At the site level, the 8728 hectare Sarawak Mangroves Forest Reserve in Malaysia supports marine fisheries worth US$2.1 million per annum and up to 3000 jobs, timber products worth US$123 217 per year and a tourist industry worth US$3.7 million per year. If the mangroves were to be damaged, all of the fisheries and timber and many of the tourism benefits would be lost, and expensive civil engineering works would be required to prevent coastal erosion, flooding and other damage.

While monetary figures such as these are often merely minimal estimates of the total economic value of natural habitats and the resources they contain, they do provide policy-makers with ammunition they can use in the kinds of battles they face in political forums. But conservation biologists also need to be aware of the danger that placing values on species or protected areas may open them up to market forces, and some policy-makers might conclude that a price tag on a resource means that it is for sale.

Use of incentives, charges and other market instruments

Regulations such as controls on hunting or the establishment of protected areas have been the mainstay of environmental policies and resource protection in virtually all countries, usually being the preferred method of control by governments. Regulations have proven their value, though often at a high cost in terms of litigation and bureaucratic interventions; but biodiversity is too complex to be managed solely by such a regulatory framework. To make further progress, the regulatory framework now needs to be complemented by market instruments such as economic incentives, producing a new mix of instruments that will be an effective means of influencing human behaviour as it affects biodiversity (Repetto, 1987; McNeely, 1988, 1993; McNeely and Dobias, 1991).

Most economic incentives to date have been used to promote resource exploitation; indeed, these might more accurately be termed 'perverse incentives' (Repetto, 1988). Subsidies which deplete biodiversity have been important to modern societies, but conservation biologists might use economic arguments to design approaches to agriculture, marine fisheries or forestry that are not coupled to commodity production which is inimical to biodiversity. For example, income support could be provided in terms of incentives for soil, water and wetland conservation, promotion of diverse crops and livestock breeds, and other measures that would conserve biodiversity.

Incentives – such as subsidies, tax differentials and other fiscal mechanisms – can be used to divert land, capital and labour towards conservation. They can ensure more equitable distribution of the costs and benefits of conserving biological resources, compensate local people for losses suffered through regulations controlling exploitation, and reward the local people who make sacrifices for the benefit of the larger public. Incentives can be attractive to policy-makers when they help conserve biological resources, at a lower economic cost than that of the economic benefits received.

Conservation biologists should also be aware of the many other market instruments that could have a significant positive impact on biodiversity. Repetto *et al.* (1992), for example, analysed a wide range of environmental charges, recreation fees for use of national forests and other public lands, product charges on ozone-depleting substances and agricultural chemicals, and the reduction of subsidies for mineral extraction and other commodities produced on public lands. Their sample of potential environmental charges would reduce a wide range of damaging activities while raising over $40 billion in revenues. Recreation fees in national forests, for example, could yield US$5 billion per year. These findings refute the argument that environmental quality can be obtained only at the cost of lost jobs and income. Instead, providing a better framework of market incentives by restructuring revenue systems can both improve environmental quality and make economies more competitive by taxing people on their consumption of resources rather than on their salaries, property and profits.

Conclusions

Since the public wants their politicians to deliver benefits, not constraints, conservation biologists advocating policy changes need to become much more politically and economically astute if they wish to have the impact they desire. The Convention on Biological Diversity provides an excellent opportunity for doing so, giving political legitimacy to issues of conservation, sustainable use and equitable sharing of benefits. It is apparent that public support is crucial to any successful conservation programme; such support will need to be based on a sound ethical footing, good information and economic benefits. Conservation biologists must build on science to demonstrate the benefits of conserving biodiversity to farmers, fishermen, ranchers, and foresters, balance the attention given to loss of biodiversity with concern for sustainable use of harvestable species, and build a broader constituency among business, the public, and academics. An effective overall strategy for mobilizing political and economic support for conserving biodiversity will:

- give management responsibility and tenure rights to the people most directly involved;
- ensure that prices fully reflect environmental costs;
- provide economic incentives to encourage individual behaviour which is in the long-term benefit of the larger society and remove incentives which promote consumption of resources;
- provide the best available science to support decision-making;
- seek a diversity of local solutions to local problems.

In short, conservation biologists need to contribute to approaches to managing biological resources which are ecologically sound, economically feasible and politically palatable.

References

Barbier, E.B., J.C. Burgess and C. Folke (1994) *Paradise lost? The ecological economics of biodiversity* London: Earthscan.

Berkes, F. (1989) *Common property resources: Ecology and community-based sustainable development* London: Belhaven Press.

Clark, C.W. (1990) *Mathematical bioeconomics*. New York: Wiley.

Costanza, R. (ed.) (1991) *Ecological economics: The science and management of sustainability*. New York: Columbia University Press.

Daly, H.E. and J.B. Cobb, Jr (1989) *For the common good: Redirecting the economy towards community, the environment and a sustainable future*. Boston, MA: Beacon Press.

Dang, H. (1991) *Human conflict in conservation*. New Delhi: Development Alternatives.

Dixon, J.A. and P.B. Sherman (1990) *Economics of protected areas: A new look at benefits and costs*. Washington: Island Press.

Glowka, L., F. Burhenne-Guilmin, H. Synge, J.A. McNeely and L. Gundling (1994) *A guide to the Convention on Biological Diversity*. Cambridge: IUCN.

Heinen, J.T. and R.S. Low (1992) Human behavioural ecology and environmental conservation. *Environmental Conservation*, **19**(2), 105–16.

Hyman, J.B. and K. Wernstedt (1991) The role of biological and economical analyses in the listing of endangered species. *Resources*, summer 1991, 5–9.

IUCN, UNEP, WWF (1991) *Caring for the earth. A strategy for sustainable living*. Gland: IUCN.

Lande, R., S. Engen and B.-E. Saether (1994) Optimal harvesting, economic discounting and extinction risk in fluctuating populations. *Nature*, **372**, 88–90.

Mathews, J. (ed.) (1991) *Preserving the global environment: The challenge of shared leadership*. New York: Norton.

McNeely, J.A. (1988) *Economics and biological diversity: Developing and using economic incentives to conserve biological diversity*. Gland: IUCN.

McNeely, J.A. (1993) Economic incentives for conserving biodiversity: lessons for Africa. *Ambio*, **22**(2–3), 144–50.

McNeely, J.A. and R.J. Dobias (1991) Economic incentives for conserving biological diversity in Thailand. *Ambio*, **20**(2), 86–90.

Morowitz, H.J. (1991) Balancing species preservation and economic considerations. *Science*, **253**, 752–4.

Myers, N. (1993) Questions of mass extinction. *Biodiversity and Conservation*, **2**, 2–17.

Pearce, D.W. (1986) *Cost–benefit analysis*. Basingstoke: Macmillan.

Pearce, D. and D. Moran (1994) *The economic value of biodiversity*. London: Earthscan.

Peters, C.M., A.H. Gentry and R.O. Mendelsohn (1989) Valuation of an Amazonian rainforest. *Nature*, **339**, 655–6.

Rappaport, R.A. (1993) Distinguished lecture in general anthropology: The anthropology of trouble. *American Anthropologist*, **95**(2), 295–303.

Repetto, R. (1987) Economic incentives for sustainable production. *Annals of Regional Science*, **21**(3), 44–59.

Repetto, R. (1988) *The forest for the trees? Government policies and the misuse of forest resources*. Washington: World Resources Institute.

Repetto, R. and M. Gillis (eds) (1988) *Public policies and the misuse of forest resources*. Cambridge: Cambridge University Press.

Repetto, R., R.C. Dower, R. Jenkins and J. Geoghegan (1992) *Green fees: How a tax shift can work for the environment and the economy*. Washington: World Resources Institute.

Schneider, R. (1992) *Brazil: an analysis of environmental problems in the Amazon*. Report 9104-BR, Latin America and Caribbean Region, World Bank, Washington, DC.

Schumacher, E.F. (1974) *Small is beautiful: A study of economics as if people mattered*. London: Blond and Briggs.

Swanson, T. and E. Barbier (1992) *Economics for the wild: Wildlife, wildlands, diversity and development*. London: Earthscan.

Tobin, R. (1990) *The expendable future: US politics and the protection of biological diversity*. Durham, NC: Duke University Press.

Wells, M. (1992) Biodiversity conservation, affluence and poverty: mismatched costs and benefits and efforts to remedy them. *Ambio*, **21**(3), 237–43.

Western, D. (1984) Amboseli National Park: human values and the conservation of a savanna ecosystem. In J.A. McNeely and K.R. Miller (eds) *National parks, conservation, and development: The role of protected areas in sustaining society*. Washington: Smithsonian Institution Press, pp. 93–100.

Westman, W.E. (1977) How much are nature's services worth? *Science*, **197**, 960–4.

Wilson, E.O. and F.M. Peter (1988) *Biodiversity*. Washington: National Academy Press.

WRI, IUCN, UNEP (1992) *Global Biodiversity Strategy: Guidelines for action to save, study, and use earth's biotic wealth sustainably and equitably*. Washington: WRI, IUCN, UNEP.

CHAPTER 5

Legislation

MICHAEL O'CONNELL

Introduction

Reversing the trend of biodiversity degradation depends greatly on changing patterns of human activity and moderating their impact. The solution partly lies in establishing a framework of common principles for human activity. Given that people rarely share the same philosophy, ideology or values regarding natural resources, conservation laws provide standards to which individuals in a society can conform. In this way, law and legislation are fundamental to the conservation of biological diversity.

Legislation to protect elements of biological diversity has existed in various forms for hundreds of years. Most historical conservation laws protected elements of biological diversity with economic or symbolic value to a society. Many regulated killing specific species of wildlife or taking highly desired plants or animals. Rulers of nations enacted laws to protect fish, game and forest resources, or to set aside land of important value for hunting and fishing. For example, European monarchs and heads of state frequently passed laws regulating their favoured wildlife or habitats.

Most contemporary conservation laws consist of prohibitions on harming particular species, authorizing purchase or setting aside of significant habitats, and regulation of activities with detrimental environmental effects. The anti-poaching laws of Central and Southern Africa are an example of species-specific laws (Leader-Williams and Miller-Gulland, 1993) while the laws resulting in the US National Wildlife Refuge System typify habitat protection statutes (Curtin, 1993). Treaties among governments are also an important element of international law protecting biological diversity, particularly where they involve heavily exploited and commercially valuable natural resources or actions of whole countries that impact upon ecological systems.

As the human species has increased in population, particularly in the twentieth century, law and legislation have become increasingly important to moderate detrimental effects. The complexity and number of laws have grown in proportion to their necessity. The United States alone, for example, has nearly 100 statutes at the national level specifically directed at conservation of flora and fauna or their habitats. Countless state and local laws supplement these, the result being a complicated and daunting web of regulation.

In the past, conservation biologists and lawmakers have paid little attention to the effect of legislation on overall biodiversity conservation. Only recently has the United States, for example, contemplated comprehensive laws specifically targeting biodiversity (Committee on Science, Space and Technology, 1988). To date, these discussions have failed to produce any significant legal results. In contrast, numerous laws and conventions have been established to protect specific elements of biodiversity, mostly at the species level. These include statutes such as the Convention on International Trade in Endangered Species of Wild Fauna and Flora (CITES) and the United States Endangered Species Act. The Biodiversity Convention of 1992 is the first international agreement created specifically to address biodiversity conservation.

The effect of law on biodiversity is both direct and indirect. In addition to laws with a conservation focus, such as those governing commerce in endangered wildlife and plants, a wide variety of laws have secondary impacts on biodiversity. These include the United States Clean Water Act, pollution control statutes, and the recent North American Free Trade Agreement. Countless other legal provisions, from local-land use zoning codes to international consumer market controls affect conservation, illustrating that the entire legal fabric of contemporary society is intertwined with human effect on biodiversity.

The subject of legislation and conservation is vast. Any attempt to discuss the entire spectrum of laws affecting biodiversity would consume volumes. This chapter is a cursory introduction to the topic of conservation law. It focuses on twentieth century laws and treaties enacted with the expressed purpose of conserving wildlife and biodiversity. It begins with a look at international laws and conventions that govern species and their habitats, including the 1992 Biodiversity Convention. As an example, the chapter examines US laws and their effect on biodiversity conservation. It then concludes with a brief discussion of the future of conservation law, its role in conservation biology and its limitations.

International treaties

Agreements between nations are among the most common means of addressing international conservation issues. The first international treaty to protect wildlife was signed in 1886 between Germany, The Netherlands, Luxembourg and Switzerland, to regulate salmon fisheries (Lyster, 1985). Treaties have become a primary means of international cooperation for conservation in the last 25–30 years.

The frequent use of international agreements does not mean they are particularly effective. Conventions and treaties have many shortcomings. Enforcement often relies on self-policing, with only good faith stopping nations prone to violation. Treaties are only binding on signatory countries. Agreements frequently must be weakened to the point of ineffectiveness for many nations to ratify them; these nations are often the most destructive to the resources in question. One example has been the continual refusal of Iceland and Norway to adhere to International Convention for the Regulation of Whaling prohibitions, despite being the primary harvesters of several species of imperilled cetaceans.

Notwithstanding their imperfections, dozens of international treaties have been signed in the twentieth century, with coverage varying from individual species to entire bioregions. The most important treaties for biodiversity appear in Table 5.1. In the absence of an overarching authority or legal framework for controlling international impacts on biodiversity, treaties are perhaps the only means of providing countries with common principles for conservation.

International treaties vary in scope. Some, such as the International Agreement on the Conservation of Polar Bears and their Habitats, signed by the five nations with territories overlapping the range of the polar bear in 1973, concern single species. Others, including the 1973 Convention on International Trade in Endangered Species of Wild Fauna and Flora (CITES), cover dozens of species. A few international agreements are geographically regional in scope, such as the African Convention on the Conservation of Nature and Natural Resources of 1968 and the Convention on the Conservation of Antarctic Marine Living Resources of 1980. Still other conventions are global, such as the World Heritage Treaty of 1971 and the Global Biodiversity Convention of 1992.

This section briefly describes four international agreements that have great bearing on the conservation of biodiversity. The scope of these conventions ranges from a multiple species-level focus, to regulation of specific habitat types, to identification of critical global resources and delineation of global conservation actions and guidelines. With the breakup of the former Soviet Union, the number of signatories to these treaties is in constant flux; this has changed the implementation of many agreements.

The Convention on International Trade in Endangered Species of Wild Fauna and Flora (CITES)

In 1973, 21 nations signed the Convention on International Trade in Endangered Species of Wild Fauna and Flora (CITES). This treaty was the culmination of several decades of concern about explosive growth in the wildlife trade business and its accompanying detriment to rare elements of biodiversity at the species level. According to Lyster (1985) CITES has proven to be one of the more successful international conservation treaties. This is largely due to its fundamental principles of sustainable trade and cooperative enforcement provisions. As of 1994, 129 nations had ratified the treaty.

CITES operates by maintaining three lists of controlled fauna and flora. These 'Appendices' determine the level of restriction placed on the trade in specific species of wildlife and plants. Appendix I contains species that are threatened with extinction by trade. International commerce is prohibited in these species, except by special permission. Species on

Table 5.1 Important international treaties

Agreement	Date	Purpose
Convention of Nature Protection in the Western Hemisphere	1940	First international agreement targeting natural diversity as a conservation goal; covered protected areas, trade, multilateral cooperation; lack of administrative structure has limited its practical application
International Convention for the Regulation of Whaling	1946	Signed by most of the major whaling nations; established the International Whaling Commission; controls harvesting and sets quotas for whale species
Convention for the Establishment of Inter-American Tropical Tuna Commission	1949	Created a multinational commission with the express purpose of regulating harvest and trade in species of tuna
African Convention on the Conservation of Nature and Natural Resources	1968	Concerned primarily with wildlife species, this convention addresses education, research, and sustainable development; no administrative structure
Convention on Wetlands of International Importance Especially as Waterfowl Habitat (Ramsar)	1971	Encouraged parties to designate wetland areas of international importance and undertake measures to ensure their conservation
Convention Concerning Protection of the World Cultural and Natural Heritage	1972	Established lists of sites of global significance and set up a monetary fund to aid in the protection of sites in danger of degradation
Convention on International Trade in Endangered Species of Wild Fauna and Flora (CITES)	1973	Agreement to restrict trade in species of flora and fauna threatened with extinction; three Appendices based on level of threat posed by trade and status
Convention on the Conservation of Migratory Species of Wild Animals (Bonn Convention)	1979	Specifically addressed species that migrate across national borders, requiring multilateral efforts to protect
Convention on the Conservation of Antarctic Marine Living Resources	1980	Agreement designed to protect and promote the rational use of Antarctic resources by establishing an oversight commission and a scientific advisory body
Biodiversity Convention	1992	First convention identifying biodiversity as a global priority and mechanisms to address conservation

Appendix II are those for which international trade poses a potential risk. Their commerce is carefully controlled through a series of permitting authorities. The third appendix is reserved for species – not found on the first two lists – that are regulated by national commerce laws in their native countries. Individual nations may request assistance from CITES parties in protecting these species by placing them on Appendix III.

The CITES system is overseen by a secretariat located in Switzerland. This authority monitors the complicated system of prohibitions and permits for trade on an international level through reports of the signatory nations. Parties to the treaty meet every two years to review, revise and amend the treaty and add or delete species on the Appendices. Each CITES party is required to establish a national Management Authority and a national Scientific Authority to implement treaty provisions and grant permits.

The Convention on Wetlands of International Importance Especially as Waterfowl Habitat (Ramsar)

Signed in 1971 and ratified in 1975, the Convention on Wetlands of International Importance Especially as Waterfowl Habitat, or 'Ramsar', was intended to slow the rampant destruction of wetlands globally. This treaty recognized the importance of fresh and saltwater wetlands as critical areas of biological productivity. It is significant as the first international treaty concerned primarily with habitat. More than 70 nations have ratified the treaty, which designates wetlands of international significance and promotes their conservation by discouraging encroachment. Areas such as the Florida Everglades and the South American Pantanal receive nominal protection from their inclusion as Ramsar sites.

Parties to the Ramsar treaty unilaterally designate wetlands of particular importance. Each party must list at least one major wetland. More than 20 million hectares of wetland systems have been identified in this manner. The treaty confirms national commitment to promote the wise use of all wetlands and protection for the listed wetlands of international significance. It encourages research into the flora and fauna of wetlands as well as their function, and designates training of wetland scientists and managers as a national priority. Cooperation among parties is also required under the convention.

Despite the clear intent of the Ramsar treaty and a few notable national successes, it has attracted criticism for its inability to bring about broad change in the way wetlands are conserved on a global scale. This is due in particular to the few legally binding provisions of the treaty, and the relatively few signatory countries. Nations with dense human populations and few remaining wetlands have been able to utilize Ramsar designation more effectively than those with larger expanses of wetlands and sparse populations.

The Convention Concerning Protection of the World Cultural and Natural Heritage (World Heritage Treaty)

First signed in 1971 and since ratified by more than 140 nations, the World Heritage Treaty addresses outstanding sites of human and biological importance. The intent of the convention was to aid the efforts of a broad array of international bodies such as UNESCO, IUCN and others by developing a list of 'crown jewels' that all parties agree are of universal importance.

The World Heritage Convention maintains two lists, one of generally important sites and another of sites in danger of degradation. Parties to the Convention meet to determine which sites receive protection. Sites must first be nominated by a host country according to a specific set of criteria and then must be ratified by all the parties to the treaty. The list of natural areas includes some of the most outstanding and diverse places on earth, such as the Great Barrier Reef, the Galápagos Islands, Serengeti National Park, and Yellowstone National Park. The listing of a site may have dramatic consequences for the host country. For example, international tourism increased by 43% in the first year following the listing of Mesa Verde National Park in the US as a World Heritage site (a World Heritage Site is depicted on Plate 3).

Unlike other treaties, the World Heritage Convention created a financial mechanism to help protect sites included on the World Heritage List. Parties contribute to the World Heritage Fund in proportion to their level of contribution to UNESCO. This means that wealthier nations supply the greater financial share of conserving globally significant resources, placing the burden on participants most able to pay. Only sites located in signatory nations receive protection funds. For a tiny country such as Ecuador, for example, the prospect of receiving assistance in conserving the Galápagos Islands far in excess of its contribution to the Fund is a strong incentive to join the Convention and select sites for inclusion on the World Heritage List.

The Biodiversity Convention

Signing of the Biodiversity Convention at the 1992 United Nations Conference on Environment and Development (UNCED) or 'Earth Summit' in Rio de Janeiro, Brazil, by 158 countries marked a watershed event for international biodiversity conservation. Building on successful elements of earlier agreements and navigating through many contentious issues, the Convention was the first global treaty created specifically to address biodiversity. The Convention contains three fundamental obligations for its signatories: conservation of biodiversity; sustainable use of natural resources; and cooperative sharing of biotechnology and the benefits of biodiversity.

Unlike most other international agreements on conservation, the Biodiversity Convention focused closely on mechanisms to finance sustainable biodiversity use and protection. The Convention designated the existing Global Environmental Facility (GEF) as the institutional structure to provide assistance in carrying out its provisions. The GEF was established in 1991 as a joint venture of the World Bank, the United Nations Development Programme and the United Nations Environment Programme (Reed, 1993). This massive trust fund is joined by more than 138 countries and funded primarily by 17 major donor nations. The United States alone has committed $2 billion. Among the goals of the GEF is to provide financing to enable sustainable development and conservation of natural resources. The Biodiversity Treaty requires a future restructuring of the GEF to more effectively meet the goals of the parties and avoid bias towards donor nations.

The Biodiversity Convention itself has four sections. These detail the objectives of the agreement, commitments and relationships of the parties, administrative procedures, and interaction between the Convention and other international laws. The Convention creates an institution known as the Conference of the Parties which meets periodically to pursue implementation. The first COP conference was held in late 1994 to examine and define specific operating principles, designate a permanent Secretariat, and establish a scientific clearing house for technical information.

Global focus on biodiversity conservation through the 1992 Convention has forced participating nations to examine critical legal and political issues. Many of these concern the imbalance of power, finances and technology between developed nations in the Northern Hemisphere and developing countries in the Southern Hemisphere. The most important are access to resources, technology transfer, and bio-safety or risk management (Barber, 1994). On the last of these the 1992 Convention did not reach resolution, leaving development of a protocol for managing risks for the Conference of the Parties.

The Biodiversity Convention was a revolutionary agreement regarding access to resources, particularly genetic material, in that it clearly acknowledges the sovereign right of national governments to control access to and distribution of the resources in their own countries. The vast commercial and market potential of tropical plant materials, for example, enables the countries of origin to share in the benefits of their conservation and sustainable use. A significant exception to this provision lies in the retroactive exemption of genetic material and resources in collections outside the countries of origin before the Convention.

On the issue of technology transfer, the Convention seeks to balance protection of proprietary biotechnologies of industrialized nations and the demand of developing nations – often on whose resources the technologies were based – that they be allowed to share them on easier terms than outright market purchase. This section of the Convention was particularly difficult to negotiate. It was cited as the primary reason why the United States initially declined to sign the Convention at the 1992 summit in Brazil. The US has since signed the agreement.

Perhaps the most important implication of the Biodiversity Convention is not its protocols or provisions, but that it marks the first international legal and political effort towards reversing the degradation of biodiversity worldwide. Although it remains to be seen how effective the agreement will prove over time, the financial and social incentives for biodiversity conservation created by the Convention promise new directions in legislation and provide a framework for future global action. Since the 1992 Earth Summit in Brazil, parties have begun planning national strategies for conservation, developing endangered species legislation, considering stronger environmental impact assessment regulations, and examining local and regional laws governing biodiversity conservation.

United States laws

The United States maintains the most extensive and complicated portfolio of conservation laws of any nation. Nearly 100 federal statutes regulate 'take' and commerce in wildlife, authorize acquisition of habitat and require consideration of the effect of economic development on species and habitat. Most were passed in the latter half of the twentieth century. Most states have also developed a set of laws affecting conservation of biodiversity, with varying degrees of overlap and coordination with federal statutes. The most significant of the federal conservation laws appear in Table 5.2.

The root of all US conservation law is the notion that elements of biodiversity belong to the public at large. The federal government may thereby regulate their use for the benefit of all. The US Congress approves legislation under its constitutional authority and these laws are carried out by designated federal agencies. According to Bean (1983), the constitutional language is general enough that interpretation of congressional action by the court system is also an important part of the legal process. US conservation law is therefore the evolutionary product of congressional authorization, federal regulation, and broad interpretation by the judiciary.

For the purposes of this text, the discussion of US conservation law will be limited to a brief introduction to a handful of broad national statutes. These include the Endangered Species Act (ESA), the National Environmental Policy Act (NEPA), the Marine Mammal Protection Act (MMPA) and the Migratory Bird Treaty Act (MBTA). Much recent attention has been paid to the potential for legislation with broader biodiversity goals in mind (Blockstein, 1988), but with

Table 5.2 Major US conservation laws

Act	Sections	Purpose
Lacey Act of 1900	16 U.S.C. Sections 701, 3371–3378	Prohibited interstate transportation and commerce in wildlife taken in violation of state law
Migratory Bird Treaty Act of 1918	16 U.S.C. Sections 703–711	Enacted treaty strictly protecting migratory birds
Wilderness Act of 1964	16 U.S.C. Sections 1131–1136	Provided for Congressional designation of strict protected areas in the national forest system
National Wildlife Refuge System Act of 1966	16 U.S.C. Sections 668dd–668ee	Consolidated federal conservation lands into a centrally administered refuge system
Marine Mammal Protection Act of 1972	16 U.S.C. Sections 1361–1407	Placed an indefinite moratorium on harvesting marine mammals and importing marine mammal products
Coastal Zone Management Act of 1972	16 U.S.C. Sections 1451–1464	Protected coastal and estuarine areas through regulation and cooperative funding for state conservation programmes
Endangered Species Act of 1973	16 U.S.C. Sections 1531–1543	Protected endangered species of wildlife and plants through regulation and prohibitions on activities
Land and Water Conservation Fund Act of 1976	16 U.S.C. Sections 4601–4611	Created federal trust fund to purchase habitats for conservation and recreation
National Environmental Policy Act of 1976	42 U.S.C. Sections 4321–4361	Created procedures to ensure that activities were not detrimental to the human environment
Alaska National Interest Land Conservation Act of 1980	43 U.S.C. Section 16	Set aside large portions of public land in the state of Alaska as wildlife refuges
Fish and Wildlife Conservation Act of 1980	16 U.S.C. Sections 2901–2911	Provided federal grants-in-aid for conservation of wildlife and fisheries

few substantive legal results. The political climate has shifted in the US such that the possibility of passing broad national biodiversity statutes in the near future seems remote. Most current US laws directly protect only species-level biodiversity.

The Endangered Species Act

The most relevant existing US law for conservation of biodiversity is the Endangered Species Act of 1973 (ESA). The goal of this law is to protect species in imminent danger of extinction and nominally the ecosystems that sustain them (16 U.S.C. 1531–1543 as amended). The US Department of Interior through the Fish and Wildlife Service and the Department of Commerce through the National Marine Fisheries Service are responsible for implementing the ESA.

The ESA has been both hailed and reviled in the US for its clarity and rigour. It strongly prohibits the 'taking' of certain protected species and provides clear penalties for violations. The law also spells out alternative processes to avoid running afoul of its prohibitions. Although it contains provisions that result in biodiversity conservation, the ESA is fundamentally a wildlife protection act. Conservation biologists and other scientists recognize its limits in protecting what we now know to be important elements of biodiversity, such as natural communities and genetic diversity. Among its shortcomings, it contains no prohibitions on taking of protected plant species by private parties.

The Endangered Species Act describes two levels of risk for species at the brink of extinction: endangered and threatened. The federal government maintains lists of these species and adds or eliminates individual species according to scientific criteria in the law. Species proposed for listing but not yet formally protected are considered 'candidates'. As of 1994,

Fig. 5.1 The Bald Eagle (*Haliaeetus leucocephalus*) is one of more than 800 species protected by the United States Endangered Species Act. (Photograph courtesy of The Nature Conservancy-US.)

more than 800 native species – including the bald eagle (*Haliaeetus leucocephalus*) (Figure 5.1) – were listed as threatened or endangered and more than 4500 candidates awaited final listing decisions. In addition, more than 300 species not found in the US appear on the lists of protection.

The ESA mandates that listing as endangered or threatened be a purely scientific determination based on several factors, including the biological status and degree of threat to the survival of a species. Once a species is listed as endangered or threatened, the ESA divides responsibility for its protection differently between federal and non-federal entities. Section 7 describes the duties of federal agencies toward protected species. It imposes both conservation duties and prohibitions. It directs federal agencies to examine their programmes and use them to further the goals of the ESA. Section 7 also authorizes specific conservation programmes for protected species.

For private or non-federal entities, the fundamental provision of the ESA is Section 9. This section prohibits 'take' of any protected species by any individual subject to the jurisdiction of the US or any of its territories. Take is very broadly defined in the ESA to include: 'harass, harm, pursue, hunt, shoot, wound, kill, trap, capture, or collect, or to attempt to engage in such conduct' (16 U.S.C. 1532(19)). This definition has been further refined through regulations to include indirect activities such as habitat modification.

The National Environmental Policy Act

The National Environmental Policy Act (NEPA) has the potential to be a critical US statute for biodiversity conservation (42 U.S.C. 4331). It was not enacted with that specific purpose, but NEPA is broadly intended to eliminate harmful effects on the human environment, among which in many cases is loss of biodiversity. The intent of NEPA is to establish procedures to examine and avoid potentially harmful environmental effects of human activity (see Plate 4).

NEPA is fundamentally a public disclosure and analysis law. It requires that all federal activities, including issue of permits for non-federal actions, be accompanied by complete investigation and publishing of all actual and potential environmental impacts. The more significant the potential effect of the activities, the more analysis must be done and the more closely the public must be involved. The intended result is that any possible adverse environmental effects will be foreseen and avoided.

The NEPA process of disclosure and analysis is straightforward. Federal guidelines for compliance were developed by the US Council on Environmental Quality, a section of the executive branch of government. NEPA guidelines state that any proposed action by a federal agency that has potential adverse environmental effects requires preparation of an Environmental Assessment (EA). The EA is a brief document that outlines the proposed activity and its potential effects, and makes a determination about whether the action is significant enough to require a more extensive Environmental Impact Statement (EIS). If the EA reveals that the action will not significantly degrade the human environment, the agency may proceed with the activity. If the EA determines that a significant impact may occur, then the agency must prepare an EIS to discover and avoid potential impacts.

The Marine Mammal Protection Act

In the late 1960s, both the legislative and scientific communities recognized severe depletion of global populations of marine mammals. Despite the existence of the International Whaling Commission and other international regulatory bodies, direct loss from hunting and incidental loss due to net fishing and other activities put many species at risk of extinction.

To address this issue, the US Congress in 1972 authorized the Marine Mammal Protection Act

(MMPA). The intent of this legislation was to centralize the authority over marine mammals and place it in the jurisdiction of the National Marine Fisheries Service and Department of Interior under a comprehensive federal programme. The MMPA not only covers cetaceans, but also includes pinnipeds, polar bears and sea otters.

The MMPA enacted an indefinite moratorium on harvesting of marine mammals by US entities to allow depleted populations to recover to the point that they would no longer require intensive protection. It also prohibited imports of internationally harvested marine mammals and products. Further, it directed renegotiation of existing treaties affecting marine mammals and restricted imports of commercial fish caught using methods resulting in excessive incidental loss of marine mammals. The only exceptions were for native Inuit and Arctic peoples and for harvesting according to international treaties. A significant exemption was also included for certain levels of indirect loss due to commercial fishing operations, an issue that continues presently, particularly with regard to dolphins and commercial tuna fishing operations.

The Migratory Bird Treaty Act

One of the first conservation statutes passed in the United States prohibiting the taking of rare and imperilled wildlife was the Migratory Bird Treaty Act. Passed in 1918, the MBTA authorized the US to implement international treaties limiting the taking of migratory birds.

Remaining largely unchanged since its authorization, the MBTA has been referred to on three additional occasions to implement international migratory bird treaties. These include the 1936 treaty between the US and Mexico on migratory birds and game mammals, the 1972 bird treaty between the US and Japan, and the 1976 treaty between the US and the former USSR concerning migratory birds. This latter treaty has yet to be renegotiated with the independent nations of the former Soviet Union.

The number of species covered by the MBTA is substantial, with little opportunity for exemption from strict seasonal prohibitions. This has led to numerous legal challenges throughout the Act's history. Questions have generally concerned the effect of seasonal take prohibitions on hunting on private property. The constitutionality of the Act has been upheld in each case. The restrictiveness of the law, however, has been counterbalanced by insufficient and in some cases non-existent enforcement.

Laws of individual states

Most of the states in the US have wildlife or conservation laws that affect biodiversity. Some, such as California, have assembled a set of statutes that in many ways closely resembles federal law. California has both an Endangered Species Act and the California Environmental Quality Act, a law similar to the federal NEPA. These laws carry out their mandate in much the same manner as federal laws do, with implementing agencies, procedural requirements, regulation of activities, and enforcement provisions.

Some state laws overlap their federal counterparts, while others have similar but non-overlapping provisions. Florida, for example, regulates wildlife and habitat through state statutes, funds an implementing agency and maintains lists of protected species, but has no designated endangered species act. Other states have no such laws, and regulate only commercially valuable species. The combination of overarching federal laws and individual, occasionally conflicting legislation in 50 states is a complicated web of environmental law that is difficult to interpret, coordinate and implement.

International elements of US conservation law

By regulating the activities of the US government or those entities subject to its authority, most laws of the United States have international implications. These generally implement the US role in multinational conventions, unilaterally affect international trade that impacts upon wildlife, or regulate the activities of international parties within US jurisdiction.

The US Endangered Species Act has several important international elements. For example, the ESA is the implementing authority for US participation in CITES. Section 8A describes the US role in CITES and Section 9 specifically prohibits trade in species listed as endangered or threatened, further reinforcing the provisions of the CITES convention. In addition, Section 7 of the ESA regulates the activities of the US government abroad where they affect protected species. Section 9 also restricts activities by non-US entities in areas subject to national authority, such as commercial fishing in territorial waters.

The Pelly Amendment to the Fishermen's Protective Act of 1967 is another element of US law that has international consequences. This amendment authorizes the US Secretary of Commerce to sanction other nations whose fisheries activities are judged to be in violation of marine conservation laws. The President of the United States may then direct the Secretary of Commerce to prohibit importation of fisheries products from those nations. The most recent use of the Pelly Amendment came in 1991, when the US cited Japan for fisheries activities that resulted in depletion of sea turtle populations.

Conclusions

Despite the existence of a large body of national and international law affecting biodiversity, the future of law as a foundation for conservation is likely to remain uncertain. In developed countries, legal and political change is slow to conform to advancing scientific knowledge. In developing nations, immediate social priorities frequently take precedence over long-term conservation needs, even where laws exist. In those countries that do have conservation laws, pressing human requirements frequently make implementation difficult or questionable.

Clearly, scientific research into the value and degradation of biodiversity far outpaces laws and regulations intended to conserve it. In the United States, for example, the controversy over reauthorization of the Endangered Species Act pits property owners who believe the law infringes on their rights with scientists and conservationists who know the law does not effectively achieve biodiversity conservation. The trend of scientific identification of conservation needs remaining ahead of law and legislation is likely to continue into the future.

Ultimately, a system of conservation based solely on the penalties and prohibitions provided by laws is doomed to failure. Laws simply supply a framework of minimum standards to which society can conform. Sustainable conservation of biological diversity will only be achieved when the deterrents of law are combined with incentives for conservation broad enough to encompass the varying needs of society, as well as education and research to better understand and focus public involvement on solutions. Nevertheless, law and legislation will always remain an essential ingredient in protecting the variety of life on earth that sustains the human species.

References

Barber, C. (1994) Focus on the convention on biodiversity conservation. In *World Resources 1994–1995*. New York: Oxford University Press, pp. 156–63.

Bean, M.J. (1983) *The evolution of national wildlife law*, 2nd edn. New York: Praeger.

Blockstein, D.E. (1988) US legislative progress toward conserving biological diversity, *Conservation Biology* **2**(4), 311–13.

Committee on Science, Space and Technology (1988) Biological diversity/prospects and problems, *US House of Representatives One Hundredth Congress, First Session, Report No. 70*, Washington: US Government Printing Office.

Curtin, C.G. (1993) The evolution of the US national wildlife refuge system and the doctrine of compatibility, *Conservation Biology*, **7**(1), 29–38

Leader-Williams, N. and E.J. Miller-Gulland (1993) Policies for the enforcement of wildlife laws: the balance between detection and penalties in Luangwa Valley, Zambia. *Conservation Biology*, **7**(3), 611–17.

Lyster, S. (1985) *International wildlife law: an analysis of the treaties concerned with the conservation of wildlife*. Llandysul, Dyfed, UK: Grotius Publications.

Reed, D. (ed.) (1993) *The Global Environment Facility: sharing responsibility for the biosphere*, vol. 2. Washington: World Wide Fund for Nature.

CHAPTER 6

Conservation strategies

DAVID W. H. WALTON

Introduction

Conservation policies are not new. The Tudors had them, the Romans had them, and even the Ancient Egyptians were concerned about the wise use of biological resources. There are, however, important differences in the driving forces behind these historical strategies and modern strategies, and the policies that flow from them. The most important of these are the damage being done to the earth by the rapidly growing human population and a recognition that biodiversity, in both species and habitat terms, matters. It was not always so.

Sustainable use of biological resources is a cultural concept embedded in the way in which land has been used by people in many countries. From the aboriginal people of Australia to the reindeer herders of Lapland there has always been a concern to ensure a future for the tribe by protecting existing plant and animal communities. That was possible in an agricultural world with a small population. The development of urbanization, intensive agriculture, technological advances and a massive increase in world population irrevocably changed that but, for a long time, the extent and importance of the ensuing damage was not realized.

By the mid-twentieth century scientific evidence was accumulating to show that new national and international initiatives were needed to tackle the loss of natural communities, the pollution of groundwater and the seas, and the consequent effects on species biodiversity. Within the United Kingdom the government had recognized the need for national nature reserves whilst at a more global level the International Union for the Conservation of Nature (IUCN) had been established to provide strategic advice at a regional level. What was missing from this was significant public interest and concern, a necessary prerequisite for governments to act quickly.

This took some little time to develop but the activities of non-governmental bodies, especially those of the increasingly well-organized groups such as Greenpeace, the World Wide Fund for Nature and Friends of the Earth, soon mobilized public support. Their approach was twofold: first to identify areas where new policies were needed (such as the whaling issue) and second to highlight the inadequacy of implementation of existing legislation. In both these approaches Greenpeace in particular used unorthodox methods to grab media and therefore public attention. Their activities reinforced the scientific initiatives that were already underway.

The formation of the new United Nations Environment Programme in the 1970s should have been an opportunity to take new initiatives worldwide, but because its objectives were not politically acceptable to many of the richest countries its activities were largely ignored. Only now is it beginning to make significant political impact after much further damage has been done. Yet in this period there has been a growing appreciation both of the complexity of the problem and that its solutions lie with no one initiative or with any single nation. Ecosystems and environmental change recognize no political frontiers, so to succeed our long-term plans must be scaled to the natural systems but implementable through the political systems. So far there is much more evidence of inertia, shortsightedness and opportunism than there is of leadership, clear thinking and determination. However, the public interest in conservation, biodiversity and sustainable management has continued to grow and it is on this that any optimism for future sustainable use and management, at all scales, should be based.

How can we best proceed? Conservation strategies must not be simply motherhood statements but clear frameworks within which defined objectives can be both explained and realized. The strategies must exist at a range of different levels of detail to cope with the markedly different needs of national government and local authority, yet within any given region all levels must be clearly related to each other. This is a difficult task in itself given both the complexity of modern society and the range of potentially conflicting interest groups who need to be consulted. Indeed, with the rate of change which seems inherent in such a topical field it is necessary to undertake regular surveys of the whole field to maintain all the elements in the correct perspective. Examples of such surveys for the United Kingdom are to be found in Warren and Goldsmith (1983) and Goldsmith and Warren (1993).

In this short chapter it is only possible to examine three illustrative examples of the problems of defining conservation strategies but in so doing the basic requirements of definition of objectives, consultation for agreement, legal enactment (where appropriate), monitoring of success and development for the future will be considered. The United Kingdom provides an example of the national case in which many of the current developments apply equally to other countries in the European Union. The Antarctic is an example of the international regional case. The final example is of a global initiative generated from within ecology itself – the Sustainable Biosphere Initiative produced by the Ecological Society of America.

The national case – United Kingdom

Objectives

In setting national objectives for nature conservation any government has to provide for a bewildering range of conflicting uses, satisfy increasing public concern about the state of the environment and yet not incur too great a financial penalty. Each political party will attach a different degree of importance to conservation but all agree that, in view of continuing public concern, it must figure in both policy making and resource allocation. It seems unlikely that the major parties in the UK will develop radically different conservation objectives, especially since much of the legislation is now derived from European initiatives. It is, however, important to recognize that the later stages – procedures for enactment, enforcement and development – can be significantly affected by the lobbying of political pressure groups, both inside and outside the government.

Table 6.1 Statutory organizations in the United Kingdom involved in conservation activities

English Nature
Scottish Heritage
Countryside Commission for Wales
Countryside Commission
Rural Development Commission
Department of the Environment
Joint Nature Conservation Committee
National Rivers Authority
Forestry Authority
Scottish Office
Ministry of Agriculture, Fisheries and Food

In the United Kingdom the statutory responsibilities for undertaking nature conservation are divided between bodies with a regional remit, such as English Nature, and national bodies with a limited responsibility for particular features, such as the National Rivers Authority. Table 6.1 lists some of the principal bodies with conservation responsibilities. The complex web of interlinking that results from this makes effective national management and planning difficult. In addition to these there are a variety of non-governmental bodies, such as the National Trust and the Royal Society for the Protection of Birds, with major national conservation interests as well as a plethora of more localized conservation trusts and activity groups.

Despite this organizational complexity all the agencies share many common elements in defining their conservation plans. English Nature can be used as an example for this. Their present strategy is based on the principles of sustainability and retention of biodiversity agreed by governments at the Earth Summit in Rio de Janeiro and comprises twelve objectives to be achieved by the year 2000 (Table 6.2).

It is in the implementation of these objectives that it becomes necessary to define more clearly what public and private interests need to be reconciled, how commerce and conservation will cohabit successfully and what compromises must be sought for financial or commercial reasons.

Elements

There are three key elements which need to be defined in the development of the strategy: the spatial

Table 6.2 Conservation strategy of English Nature to the year 2000

- Maintain and enhance characteristic plant and animal communities, and natural features
- Achieve the commitment of owners and managers to the maintenance of England's biological diversity within SSSIs
- Maintain a range of sites representative of the diversity of rock type, landforms and geological history of England
- Increase populations and distributions of specified statutorily protected species
- Stabilize population distributions and increase the numbers of defined national level indicator species
- Achieve wider acceptance of the role of personal stewardship
- Establish a strong constituency of support for nature conservation, and for English Nature
- Increase opportunities for people to experience wildlife and natural features
- Acceptance of sustainability as a fundamental principle
- Ensure that comprehensive environmental appraisals and audits become integral to the decision-making process
- Seek widespread acceptance of the environmental quality standards
- Ensure that statements of nature conservation objectives are integrated with other objectives within all relevant Government plans

Source: English Nature (1993).

elements, the legal elements and the affected user community. The spatial element is the geographical remit of the strategy (global, regional, national, local). The legal element involves any existing national or international legislation which affects species or habitats. The users must include both those who are actually involved in implementing the conservation strategy and the wider community, e.g. farmers, industry, etc., whose activities are likely to be affected by the implementation of the strategy.

In planning any strategy these three elements have to be related in both time and space. This is true as much for the user interests as it is for the actual species and sites. For example, in consultation exercises it can be crucial to make sure that if the object of protection is of international importance it is clearly identified as internationally important by government departments. Conversely, if a site is of regional or local importance the appropriate levels in the bureaucratic chain are likely to be much lower. This may seem obvious but there have been many instances of battles lost because the wrong level in the official system was targeted.

It follows from this that the various official levels, from district councils up to the national government, are all required to develop separate plans for conservation but that all the components should be self-compatible within a national framework. In the past the national framework was provided by a single government agency – the Nature Conservancy Council. The devolution of responsibility to national conservation agencies (English Nature, Scottish Heritage, Countryside Council for Wales), with only the Joint National Conservation Committee to provide integration, has been seen by many as an indication of diminishing interest in providing the necessary strategic coherence centrally. Now there is a further problem. Supranational legislation, derived from the European Union, is rapidly becoming the determinant of many key parts of conservation strategy and UK activities are having to be remoulded to fit in with this.

Users

Defining the user community can be both complicated and contentious. At a local level the owners and managers of a site are normally easy to identify and must be consulted. Once beyond this level the interested user community becomes more difficult to identify and consult. Often there are statutory or elected bodies who must be consulted and might, at least in some instances, be assumed to fairly reflect informed public opinion. These clearly include District and County Councils, National Park Planning Boards, the National Rivers Authority, and other more local bodies such as Drainage Boards where appropriate. In both local and national policy development there are a wide range of special interest groups who claim to represent important user elements, e.g. the Farmers' Union, the Confederation of British Industry, Greenpeace, etc., or are able to provide specialist advice and information, e.g. the British Ecological Society, the British Trust for Ornithology, the Royal Town Planning Institute, etc. Collecting the views and suggestions from potential users is a difficult task; incorporating these into the final policies may not prove possible.

The elements of the policy can be viewed in a number of different contexts. For example, habitat-based conservation has been used as a key feature in many strategies partly because of the ease with which conservation units could be described both legally and to the public. This has a number of important drawbacks, as described in Chapter 12, but seems likely to remain the principal tool of conservation strategy. The history of Sites of Special Scientific Interest (SSSIs) provides a useful example of the development of habitat conservation in terms of both its changing role in conservation policy and the way in which protected sites might be currently viewed by various interested groups.

In the original parliamentary report on nature conservation, written by J.S. Huxley in 1947, the sites that later became SSSIs were a very minor element, described only as 'many hundreds of small sites of considerable biological or other scientific importance'. It was assumed that landowners would protect them automatically once their value and interest had been explained. This was not easy to do and little headway had been made before the establishment of the Nature Conservancy in 1949 brought a new impetus to conservation. These sites were now categorized and individually described over the following decade or so. This official designation meant that they could be given legal protection from development and damage. By 1980 policy had shifted further and SSSIs were proclaimed to be a major element in site conservation and their management a top priority for the Nature Conservancy. Many farmers and other landowners had by now come to recognize the importance of SSSIs but there was increasing pressure from mineral extraction, forestry and developers on some sites. By the 1990s policy had shifted again as it became clear from scientific evidence that SSSIs were at best only a partial answer to conservation, since many were too small to maintain wildlife integrity in an increasingly hostile agricultural desert and with limited protection from development damage. By now changes in European agricultural policy limiting crop production were feeding back into farming, changing the range of land-use options available to the farmer. There is now recognition of the need to manage larger areas for habitat protection as well as population viability, in particular to manage the surrounding estate in a supportive manner and to redress the problems of atmospheric pollution, etc.

Recognizing the importance of SSSIs, how then are they viewed by different groups? The scientists see the SSSI as a key but inadequate element in any successful conservation framework, providing a refuge from direct human impacts for a carefully selected sample of flora and fauna. Loss or damage to such areas is a loss of biodiversity and a loss to future scientific investigations. Developers on the other hand see it as an expendable or transplantable resource of little other significance if it stands in the way of a major road or housing development. A third group with potentially conflicting interests could be farmers with land adjoining the SSSI. In order to protect its value it might be necessary to limit their use of sprays, provide access routes across their surrounding land, conclude management agreements to provide for buffer zone protection around the site, etc. All of this may prove difficult to achieve since not all farmers are convinced of the importance or value of scientific reserves which may harbour species they regard as pests.

These conflicts over use are not new but are now taking on a new and urgent importance as pressure on the environment grows. With research increasingly indicating that biodiversity and community structure cannot be protected adequately by a series of relict ecosystem fragments surrounded by the wastelands of modern agriculture, forestry and urban development (see also Chapter 14), new policies are needed to meet new conservation needs.

How is policy developed to provide the framework for implementation? Typically there are consultation exercises at various scales. At a governmental level these might include interdepartmental working groups to establish where there may be conflicts of interest between, for example, the Department of the Environment and the Department of Transport over road-building policies. Equally important at the more local level are public consultations through local councils. This has been used very successfully in urban areas to produce agreement on prioritization of areas for conservation initiatives and the management guidelines to be applied to these areas by both official and amateur groups. Figure 6.1 illustrates this process for London. It is at this consultation stage that key decisions are often taken over how rigorously conservation policies can be maintained when faced with the conflicting demands from wealth creation, convenience or political opportunism.

Any strategy is only as good as its implementation. And it is here, at the interface with people, that the most difficult decisions have to be made. The government's attitude to conservation and environmental management influences all of the legal and economic instruments used in implementation of conservation and environmental management, i.e. directives, regulations, subsidies, etc.

We can consider the effectiveness of a strategy in a different way – by looking at the way in which key elements of it are derived and implemented. In this instance we will take three examples which cover habitat conservation (SSSIs) and land use (Environmentally Sensitive Areas and Nitrate Sensitive Areas).

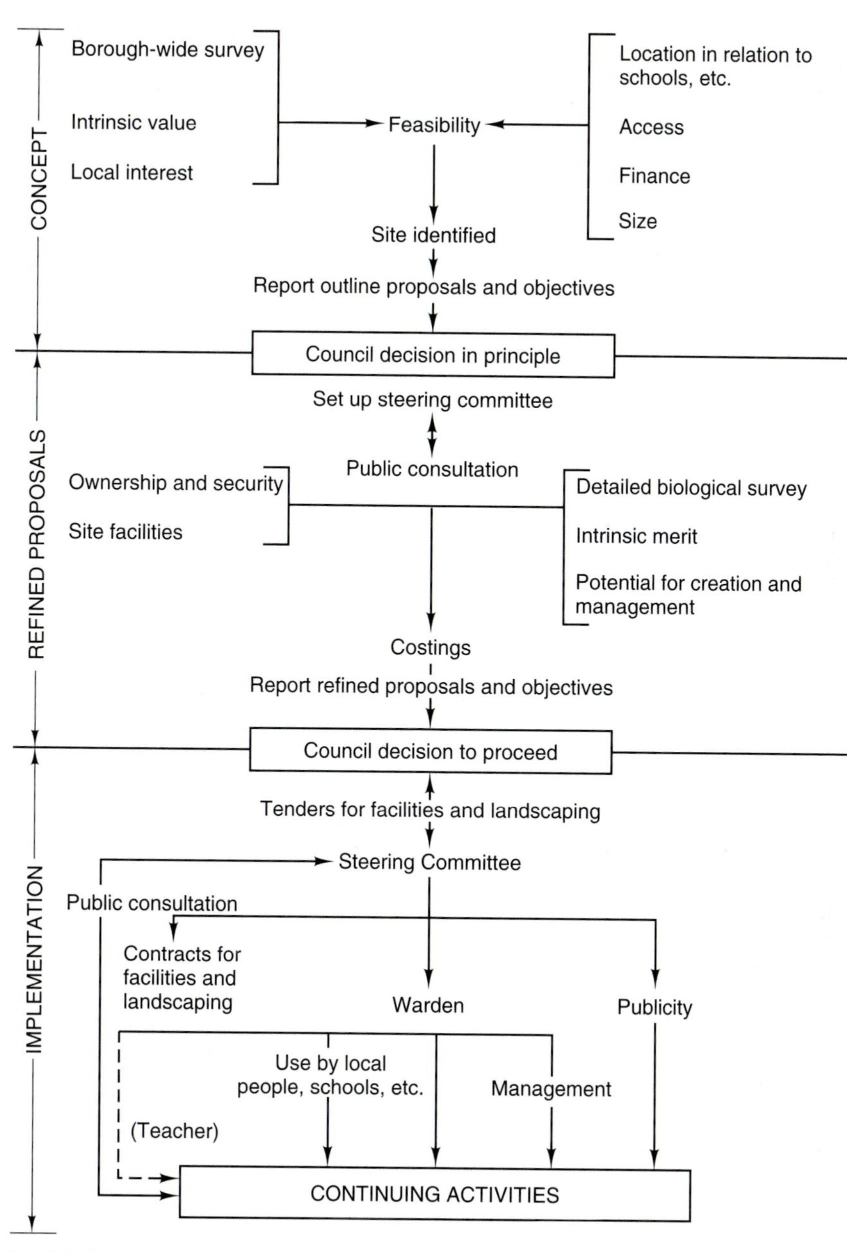

Fig. 6.1 Consultation steps involved in setting up a local nature reserve in London (GLC, 1985).

Habitat conservation

For many years Britain has had a system for identifying and designating Sites of Special Scientific Interest to ensure their long-term conservation. The sites comprise a basic minimum of habitats which must be protected for future generations. The last UK legislation to cover these protected areas was the Wildlife and Countryside Act 1981. This is now being superseded by the implementation of the European Habitat Directive.

At present in England there are 3707 SSSIs with a total area of about 840 000 hectares – almost 8% of the land area of Great Britain – with around

18 000 owners and managers (English Nature, 1993). Of these 80% were designated because of their biological interest. It might be expected that considerable care would be taken of such an important sample of our natural heritage. But a recent report (Rowell, 1991) illustrates only too clearly how extensive the damage is despite the strategy, policy and laws. Since 1981 about 5% of the SSSIs are recorded as being damaged in any one year, but this is known to be an underestimate because of a lack of systematic recording. A detailed assessment of one-third of SSSIs in England in 1990 showed that 40% were damaged and a further 21% were under threat. The majority of damage is short term and probably could be rectified within 15 years with appropriate management. However, inappropriate or inadequate management is much more widespread than published figures suggest. A major factor in long-term damage is activities requiring planning permission such as road building and extraction of peat or minerals.

Most disturbingly, most management agreements merely prevent an intentional damaging operation rather than providing for positive management. Indeed, according to the report, it is difficult to show from available data that management agreements provide any clear conservation benefits. In addition there is a reluctance to use compulsory purchase orders to protect endangered sites and an apparent inability to mount prosecutions for damage because of a failure to detect it early enough.

In response to this, English Nature has started a quarterly newsletter for those involved in SSSI management to improve communications, developed a Wildlife Enhancement Scheme and a Reserve Enhancement Scheme to encourage land management that favours wildlife, and tried to emphasize positive management in site agreements rather than paying the owners simply not to damage the sites.

Land use

In a small and crowded country such as the United Kingdom conservation strategies have to accommodate disparate and often conflicting uses. Large areas cannot be reserved simply for the conservation of the native flora and fauna, as has been done in some other countries. In the United Kingdom conservation must be more complicated, but with probably the best researched flora and fauna of any country the UK should be in the best position to develop new conservation initiatives.

In developing both the strategy and its implementation, initial attention needs to focus on the expectations both of the public at large and of the wide range of special interest groups. A second element in the policy development is the need to recognize the importance of scale – local, regional, national and international – at all of which there are both interlinked and separate demands.

In case all of this so far sounds too theoretical, let us come down to examining some examples of how a strategy and its policy elements have accommodated differing interests.

Farming constitutes the principal land use in the UK. Its legitimate activities must be considered as part of the strategy. The range of farming types is wide, from low-density grazing of upland areas by sheep, through livestock (including chickens, dairy herds and high-density pig farms) to high-intensity arable farms growing a very limited range of crops. Major elements of concern to conservation are intensity of grazing, fertilizer use, the effects of slurry and manuring, the use of herbicides and pesticides, and the extent of the removal of natural vegetation from farms and changes in traditional farming practices which affect the landscape.

In the last few years there has been a considerable expansion in the number of financial incentives available to farmers, with the principal objective of improving environmental management or limiting the damage to natural ecosystems from intensive agriculture. The nature and purpose of the payments varies considerably but some of the important forms of subsidy are:

- incentives for traditional land management, for example in the Environmentally Sensitive Areas (see Box 6.1);
- compensation payments, for example for reduction in fertilizer use (Nitrate Sensitive Areas – Box 6.2) or taking land out of production (Set Aside);
- incentives for restoring habitats and improving public access to them, for example in the Countryside Stewardship Scheme;
- incentives for conversion to more acceptable crops, for example the Farm Forestry initiative.

These initiatives underlie the clear linkage between economics and ecology in implementing the objectives of a strategy. Ecological understanding is not sufficient on its own. To be effective in a complex system of multiple land use the economic costs of each level of conservation must be clearly understood. In this field there is considerable scope for

Box 6.1 Environmentally Sensitive Areas (ESAs)

These were introduced in 1987 by the Ministry of Agriculture, Fisheries and Food to encourage farmers to help safeguard areas of the countryside where the landscape, wildlife or historic interest is of national importance. Farmers in each ESA join the scheme by entering into a management agreement, normally for a period of 10 years. In return for adopting agricultural practices which will help to protect and enhance the environment, the farmer receives an annual payment for each hectare entered in the scheme. There are usually three tiers of entry, with higher payments and greater environmental benefits for the higher tiers. Typical restrictions are a ban on converting grassland to arable, limits on fertilizer application and the use of herbicides, maximum stocking levels and the timing of hay cuts. Typical inclusions within the higher tiers are requirements to maintain traditional buildings and walls, hedge planting, and the creation of flower-rich meadows. Twenty-two areas have now been designated in England (see figure) and Wales, another five in Scotland and two in Northern Ireland. The Ministry is required to monitor the effects of the scheme and review the working of each at regular intervals to adjust, if necessary, the management requirements in each tier to achieve the conservation objectives.

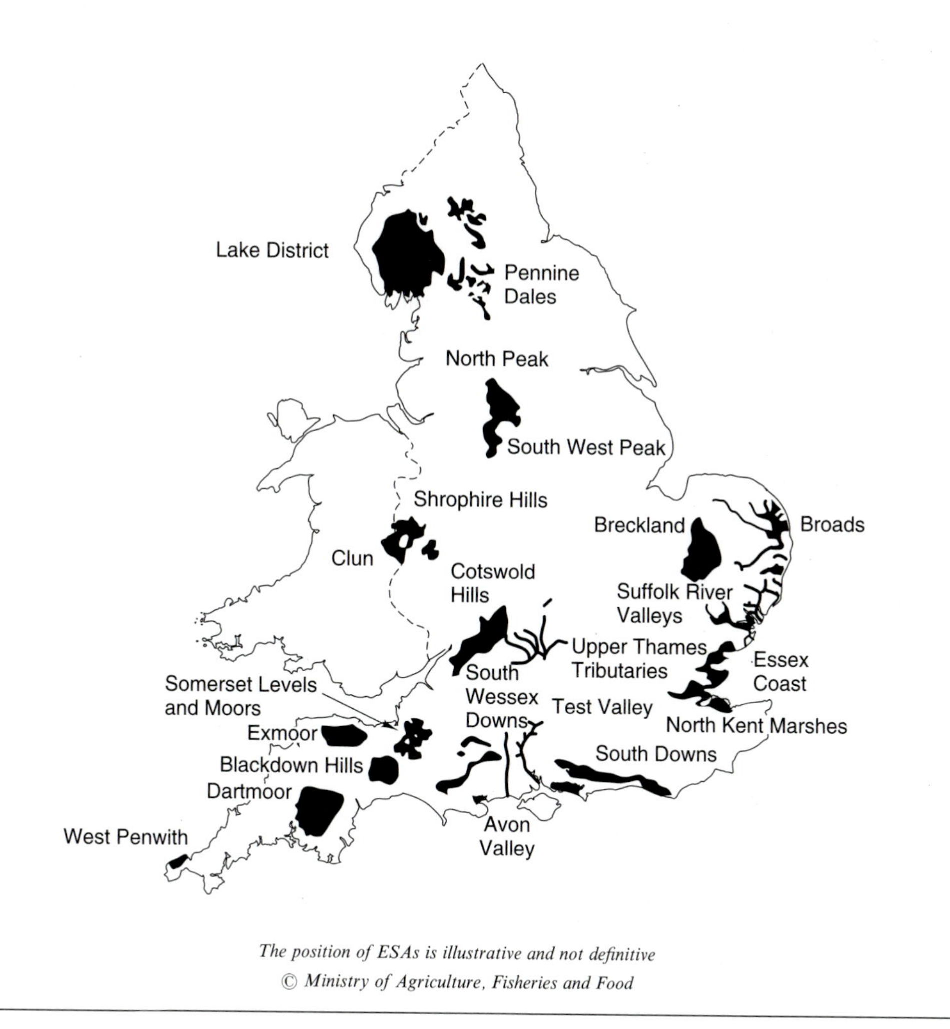

The position of ESAs is illustrative and not definitive

> **Box 6.2 Nitrate Sensitive Areas**
>
> Rising levels of nitrate in groundwater are a significant problem in many parts of Europe. From an ecological point of view, elevated levels have disastrous effects on freshwater bodies, encouraging algal blooms and significant loss of biodiversity. The European Commission issued a directive in 1991 to protect water supplies against pollution by nitrates from agricultural sources. By December 1993 every member state was required to designate 'vulnerable zones' where nitrate concentrations in water exceeded 50 mg/l, and by 1995 compulsory action plans must be established for these zones. In the UK the Ministry of Agriculture, Fisheries and Food established ten Pilot Nitrate Sensitive Areas covering three types of aquifer: sandstone, chalk and limestone. Under the scheme farmers are paid to limit or stop applications of fertilizer and manure, convert arable to low intensity grassland and restrict ploughing of existing grasslands. In a further nine areas MAFF are attempting to achieve a reduction in nitrate leaching by means of an intensified programme of land management advice to farmers.

development as economists have yet to fashion adequate instruments to deal effectively with the values of ecosystem integrity, biodiversity and sustainable management.

The international case – Antarctica

Established in 1948, IUCN (now called the World Conservation Union) has provided a unique link between the non-governmental bodies, government agencies and sovereign states. As an international and independent organization it attempts to provide both leadership in and a common approach to world conservation. Its three principal objectives are:

- to ensure the conservation of nature and especially of biological diversity, as an essential foundation for the future;
- to ensure that where the earth's natural resources are used this is done in a wise, equitable and sustainable way;
- to guide the development of human communities towards ways of life that are both of good quality and in enduring harmony with other components of the biosphere.

Its membership comprises 64 sovereign states, 98 government agencies, and over 570 non-governmental organizations. In this remarkable mix of members 117 countries are represented. Organized in various Commissions, the World Conservation Union is a major part of the international framework of UN-linked agencies which provide the opportunities for both assessing the needs and initiating the actions needed for global conservation and environmental management. At its General Assembly, held every three years, the diverse constituents of the Union come together to discuss new policies and conservation initiatives.

During the 1970s it became increasingly obvious that population growth, technological developments and industrial expansion were creating conservation problems on a much wider and more fundamental scale than previously. Governments were unable to see beyond their national interests, so little cross-boundary agreement was evident. Even within countries there was little evidence of policies for resource exploitation, environmental management or even structured decision-making in which conservation had any role to play. Conservation is by its very nature a cross-sector activity and bureaucracies are never good at dealing with such difficult areas.

Against this background IUCN set itself the task of defining a World Conservation Strategy which would provide the overarching framework within which national and regional policies could be developed. Published in 1980, the World Conservation Strategy set out three fundamental principles as the basis of conservation activity at all levels:

- The essential ecological processes and life-support systems must be maintained.
- Genetic diversity must be preserved.
- Any use of species and ecosystems must be sustainable.

These can be seen as the essential 'motherhood' statements so beloved of politicians. For them to be of any use they needed to be converted into practical principles that would initiate effective action. To do this the World Conservation Strategy proposed the compilation of regional and national strategies, allowing them to be interpreted in terms of the diversity of environmental, economic and political situations worldwide. It was suggested that each of these sub-strategies should be structured around four objectives:

1 A definition of the objectives for living resource conservation

2 The priority attached to each objective
3 How greater public participation in planning and decision-making could be developed
4 The deliberate adoption of pro-active cross-sectoral policies rather than reactive sectoral ones

One such regional conservation strategy is that for the Antarctic. Although there are a number of territorial claims to the Antarctic by the United Kingdom, Chile, Argentina, Norway, France, New Zealand and Australia, the ratification of the Antarctic Treaty in 1961 set these aside in favour of an international regime for the whole continent. Since 1961 the Antarctic has been governed through periodic meetings, called Consultative Meetings, between those countries with active scientific research programmes in the Antarctic (Triggs, 1987). The original treaty is concerned with establishing the Antarctic as a nuclear-free area, ensuring no military development and proclaiming the continent's main use to be for scientific research (Walton, 1987). The only sentence concerned with conservation is simply a recognition of the need to consider it!

IUCN had recognized the need for specific initiatives in Antarctic conservation as long ago as 1960. Initially the Antarctic Treaty countries met this by accepting the 'Agreed Measures for the conservation of the Antarctic flora and fauna' at the meeting in 1964. Several Treaty countries then decided to provide further Recommendations on other aspects of conservation and environmental management. Over the next two decades a considerable number of conservation proposals on protected areas, protected species and management of pollution were agreed at Treaty meetings but there was still no formal framework or strategic overview linking the disparate units.

Antarctica was mentioned again at the IUCN 1978 General Assembly; there, attention was shifted to the marine ecosystem. Increasing concern about marine resource depletion in the Southern Ocean was expressed at the Assemblies in 1981, 1984 and 1988. Finally, in 1990, a comprehensive resolution on Antarctic conservation was sent by IUCN to the Antarctic Treaty Meeting in Chile at which the Protocol for Protection of the Antarctic Environment was being discussed (IUCN, 1991).

Meanwhile, a series of parallel activities involving both IUCN and SCAR (Scientific Committee for Antarctic Research), the principal scientific non-governmental body for the Antarctic, had been underway for some years. With the acceptance of the 'Agreed Measures...' SCAR began to work for a more comprehensive range of conservation measures within the Antarctic Treaty. In due course this resulted in the development of a protected areas system, the nomination of protected species, specification of environmental impact procedures, agreement on the management of marine resources, and now the implementation of a comprehensive range of controls on waste management and marine pollution.

Those outside the Treaty meetings could see the need for a strategy document linking these elements. During the 1980s IUCN brought together conservationists and SCAR scientists to develop an Antarctic conservation strategy. This was a more unusual instrument than the other regional strategies since the Antarctic is an international area and all the applicable legislation is also international – that agreed at the Antarctic Treaty Consultative Meetings (see Box 6.3).

The continent has no indigenous population and no industries except fishing and tourism, but a key role to play in scientific research of global importance. There appeared to be little in the way of the types of

Box 6.3 The Antarctic Treaty

This international treaty, signed in 1961, runs indefinitely and applies to all land and ice shelves south of 60°S but not to the seas. It sets aside all national claims to sovereignty, bans all military activities and nuclear waste disposal, gives complete freedom for scientific investigations and allows any country to inspect the activities of any other. Major additions to the Treaty during the last 30 years have included conventions for the protection of Antarctic seals and for the conservation of Antarctic marine living resources. Most recently agreement to the Protocol for the Protection of the Antarctic Environment has provided a very high level of protection for Antarctica, including a ban on mining for at least the next 50 years.

Those countries with active scientific research programmes in Antarctica are called 'Consultative Parties' whilst those countries which simply support the principles of the Treaty are called 'Acceding States'. The Treaty meets every year in a different country and forms the only effective international authority governing the Antarctic.

At present there are 26 Consultative Parties and 14 Acceding States. Between them they represent about 80% of the world's population and include both developed and less developed countries.

conflicting interests often so prominent in other parts of the world but there was the extra dimension of gaining agreement to the strategy from the 26 countries operating in the Antarctic.

The strategy identified ten key issues where action was required to strengthen the existing policies:

1 Conservation principles and goals
2 Legal instruments
3 Institutional mechanisms
4 Public information and education
5 The management of science
6 Logistic facilities
7 Protected areas
8 Tourism
9 Marine living resources
10 Minerals activities

Many of these specific concerns were addressed in the Protocol on Environmental Protection to the Antarctic Treaty (Foreign and Commonwealth Office, 1992) which was agreed just after the Strategy document was completed. Indeed, in many respects the Protocol *is* the new conservation strategy for the Antarctic, embracing most of the elements highlighted in the IUCN document.

The Protocol not only comprises the framework strategy for conservation but also has a series of Annexes – on environmental impact assessment, waste management, marine pollution, protected areas, conservation of flora and fauna – which define how the general objectives should be implemented in each of the major fields and how environmental damage should be monitored. Thus in one agreed international document we have the strategy for Antarctic conservation, its application to the various fields of environmental management and species protection and a requirement to report on progress in these fields – and thus assess successes and failures.

Legal implementation of the Protocol requires each country to pass national legislation embodying its major points, so that citizens of all the countries active in the Antarctic are required to undertake environmental management and conservation in the same way. This rewriting of the Protocol into an acceptable form for Japan, France, Australia, South Africa, Russia, etc., is a major undertaking and its completion for all 26 countries will probably take several more years. It does, however, show that international agreement for scientific conservation is not only possible but politically important for many countries.

The ecological case

In 1988 the Ecological Society of America (ESA) initiated discussions on the research priorities for ecology to the end of the twentieth century. Recognizing that many of the environmental problems of the world are fundamentally ecological in nature, the ESA concluded that 'investigator initiated, peer-reviewed basic research is the foundation on which informed environmental decisions must be based'. Resources for research are, however, finite so whilst identifying the problems it was also necessary to prioritize them. The resulting document, 'The Sustainable Biosphere Initiative: an ecological research agenda' (Lubchenko *et al.*, 1991) can be considered as the first global strategy document generated entirely by ecologists and directed at the key areas of global change, conservation of biodiversity and sustainable management of ecosystems. Its emphasis is emphatically scientific, examining the intellectual challenges to ecologists in each of the key areas, but it recognizes that success depends as much on education and informed decision-making as it does on progress in research. The ESA sees the strategic areas of this initiative as applying internationally as well as nationally. Their establishment of a Sustainable Biosphere Initiative (SBI) Office is a major step in moving from strategy to implementation, and provides the essential link between research, education and application that underlies all strategies (Figure 6.2).

Conclusions

The words ecology, conservation and strategy have all too often been misused, especially by politicians. Yet it is these same politicians who, at the end of the day, agree to a particular conservation strategy and its vital underpinning of ecological information and research. The better informed our politicians are about the real priorities in conservation, the more sensible will be the objectives, the implementation and the resourcing. The future requires just as much effort as the past in educating the policy makers.

A good strategy must fulfil the following conditions:

- consultation with all interested parties
- clear objectives
- practical implementation

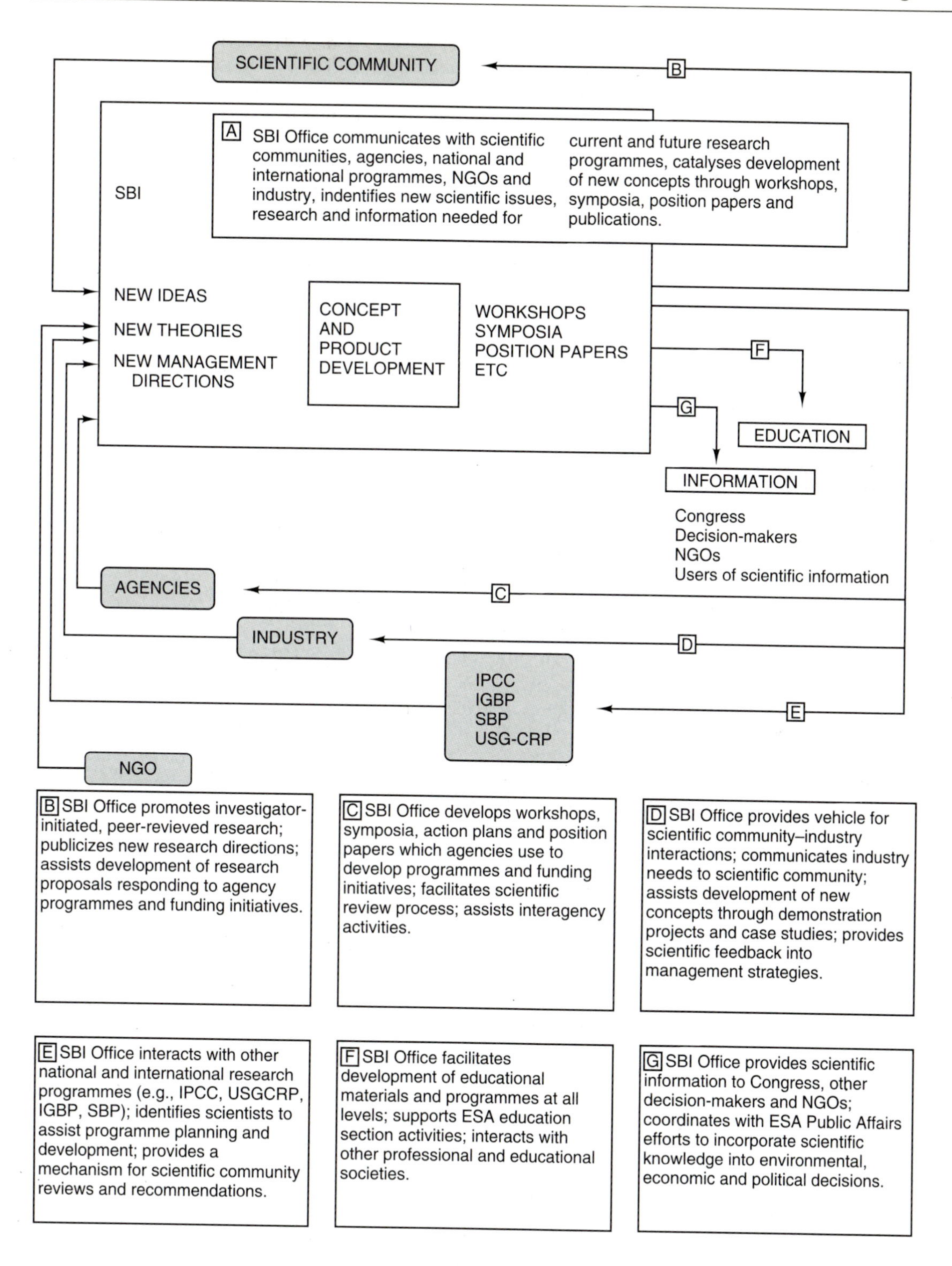

Fig. 6.2 The Sustainable Biosphere Initiative Implementation Plan (Dennis, 1992).

- potential for development
- a time scale for review
- methods for assessing success and failure
- general support from the user community

Nature conservation is an activity undertaken by people for people. As the primary custodians of the planet we have both a self-interest in protecting our own environment and a need to enhance the quality of our lives. Having the right strategy is fundamental to all conservation at global, regional, national and local levels. Achieving it requires the recognition that nature matters to everyone and that, although regulation and legislation will be needed, success depends upon the development of active partnerships between potentially conflicting interest groups. The international agreements in Rio de Janeiro on the importance of sustainability and the conservation of biodiversity offer considerable opportunities for the development of strategies that protect the needs and opportunities of future generations without compromising the requirements of today.

There are considerable hurdles to be overcome both inside and outside ecology. Within the science there is a need for a more concerted agreement on priorities, and a more realistic recognition by ecologists of economic realities. Outside there is a pressing requirement for better and more accessible information flow to decision makers and more active development of a constructive dialogue between scientists and politicians.

If politics is the art of the possible, too often conservation must seem to be the science of the impossible. This chapter illustrates that this is increasingly not so. Optimism, education and the application of science can together develop the conservation strategies we need for the future.

References

Dennis, D. (1992) Idealism to realism: applying ecological science to the Society's environmental problems. The Sustainable Biosphere Initiative Project Office creates implementation plan. *Bulletin of the Ecological Society of America*, **73**, 219–21.

English Nature (1993) *Strategy for the 1990s*. Peterborough: English Nature.

Foreign and Commonwealth Office (1992) *Protocol for the Protection of the Antarctic Environment*. London: HMSO.

GLC (1985) *Nature conservation guidelines for London*. Ecology Handbook No. 3. London: Greater London Council.

Goldsmith, F.B. and A. Warren (eds) (1993) *Conservation in progress*. Chichester: Wiley.

IUCN (1991) *A strategy for Antarctic conservation*. Gland: IUCN.

Lubchenko, J. *et al.* (1991) The Sustainable Biosphere Initiative: an ecological research agenda. *Ecology*, **72**, 371–412.

Rowell, T.A. (1991) *SSSIs: a health check*. London: Wildlife Link.

Triggs, G.D. (ed.) (1987) *The Antarctic Treaty Regime: law, environment and resources*. Cambridge: Cambridge University Press.

Walton, D.W.H. (ed.) (1987) *Antarctic science*. Cambridge: Cambridge University Press.

Warren, A. and F.B. Goldsmith (eds) (1983) *Conservation in perspective*. Chichester: Wiley.

CHAPTER 7 Conservation education

MARK N. BOULTON and DAVID KNIGHT

Introduction

> If you are thinking one year ahead, plant rice
> If you are thinking ten years ahead, plant trees
> If you are thinking one hundred years ahead, educate the people
>
> Chinese proverb

For many conservationists the major achievements of the United Nations Conference on Environment and Development (the Earth Summit) were the general acceptance of the conventions on climate change and biodiversity and the development of a wide-ranging programme of action laid out in the forty chapters of Agenda 21. Unfortunately neither international agreements nor improved legislation are, by themselves, likely to prevent the continuing deterioration of the ozone layer, the increase of greenhouse gases in the atmosphere or the steady decline in habitats and numbers of animal and plant species. There is still a widespread lack of awareness regarding the interrelated nature of human activities and the environment.

'Environmentally sound' development, the ultimate goal of the Earth Summit, depends not only on a more thorough understanding of the human species as an integral part of nature but also on an acceptance of the need for every citizen to develop a more sustainable pattern of daily living. Education is absolutely fundamental to that process. What the Earth Summit also showed was that issues relating to nature conservation are fundamentally linked to those of environment and development. Nature conservation education cannot be separated from environmental education.

Although environmental education was overshadowed in Rio by high profile issues, it is significant that the word *education* actually appears in Agenda 21 more than 600 times, second only to the word *government*. In Chapter 36 of this important document UNCED clearly identified education, public awareness and training as a vital cross-sectoral issue:

> Education is critical for promoting sustainable development and improving the capacity of the people to address environment and development issues. ... Both formal and non-formal education are indispensable to changing people's attitudes so that they have the capacity to assess and address their sustainable development concerns. It is also critical for achieving environmental and ethical awareness, values and attitudes, skills and behaviour consistent with sustainable development and for effective public participation in decision making.
> (Agenda 21, Chapter 36, UNCED, 1992)

Early human societies once lived in close relation with their surroundings; true, there were far fewer individuals, but as studies of aboriginal and isolated rural peoples today demonstrate, such societies were very conscious of their dependence on nature and natural resources (Cohen, 1977). Early civilizations in Asia maintained close and harmonious links with nature, traditions which are still clearly evident in Buddhist monasteries today. As we approach the end of the second millennium, the legacy of agricultural and industrial revolution, the inventions of modern science and constant advances in information technology have changed the face of human society and in so doing have severed the umbilical cord which once linked *Homo sapiens* to the biosphere.

The emerging environmental issues of the last few decades have fostered the need to establish environment and development literacy. There is an urgent need for every person on the planet to be aware that it is the cumulative result of the activities of well over five billion of us which threatens the long-term viability of all life on earth, and for the citizens of western nations to acknowledge their proportionately greater contribution to those threats. This is the ultimate challenge of environmental education.

A brief history of environmental education

In Britain, one of the earliest focuses for environmental education occurred in 1943 with the formation of the Council for the Promotion of Field Studies, from which grew the Field Studies Council (FSC) and a network of field centres around the country. Wheeler (1981) described the FSC thus: 'It was a major innovation in the practice of environmental education and it heralded an upsurge of educational activity related to the countryside.'

The need for environmental education grew more urgent in the 1960s in response to increasing evidence of environmental degradation. The World Wildlife Fund (now the World Wide Fund for Nature, WWF), launched in 1960 to raise the level of public concern about disappearing species, was soon followed by the publication of Rachel Carson's *Silent Spring* (1963) which drew attention to the potential dangers of excessive use of pesticides. Just a few years later a delegation from Sweden raised some of these growing concerns within the forum of the United Nations, an initiative which was to lead to one of the first great milestones in the conservation movement, the Conference on the Human Environment held in Stockholm in 1972 (United Nations, 1973).

This also proved to be a significant landmark for environmental education since the importance of education and training was recognized as essential (Conference Recommendation 19) and there was agreement on the need for member states to establish an interdisciplinary international programme of environmental education both in and out of school with the support of UNESCO (Recommendation 96). It was these recommendations which almost certainly led to the establishment of the International Environmental Education Programme, a comprehensive long-term programme developed jointly by UNESCO and UNEP.

Original worries about the loss of species and habitats soon embraced the wider concerns of pollution and population growth and these in turn led to a recognition that people themselves are an inseparable part of the total environment. It was the Belgrade Charter in 1975, arising from a ten-day UNESCO–UNEP workshop on environmental education, which pointed out the need for the establishment of harmony between humanity and the environment and called for the eradication of hunger, poverty, illiteracy, pollution and exploitation. The Belgrade Charter laid the first real foundation stone for environmental education, stating that the process should lead to 'the development of a world population that is aware of and concerned about the environment and its associated problems and which has the knowledge, skills, attitudes, motivation and commitment to work individually and collectively towards the solution of its current problems and prevention of new ones'.

Perhaps the most comprehensive, widely accepted and enduring description of environmental education was that adopted by the Intergovernmental Conference on Environmental Education held at Tbilisi in the former USSR in 1977. Building on the foundations of Stockholm and Belgrade, Tbilisi redefined and extended the importance of environmental education, putting forward a total of 41 detailed recommendations, many with a strong emphasis on the need for closer links between education and everyday life.

The Tbilisi conference felt that the objectives of environmental education should be as follows:

- to foster clear awareness of, and concern about, economic, social, political and ecological interdependence in urban and rural areas;
- to provide every person with the opportunities to acquire the knowledge, attitudes, commitment and skills needed to protect and improve the environment and;
- to create new patterns of behaviour of individuals, groups and society as a whole towards the environment.

Three years later the International Union for the Conservation of Nature and Natural Resources (IUCN) launched *The World Conservation Strategy* in partnership with UNESCO, UNEP and WWF. This important document put forward a much more planned approach to addressing environmental issues and also identified lack of awareness, especially in western societies, emphasizing the need to develop environmental education to raise public awareness on major environmental issues. It was one of the first publications to develop the concept of sustainable development (IUCN *et al.*, 1980).

The importance of education was underlined in *Our Common Future*, the report of the World Commission on Environment and Development (WCED, 1987). 'The changes in human attitudes that we call for', it declared, 'depend on a vast campaign of education, debate and public participation'. Following the launch of *Caring for the Earth*, the second version of IUCN's World Conservation Strategy (IUCN *et al.*, 1991), and the establishment of the Commission on Sustainable Development as an outcome of UNCED, the term environmental education began to be challenged by new terms such as 'development education' and 'education for ecologically sustainable development'. IUCN defines education for sustainability as 'a life long process of learning that develops human ability to participate in creating a future that is fair, just and ecologically sustainable'.

Today there are encouraging indications of this more holistic view; poverty, health, education for all, the rights of indigenous peoples, the role of women and the importance of cultural diversity are all increasingly recognized as coming within the remit of environmental education.

Why do we need environmental education?

If we are to move significantly towards UNCED's goal of achieving a sustainable international community we will need a 'global education effort to strengthen attitudes, values and actions that are environmentally sound and that support sustainable development' (Agenda 21, Chapter 35, UNCED, 1992).

- We need to increase awareness of the interdependence of environment and development.
- We need to increase the level of knowledge and skills necessary if individuals are to be able to sensitively manage the environment.
- We need to foster critical thinking and thoughtful decision making which will encourage people to recognize the frequent conflicts of interest between environment and development and help them reach sensible conclusions.
- We need to encourage a sense of personal responsibility towards the sustainable use of the earth's natural resources.

Environmental education in the formal sector

Perhaps the most obvious focus for environmental education should be the formal education sector. Around 800 million young people attend school worldwide, providing a vital opportunity for developing environmental awareness. However, even such apparently fertile ground is fraught with problems.

In the developed world a host of subjects are already competing for position in overcrowded curricula. There has long been a debate as to whether or not environmental education should be taught as a specific subject or infused across the entire curriculum, especially through links with science, geography, general studies and religious education. It is generally and perhaps correctly given a cross-curricular approach but this can sometimes lead to marginalization, with the bias on core or examination subjects. Despite this the cross-curricular approach is generally more acceptable and takes account of the premiss that if it is to be really effective, environmental education must address and affect all aspects of life.

In England environmental education is one of five cross-curricular themes officially recognized and documented (NCC, 1990). In planning for environmental education through national curriculum subjects the National Curriculum Council envisaged environmental education as having three linked components:

- education about the environment (knowledge)
- education for the environment (values, attitudes and positive action)
- education in or through the environment (a resource)

Under the National Curriculum the objectives for environmental education include gaining skills such as numeracy, communication and problem-solving skills, gaining knowledge and positive attitudes about the environment. Table 7.1 lists the latter two objectives. It is interesting that the concept of conservation is not explicitly stated here, but is obviously implicit in many of the points.

Organizing environmental education as a cross-sectoral subject is by no means easy and the end result is that environmental education teaching can be ineffective or neglected. Teachers themselves have often had little training in environment and development and feel they lack the necessary skills to present the subject with confidence.

Palmer and Neal (1994) believe it is possible for schools to positively promote environmental education even with the problems outlined above. This involves the commitment of school governors, managers and subject teachers, the creative use of the local environment, including the school grounds (Young, 1990), the appointment of an environmental

Table 7.1 Knowledge and attitudes about the environment that school pupils should gain from environmental education as a cross-curricular subject within the National Curriculum

Knowledge

- The natural processes which take place in the environment
- The impact of human activities on the environment
- Different environments, both past and present
- Environmental issues such as the greenhouse effect, acid rain, air pollution
- Local, national and international legislative controls to protect and manage the environment; how policies and decisions are made about the environment
- The environmental interdependence of individuals, groups, communities and nations – how, for example, power station emissions in Britain can affect Scandinavia
- How human lives and livelihoods are dependent on the environment
- The conflicts which can arise about environmental issues
- How the environment has been affected by past decisions and actions
- The importance of planning, design and aesthetic considerations
- The importance of effective action to protect and manage the environment

Attitudes

- Appreciation of, and care and concern for, the environment and for other living things
- Independence of thought on environmental issues
- A respect for the beliefs and opinions of others
- A respect for evidence and rational argument
- Tolerance and open-mindedness

Source: NCC (1990).

education coordinator and the use of environmental auditing techniques in an educational context (NAEE, 1992).

In developing countries the problems are much more fundamental. Access to even the most elementary education is often restricted, especially for girls, and schools lack all but the most basic facilities. Teachers themselves have limited training and minimal resources, and are usually very poorly paid; emphasis is inevitably on the three R's. In Africa it has been acknowledged that where school curricula did exist they were rigid and compartmentalized and unsuitable vehicles for environmental education (UNESCO, 1977).

Gould (1993) argues that formal education and training policies have an important role to play in enhancing the quality of the human resource base in the context of economic development. Whilst accepting the controversy surrounding this approach Gould believes that environmental education, with its focus on local problems and how they may be managed, provides a strong basis for a curriculum of general relevance. Such an approach has been adopted in Ethiopia, with many critical environmental problems to tackle (Fitzgerald, 1990).

Extra-curricular environmental education

Whilst there are indications that more attention is now being given to the incorporation of environmental education into school curricula, the evidence is that much more progress has hitherto taken place in the field of extra-curricular activities. Wildlife or Nature Clubs are now to be found in developed and developing countries alike. Watch Groups and Young Ornithologists' Clubs flourish in the United Kingdom, providing opportunities for youngsters in primary and secondary schools to carry out practical conservation activities.

Well organized research projects have involved children in carrying out studies on the environmental effects of acid rain (Thomson, 1987) or encouraged the collection of data on the ecology of species such as bumblebees (Fussell and Corbet, 1991). Older students are able to join groups like the British Trust for Conservation Volunteers and take part in active habitat management such as the restoration of ponds or the planting of trees.

The Earth Education movement, brainchild of Steve Van Matre in the USA, has played a major role in bringing young people into a much closer relationship with nature (Van Matre, 1990) and a range of activities modelled on the American programme is now popular in the UK.

Successful out-of-school activities are now well established in developing countries. The Wildlife Clubs of Kenya, established in the country's secondary schools in 1968, have become a model for similar activities in many other parts of Africa, including Tanzania, Uganda, Botswana and Malawi, and new groups are springing up elsewhere. Regular activities include the establishment of school museums, vegetable gardens, tree plots or fish pond construction whilst on a national basis Wildlife Clubs take part in rallies, competitions, workshops and visits to national parks (Withrington, 1977).

A similar initiative took place in primary schools in Zambia in 1972 with the establishment of the Chongololo Clubs. Chongololo is the local name for the large millipedes which emerge during the rainy

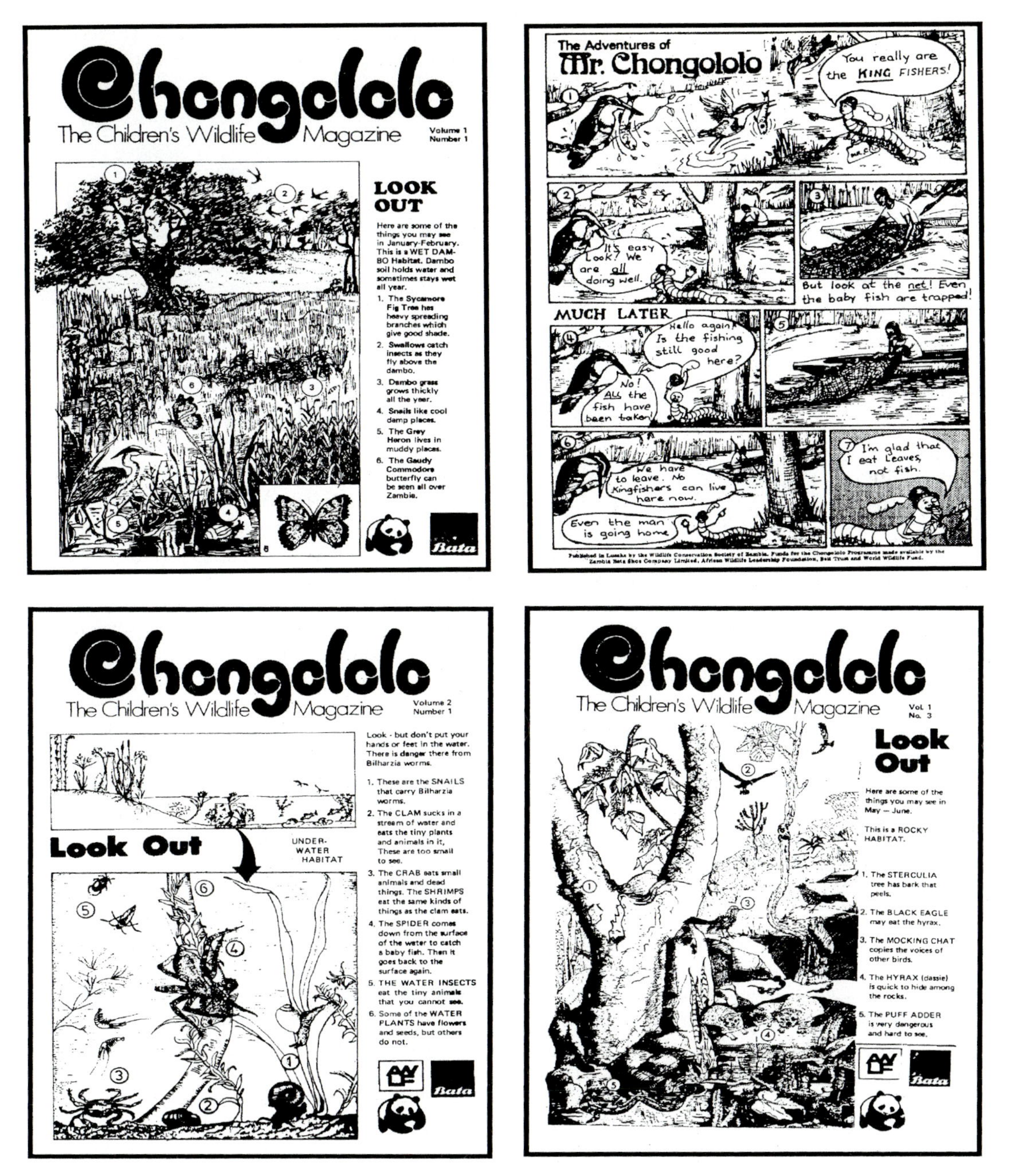

Fig. 7.1 Examples of Chongololo Club materials.

season and symbolize the clubs' concern for their local environment and their immediate surroundings. Organized by the Wildlife Conservation Society of Zambia, which received the coveted UNEP Global 500 Award for its innovative educational work, the Chongololo Clubs produce a regular magazine on environmental activities, including a special additional complementary booklet for use by teachers (Figure 7.1). These materials have proved such a valuable resource in Zambia's schools that the Ministry of Education now actively encourages and supports this work which has spread into secondary schools. A much wider audience is reached through the Chongololo Clubs of the Air, a popular weekly radio broadcast enthusiastically listened to by thousands of adults as well as children.

In India WWF has played a pioneering role in establishing Nature Clubs and running Nature Camps. Regular camping programmes have proved to be an effective way of harnessing the natural enthusiasm of young people and in inculcating an appreciation and respect for nature. Whilst such programmes, including the Eco-Clubs more recently developed by the Indian government, are to be strongly encouraged and actively supported, they must not in the longer term be regarded as substitutes for mainline curriculum-based teaching.

Organizational resources for environmental education

There are many informal environmental education programmes reaching not only children but the public at large using the resources of organizations. Zoological and botanical gardens provide an excellent opportunity for broad-based environmental education and the Botanical Gardens Secretariat as well as the International Association of Zoo Educators are constantly developing and improving their outreach programmes (see Chapter 15). Museums too have well-developed educational programmes not only related to their exhibits and displays but often through special education programmes provided for children during the school vacations.

National parks and other protected areas also have an important role to play. Many have a special information or conservation centre which forms the focus for a range of educational activities. Interactive displays, regular audiovisual shows, and printed materials help to educate the casual visitor or tourist whilst specially arranged programmes are frequently arranged for school groups. The Wildfowl and Wetlands Trust, the late Sir Peter Scott's centre for research and conservation of waterfowl, has excellent interpretative and education programmes. Well-prepared and thoroughly tested educational resource materials help teachers make optimal use of the facilities when they visit the Trust's headquarters at Slimbridge in Gloucestershire with their classes, and the Trust also offers additional materials for use before and after their visit. Much of the WWT's work, which encompasses innovative programmes for people with handicaps including the visually impaired, has recently been extended through the Wetlands Education International Programme which has been assisting the development of similar centres in Hong Kong and the United States.

Fig. 7.2 African students in the field at Lake Naivasha.

Forest areas too have become the focus for the development of new Nature Centres such as the Seloliman Centre in Eastern Java which brings together students, educators, farmers, government, business, women's groups and the general public to learn about the environment. Their courses have included seminars to promote the integration of environmental education into school curricula and workshops for local villagers to learn skills and methods of environmentally friendly living. In Kenya, part of the home of the late George and Joy Adamson of *Born Free* fame, on the shores of Lake Naivasha in the Rift Valley, has been developed into the Elsa Field Studies Centre (Figure 7.2). The residential programme conducts courses for a wide range of target groups from school children to game wardens and from school teachers to government officials and is supported by the Elsa Conservation Trust.

Specialist training in communicating conservation also takes place at the UK-based International Centre for Conservation Education in the Cotswolds. Over 300 people from more than 50 developing countries, many themselves trainers, have attended specialized courses carefully tailored to their particular needs. Strategic planning, tutorials and seminars, study visits and practical communication techniques – audiovisual production, graphic design and role play – help to strengthen existing skills and to develop new ones.

Environmental education and the arts

Some public awareness programmes have chosen to use the arts including music and drama, street theatre and puppetry as innovative and effective means of communicating conservation. *Yanomamo* is a WWF musical promoting conservation of tropical rain-forests and has been successfully performed by children in many parts of the world; *Help Save the Planet*, presented by young people from Wisconsin, USA, was given a gala performance before the United Nations in New York and to the Eco-Ed Conference in Canada in 1992.

Street theatre has been used by the Centre for Environment Education (CEE) in India to involve local communities in graphically portraying their environmental problems, and groups in rural parts of Zambia have used the same technique to put over such problems as poaching and deforestation. In many parts of West Africa, puppets or masks have been used to promote the message of sustainable development. Role play and environmental games provide ways in which learning about conservation can be not only extremely effective but also great fun!

The role of the media

There is no doubt that the media have played a major role in raising public awareness of global conservation issues in recent years. Much of the reporting in newspapers and magazines as well on radio and television, however, has tended to be somewhat depressing and rather negative. With the advent of satellite TV, the general public in an increasing number of countries has been deluged with programmes on acid rain, holes in the ozone layer, forest destruction and the dangers facing a host of species from elephants to rhinos, and from tigers to whales.

Burgess (1992) considers one effect of these messages is for ordinary people to feel shut out and cut off from the pleasures of contact with nature as they interpret the damage caused by humans to plants and animals to mean that people and nature are mutually exclusive. However, Burgess also believes that the traditional wildlife documentary film shot and edited in such a way as to be 'brimming with animals' will also be increasingly challenged by viewers, as awareness of habitat destruction locally and around the world increases.

Traditionally, film companies have spent long periods filming natural history programmes in national parks and wildlife reserves in Africa and elsewhere, yet frequently neglected to provide copies of the finished film for use in the host country. Fortunately a number of television producers have begun to make available supporting resource materials which not only provide additional information on the programme content but also suggest practical ways in which the viewer may be able to help.

A particularly encouraging and successful initiative has been the establishment in 1984 of the London based Television Trust for the Environment which has not only commissioned an extensive range of television programmes but is increasingly making them available to TV networks in developing countries.

But it is not just the professional producers who have been communicating conservation through film and video. The new generation of compact lightweight video cameras are being used by enthusiastic amateurs and even groups of school children and village communities to record situations, stories and strategies for addressing local environmental problems. Played back on battery-operated (often solar-powered) portable TV screens, such programmes have often captured the initial interest of local people and engendered the enthusiasm necessary for new projects to be accepted by local communities.

Working with local communities

Many rural communities are still directly dependent on their immediate environment for their daily needs including land on which to grow their crops, timber for building construction and fuel and water for drinking, washing and irrigation. In areas of growing populations, demand frequently exceeds supply and there is progressive environmental deterioration. What is more, the natural resources of such regions – land, forests and wildlife – are frequently subject to commercial exploitation with little if any benefit accruing to local people who are frequently ousted from, or given restricted access to, areas which traditionally contributed to their well-being. The total exclusion of local people from national parks or game reserves or the prohibition of local hunting in game management areas earmarked for international hunters are two widespread examples.

It is only in recent years that serious attention has been given to what have become known as

Community Conservation Programmes. Perhaps the best known is the Camp Fire programme in Zimbabwe. Camp Fire is the acronym for Communal Areas Management Programme for Indigenous Resources and represents an innovative approach to rural development. Camp Fire promotes natural resource utilization, including wildlife, as an economic and sustainable land-use option in the interests of both the conservation of natural resources and the relief of human poverty. It seeks to establish and strengthen institutions at village level, encouraging rural communities to use their natural resources on a sustainable basis and to manage the revenues derived from their activities for the benefit of their communities. Whilst such activities do encompass education and public awareness, their most important role is that of helping to reunite communities with their cultures and traditions and to develop programmes of sustainable rural development.

Similar initiatives have been taken in IUCN's Mount Elgon Conservation and Development Project in Uganda. Breakdown of law and order in the late 1970s resulted in progressive encroachment of the mountain's tropical montane forest reserve by farmers and small-scale commercial loggers, resulting in the destruction of around one-third of the 118 000 hectare reserve.

The problem was addressed initially by means of an armed environmental task force which, like traditional anti-poaching units, physically moved people out of the reserve. To counterbalance this rather heavy-handed approach a rural Extension and Education Unit (EEU) began to take on a more positive role and to actually involve local communities living near the forest. A combination of posters, video and other visual aids, often in comic strip form, together with a mobile Parish Roadshow which often attracted audiences of 2000 people, were used to demonstrate the importance of the forest, especially in terms of the protection it provided to the local water supply. Courses on fuelwood conservation and developing tree nurseries were provided for local women's groups and workshops arranged for school teachers.

An in-forest exploration centre is planned which will help school children to understand the ecological processes at work in the forest. Through such approaches the project hopes to ensure the conservation of biological diversity, to safeguard the quality and quantity of local water supplies and to encourage sustainable use of the land surrounding the forest.

The ethical dimension

In 1986 the World Wide Fund for Nature convened a major international conference in Assisi, Italy, home of St Francis, who is often referred to as the Patron Saint of Conservation. Leaders of most of the world's major religions met for three days to consider their roles and responsibilities in promoting the message of conservation and the wise use of natural resources to their respective communities (WWF, 1986).

Since then religious communities in many parts of the world have given greater prominence to the stewardship of natural resources and greater attention has been given worldwide to the ethical dimension of conservation. The Social Responsibility Division of the Methodist Church in the United Kingdom has developed a Trees for Life project in West Africa, a programme in which an experienced agro-forester works with local church communities to establish tree nurseries and tree-planting schemes and encourage improved horticultural and agricultural practices. It also supports a permaculture project for local communities in Nepal.

Conservation on wheels

The concept of mobile information units is not new. For many years vehicles equipped with film projection equipment have visited outlying towns and villages in many developing countries. Specially fitted four-wheel drive vehicles operated by the Ministry of Information, Health or Agriculture have used film shows to increase public awareness about important issues.

An interesting development of this theme is demonstrated by the mini-mobile units (Figure 7.3) developed by WWF's Special Project for Conservation Education in Developing Countries – now ICCE. Based on the economical Renault Fourgonette van, units equipped with a small generator, film and slide projectors, roof mounted screen and public address system were used in a number of African conservation education programmes, notably in Rwanda and Zambia (Cowan and Stapp, 1982). Operated by volunteer educators from the British Voluntary Service Overseas or the American Peace Corps, these vehicles travelled thousands of kilometres disseminating information through films,

Fig. 7.3 ICCE mobile unit.

slide-tape and live lectures. Working with local African counterparts, the Rwandan unit prepared a range of special educational programmes in Kinyrwandan about the threatened Mountain Gorillas which were often shown to audiences of more than 2000 people. The Zambian mobile unit, operated by the Wildlife Conservation Society, carried out a series of training programmes throughout the country involving local government officials, community leaders and school teachers.

Communication resources for environmental education

Most organizations with an interest in environmental education are involved with the development, use and distribution of a wide range of resource materials. Traditionally emphasis was placed on printed materials such as books, pamphlets and posters to promote the work of the organization itself or to campaign about specific issues such as poaching or deforestation. Some were also prepared for use in the formal sector.

In the early 1970s, a major United Nations project in Zambia, the Luangwa Valley Conservation and Development Project, cooperated with the Ministry of Education's Curriculum Development Centre to prepare and publish two primary school books addressing many of the country's conservation problems. Since they were also designated approved English reading texts, they were actually distributed as official text books and more than 200 000 were distributed throughout Zambian schools. This proved to be a particularly cost-effective way of reaching large numbers of Zambian primary children.

Some interesting programmes have also been carried out in the field of adult literacy where materials developed to raise the educational literacy of adults were based on conservation issues of relevance to the communities concerned. Adekoya (1991) describes how oil companies in Nigeria are including environmental protection and conservation in the curricula of community development and adult literacy programmes they are administering.

With the development of information technology and the increasing availability of computers, photocopiers, copyprinters and small offset-printing machines, many organizations are now developing their own in-house publishing facilities. In the most basic form this may be little more than simple hand-generated text and artwork and a hand-operated office duplicator. Such basic technology is being used to generate simple educational leaflets for use in the local community. Modern personal photocopiers require minimal maintenance and can often be used very successfully to prepare printed materials in remote field conditions. The excellent educational publication *Walia*, published by the IUCN project in Mali, was originally produced on a small personal photocopier until eventually superseded by a small offset-printing press which was itself operated by staff at the project site (Du Roy, 1991).

At the other end of the scale, the Centre for Environmental Education (CEE) in Ahmedabad, India, runs a full-scale publishing service within the organization and its team of artists, designers and printers are able to produce an impressive range of full-colour posters, books and displays.

Printed posters are of course widely available covering a great variety of topics and carrying a host of different messages. Whilst many are attractive and colourful, far too little emphasis has been placed on how effective they are in actually getting the message across. A particularly valuable contribution to improving communication techniques was made by the Swedish researcher Andreas Fuglesang (1973). He investigated the relative merits of using different styles of visual presentation – photographs, photo cut-outs, drawings, and silhouettes – and demonstrated that the choice of medium and the simplicity of the illustrations were important considerations in determining the effectiveness of his health education programmes.

Today there is much emphasis on multimedia education and communication. Reference has already been made to the use of film and video material by the media and to mobile conservation units. Increasingly sophisticated electronic media, notably CD-ROM technology, is enabling large amounts of information about the environment – text, maps, graphics, photographs and even clips of moving video – to be incorporated on a single disk which can be accessed by computer. Such information can be presented interactively, providing the user with a much greater level of interest.

National environmental education strategies

One of the major difficulties limiting the effectiveness of environmental education today is the frequent lack of any clearly defined mechanism for national coordination. Whilst most countries have a range of important individual initiatives which are often very effective, they rarely form part of a strategic plan. This often results in duplication of effort, lack of adequate resources and the neglect of important target groups.

There has been considerable progress over the past decade or so in the preparation of National Conservation Strategies and National Environmental Action Plans. However, there are relatively few examples of National Environmental Education Strategies. Amongst the most notable exceptions in this field are those of the State of Victoria in Australia and of Scotland in the United Kingdom.

Victoria's 'Learning to Care for Our Environment' programme was launched as a joint venture involving the state government and NGOs (Victorian Environmental Education Council, 1992). It involves an eight-year plan which sets out to enable all Victorians to develop the awareness, understanding, skills and attitudes necessary for meeting Victoria's environmental needs both now and in the future, consistent with social, economic and other interrelated needs. The strategy clearly defines a series of aims and specific actions which are fundamental to achieving the purpose of the strategy. Whilst emphasizing that everyone has a part to play at the personal level, the strategy goes on to enumerate the potential roles of particular sectors of society and suggests specific focuses for action.

Learning for Life: National Strategy for Environmental Education Inside Scotland (Scottish Office, 1993) was the outcome of a two-year working party set up by the Secretary of State as a response to Agenda 21 which calls upon each nation to develop a national environmental education strategy. The Scottish report was built on two main premisses: that education aimed at improving personal behaviour towards the environment is essential for social, cultural, economic and industrial survival; and that environmental education is a sustained learning experience important to everyone at every stage and in every aspect of life. The report reviews the development of environmental education in Scotland and illustrates a wide variety of approaches, projects, techniques and initiatives which are being made in almost all sectors of society. From these diverse activities, *Learning for Life* formulates the basis of a national strategy identifying key needs and priorities and proposing seven crucial areas for action: policy, curriculum, training, research, information, networking and review. It concludes by enumerating a series of clear recommendations to the Secretary of State and appropriate government departments.

In order to provide practical assistance to those involved in planning, producing or strengthening of national policies and strategies for education, the Commission on Education and Communication of IUCN published a series of guidelines entitled *Education for Sustainability* (IUCN, 1993). It is hoped that this will assist many other nations in creating national environmental education strategies which suit their own particular circumstances.

Conclusion

The conservation of the biological resources of the planet can only come about with an informed and educated world citizenry who are able to place biological conservation into a social, political and economic context at global, national and local levels. Individuals must be able to relate their own requirements for a fulfilling life with the constraints placed upon the expression of that lifestyle if truly environmentally sustainable development is to be achieved.

Conservation education embedded within environmental education at primary, secondary and tertiary levels of formal education and in extra-curricular activities for children and adult education programmes in both First and Third World countries offers perhaps our only hope for the future.

References

Adekoya, A. (1991) Environmental education in adult literacy programmes in Nigeria. *Adult Education and Development* **37**, 69–75.

Burgess, J. (1992) The nature of the box. *BBC Wildlife* **10**(10); 52–3

Carson, R. (1963) *Silent spring*. London: Hamilton.

Cohen, M.N. (1977) *The food crisis in prehistory: Overpopulation and the origins of agriculture*. New Haven: Yale University Press.

Cowan, M.E. and W.B. Stapp (1982) *Environmental education in action. V: International case studies in environmental education*. Columbus, Ohio: ERIC Clearinghouse for Science, Mathematics and Environmental Education.

Du Roy, N. (1991) *Walia: the approach, practical guidelines/ Schools Environmental Education Project, Walia*. [English translation from French by C. Norton] Mopti, Mali: Walia; Gland, Switzerland: IUCN Sahel Programme.

Fitzgerald, M. (1990) Education for sustainable development. Decision making for environmental education in Ethiopia. *International Journal for Educational Development*, **10**, 289–302.

Fuglesang, A. (1973) *Applied communication for developing countries*. Uppsala, Sweden: Dag Hammarskjold Foundation.

Fussell, M. and S.A. Corbet (1991) Bumblebee habitat requirements: a public survey. *Acta Horticulturae* **288**, 159–63.

Gould, W.T.S. (1993) *People and education in the Third World*. London: Longman.

IUCN (1993) *Education for sustainability*. Discussion document prepared by the IUCN Commission on Education and Communication. Gland: IUCN.

IUCN, UNEP, WWF (1980) *The World Conservation Strategy. Living resource conservation for sustainable development*. Gland: IUCN.

IUCN, UNEP, WWF (1991) *Caring for the earth. A strategy for sustainable living*. Gland: IUCN.

NAEE (National Association for Environmental Education) (1992) *Towards a school policy for environmental education: An environmental audit*. Walsall: NAEE.

NCC (National Curriculum Council) 1990 *Curriculum guidance 7: Environmental education*. York: NCC.

Palmer, J and P Neal (1994) *The handbook of environmental education*. London: Routledge.

Scottish Office (1993) *Learning for life: National strategy for environmental education inside Scotland*. Edinburgh: Scottish Office.

Thomson, C.H. (1987) The acid drop project: pollution monitoring by young people. *Biological Journal of the Linnean Society*, **32**, 127–35

UNCED (1992) *The global partnership for environment and development: A guide to Agenda 21*. Geneva: UNCED.

UNESCO (1977) *Regional Meeting of Experts on Environmental Education in Africa. Brazzaville, People's Republic of the Congo, 11–16 September 1976*. Paris: UNESCO.

United Nations (1973) *Report of the United Nations Conference on the Human Environment, Stockholm, 5–16 June 1972*. New York: United Nations.

Van Matre, S. (1990) *Earth education: A new beginning*. Greenville, WV, USA: Institute for Earth Education.

Victorian Environmental Education Council (1992) *Learning to care for our environment: Victoria's environmental education strategy*. Melbourne: Victorian Environmental Education Council.

WCED (World Commission on Environment and Development) (1987) *Our common future*. Oxford and New York: Oxford University Press.

Wheeler, K.S. (1981) A brief history of environmental education in the UK. In Department of Education and Science, *Environmental education. A review*. London: HMSO, pp. 22–3.

Withrington, D.K.J. 1977 Environmental education programmes for out-of-school youth. In UNESCO, *Trends in environmental education*. Paris: UNESCO.

WWF (World Wide Fund for Nature) (1986) *The Assisi Declarations. Messages on man and nature from Buddhism, Christianity, Hinduism, Islam and Judaism*. London: WWF.

Young, K. (1990) *Learning through landscapes: Using school grounds as an educational resource*. Winchester: Learning Through Landscapes Trust.

CHAPTER 8

Biological conservation and site classification; the importance of abiotic variables

D. GRAHAM PYATT

Introduction

Conservation in practice involves politics, economics, legislation, policy and education. Also, problems in biological conservation cannot be properly confronted without a prior consideration of the abiotic environment, because this sets limits for the level and kind of biodiversity present. Mapping of biological communities and habitats usually requires a climatic and soil map if 'ecological regions' are to be established.

This chapter is limited to terrestrial ecosystems where the abiotic factors can be summarized as climatic and edaphic. Indeed, an ecosystem may be defined as a combination of the climate, the soil and the plant and animal community (but note comments in Chapter 13). The climatic and soil factors taken together define the site. The ecological quality of the site sets limits on the potential composition of the plant and animal community, although, for a variety of reasons, many different communities can exist on a given site type. One of these factors is time itself, as it can take tens or even hundreds of years for a community to mature. On many site types there is a succession of different communities leading towards a climax community. If the soil has no obvious limiting factors, i.e. it is not very dry or very wet or very infertile (a so-called zonal site), then the development of a climatic climax community is possible, given sufficient time and freedom from natural catastrophe or human intervention. Of course it is not possible completely to separate the influences of climate, soil, vegetation and animals, as there are various feedback processes. The soil has its own flora and fauna, the activities of which enable the soil to function as the medium for plant growth. These processes are in turn influenced by the supply of organic matter, mainly as leaf-fall and dead roots. The term 'site diagnosis' is used for the process of determining the ecological quality of the site in terms of climate and soil conditions.

Ecosystems defined in this way can occupy very small areas, typically from tens of square metres to several hectares. It is obvious that such small ecosystems cannot be considered to exist entirely independently of each other. There are transfers between ecosystems of many kinds, including soil water and its dissolved nutrients, organic matter and its nutrients through the feeding and excreting of animals and the fall of leaf litter.

Some species will be restricted to individual ecosystems because *in competition with other species* they can only survive in a narrow range of site conditions. These species are said to have narrow ecological amplitude. Looked at in another way, the presence of one of these species may be a reliable indication of specific site conditions. Other species have a wider ecological amplitude and may be found on a number of adjacent or separate ecosystems within the wider landscape. In practical conservation it will often be necessary to deal in multiple, interdependent ecosystems (see specific examples in Chapter 16). This is merely a demonstration of the importance of scale. From the whole planet at one extreme to the individual plant or animal or even the individual gene at the other extreme, there is a continuous series of scales of increasing detail.

World climate and vegetation

The relationships between the major vegetation zones of the world and climate have been described by many authors since the late nineteenth century, of whom Köppen is probably the best known (Griffiths, 1966; Trewartha and Horn, 1980). These workers used

Box 8.1 The Köppen system

A: Tropical rainy climates
Temperature of coolest month >18°C. With lower temperatures than this sensitive tropical plants do not thrive.

Af: Tropical wet climate: rainfall of the driest month > 60 mm. This is the typical hot wet climate of the rainforest.
Aw: Tropical wet-and-dry climate: at least one month with rainfall <60 mm. This characterizes the hot savanna zone.
Am: Tropical monsoon climate: intermediate between Af and Aw, there is a short dry season but the soil remains moist enough to support rainforest.

B: Dry climates
There is an excess of potential evaporation over precipitation.

BW: Arid or desert climate.
BS: Semiarid climate or steppe.

C: Mesothermal climates
Mild temperate rainy climates, with temperature of coldest month between 18°C and −3°C, that of warmest month >10°C.

Cf: No distinct dry season: driest month >30 mm. Temperate forest, broadleaves or conifers.
Cw: Winter dry: wettest month (in summer) is more than 10 times wetter than driest month. Monsoon and upland savanna.
Cs: Summer dry: wettest month (in winter) is more than 3 times wetter than driest month. Driest month has <30 mm precipitation. The Mediterranean type of climate. Sclerophyllous shrub woodland, including evergreen broadleaves and some conifers.

Mesothermal climates are further subdivided:

a: Temperature of warmest month >22°C. Warm mesothermal.
b: Temperature of warmest month <22°C. Cool mesothermal.
c: Less than 4 months warmer than 10°C. Very cool mesothermal, reaches the polar limit of the forest in an oceanic climate.

D: Microthermal climates
Mean temperature of the coldest month below −3°C, that of the warmest month >10°C. The summer temperatures are supposed to coincide with the poleward limit of forest. Microthermal climates have frozen ground (permafrost) and snow cover of several months' duration. They roughly coincide with the northern hemisphere boreal zone of predominantly pine–spruce–birch forest.

Df: With humid winters.
Dw: With dry winters.

E: Polar climates
Mean temperature of the warmest month <10°C.

ET: Tundra climate: mean temperature of warmest month between 0°C and 10°C.
EF: Perpetual frost: mean temperature of all months <0°C.

vegetation patterns to decide which critical or threshold values of climatic variables to choose to draw up a climatic classification. This can be illustrated by considering the definitions of the main climatic classes of the Köppen system (Box 8.1).

The following small selection of examples from Europe and Canada illustrates how different classifications of terrestrial ecosystems deal with the relationships between climate, soil and biotic communities, and how a system might be constructed for the particular requirements of one region, Britain.

Scandinavia

Most of Scandinavia falls within Köppen's microthermal zone, except for the mesothermal southern parts of Sweden and Norway (Cfb) and the extreme coastal strip of Norway (Cfc). Forests and mires are the two main groups of ecosystems below the tree line.

The Finnish classification of forest types was developed by Cajander (1926, 1949) and has changed little since. In this system the composition of the ground vegetation is considered to express site quality. The relationship between site quality and the tree species composition is less clear and frequently affected by forest management. The forest types are thus effectively site types and can be identified irrespective of the condition or age of the forest stand. The site types were conceived as lying on a unidimensional gradient of site fertility, from poor-and-dry to moist-and-rich.

Later workers in Finland have shown that climatic differences between regions affect the composition of the ground flora. Different regions have similar sets of forest types but with slightly different combinations of indicator species. Also it has been shown that a relationship between tree species and site types needs to be recognized. Thus, spruce (*Picea abies*) ecosystems generally have higher nutrient levels than pine (*Pinus sylvestris*) ecosystems. The shade cast by the forest canopy at its different stages of growth affects the ground vegetation, and spruce normally casts greater shade than pine or birch (*Betula* sp.). Kuusipalo (1985) has shown that the Cajander system is effective in predicting the nutrient regime of the humus layer. He also provides a useful review of the historical development of forest site classification in Finland.

Mires can be classified into various kinds of fens, raised bogs, blanket bogs, aapamires and so on (Moore and Bellamy, 1974). The distribution of these main groups, in Finland and indeed in the whole of Europe, appears to be related to the warmth or 'continentality' of the climate. In Finland and Sweden, growing season warmth measured as the accumulated temperature in degree-days above a threshold of 5°C has been recognized as an important climatic variable limiting forest yield. In Scandinavia in general this increases as latitude decreases.

In the transition between forest and treeless mire there are many examples of wooded mire or scrub mire. Many mires show a concentric zonation from a treeless centre through a zone of scrub to a tall wooded fringe as the peat thins and becomes less wet. For forestry purposes in Finland there are three main groups of mire: treeless mires, pine swamps and spruce swamps. The treeless mires are so because they are excessively wet, and can be divided into relatively fertile fens and infertile bogs. Pine swamps are generally less fertile than spruce swamps. Fertility is measured by a combination of acidity and nitrogen availability and expressed in the species composition of the vegetation. Some 28 types of peatland are recognized in this way and arranged on a simple 'ecogram' or grid with axes of wetness and nutrient status. Neither axis is quantified or divided into classes.

Central Europe

Apart from the alpine zone, this part of Europe was almost completely forested before man began extensive settlement. Only saltmarshes, windswept coastal dunes, the wettest and most nutrient-poor mires, bare rocks and avalanche tracks would have been naturally treeless. In particular the extensive grasslands and heaths are man-made. Western and Central Europe fall within Köppen's mesothermal zone, mostly Cfb, with Csa around the Mediterranean Sea.

Ellenberg (1988) has used a grid with axes of annual rainfall and mean annual air temperature to illustrate different climate types within the Cfb subzone. He also used an ecogram, with axes of soil moisture regime and soil acidity, to illustrate the ecological distribution of different types of woodland. In the submontane belt in a temperate sub-oceanic climate, beech (*Fagus sylvatica*) is the dominant species in climatic climax forests, i.e. on zonal soils, with oaks (*Quercus* spp.) and other species being dominant only on wetter, drier or the most acid soils. A third climatic variable, oceanicity (or its converse continentality), is illustrated for Europe as a whole using the distribution of vascular plant species.

An important feature of Ellenberg's (1988) approach to vegetation ecology is the allocation of indicator values on a scale of 1 to 9 to all plant species. These indicate the ecological amplitudes of the species when growing in competition with others. The latest version (Ellenberg *et al.*, 1992) lists indicator values for light, temperature, continentality, moisture, acidity and nitrogen for over 2000 species. Indicator values are omitted for plants which show no clear preference for a particular factor. The indicator values can be used in reverse, to calculate approximate site values for moisture, acidity, etc., by applying the respective values to the species present on the site. Weighting by frequency of occurrence or cover fraction (abundance) is advisable. Groups of plants with similar ecological amplitudes characterize particular parts of the ecogram and short-lists are provided, named after a characteristic member. In the Ellenberg (1988) classification the use of numerical indicator values for each ecological factor is given more prominence than is the quantification of the axes of the ecogram.

France

A compendious ecological flora for the forests of France covering more than 2000 vascular plants and bryophytes has recently been produced by Rameau *et al.* (1989, 1993). France is divided into three geographic/climatic/ecological regions (plains and collines, mountains, Mediterranean) and there are

Table 8.1 Forest zones in the plains and mountain regions of France

Zone	Mean annual temperature (°C)	Length of growing season (days)	Elevation range (m), in NW Alps
Permanent snow	<0	0	>2500
Alpine	0–2	<100	2000–2500
Subalpine	2–4	100–200	1300–2000
Montane	4–8	>200	600–1300
Colline (hills)	8–12	>200	<600
Plains	>12	>200	not stated

Source: Adapted from Rameau *et al.* (1993).

separate publications for each. The descriptions of these regions are incomplete but a zonation with elevation is given in quantitative terms as Table 8.1.

The ecological amplitude of each species is illustrated on a grid very similar to that of the ecogram of Ellenberg, but numerical indicator values are not provided. The ecogram has eight classes along the soil moisture gradient and six classes along the soil acidity gradient (referred to as the trophic gradient), although these are not defined quantitatively. Among supporting information supplied for each species are the geographical distribution and the ecological amplitudes for climate, light and humus form. Humus forms are directly related to the trophic gradient. Thus mor humus is found only on very acid soil, moder humus on very acid to acid soil, and various forms of mull humus on rather acid through to calcareous soil. Indicator groups of species are described as occupying particular, often somewhat overlapping, portions of the ecogram. In addition, two groups of species indicative of high availability of soil nitrogen are given. Each species is allocated to one or more of these groups, and short-lists of common indicator species are provided. Sites will normally carry species from several adjacent indicator groups, hence advice is given on how to judge the position of the site on the ecogram from the mix of groups present.

Belgium

In his classification of forests and forest sites, Noirfalise (1984) subdivides the elevation range of Belgium into four zones and uses the temperature of 10°C as the threshold above which to sum the day-degrees of warmth (Table 8.2). He also evaluates the water

Table 8.2 Forest zones of Belgium

Zone	Elevation (m)	No. of days >10°C	Day-degrees >10°C
Low Belgium	0–100	155–160	2400–2500
Middle Belgium	100–300	150–155	2200–2300
Low and middle Ardenne	300–500	146–149	1900–2000
High Ardenne plateaux	500–700	140–145	1800–1900

Source: Adapted from Noirfalise (1984).

balance of forest sites in terms of the balance between rainfall and potential evaporation given by the cumulative soil water deficit. Although he does not use an ecogram to represent variation in soil quality, Noirfalise (1984) recognizes seven indicator groups of plants with respect to soil water regime and five indicator groups with respect to soil nutrient regime ('indifferent' species are placed in a sixth group). As in Rameau *et al.* (1989, 1993) most of the nutrient regimes are concerned with soil acidity, but there is also a class of nitrate-loving plants.

In the *Ecological File of Tree Species* (Ministère de la Région Wallonne, 1991a,b) five climatic variables are considered of importance for the choice of tree species for the region of Wallonia in Belgium. These are the mean annual temperature, mean growing season temperature, annual precipitation, growing season precipitation, and length of growing season. Soil quality is presented as an ecogram with four classes of soil moisture regime and four classes of soil nutrient regime. The nutrient classes are intended to be thought of as general trophic regimes, although they are defined quantitatively only by pH ranges. Some classes are then subdivided and others joined to form 17 site types or ecological groups as rectangular portions of the grid. The ecological groups are identified by lists of indicator plant species. This set of site types is used to delineate suitable sites for each of 29 tree species.

Canada

Most of the Canadian provinces have developed site classifications for forests (by far their most important land use). From the 1920s onwards Canadians followed the Cajander methodology (page 81), finding many similarities in the boreal forests of

Canada and Finland. In the 1960s in Ontario increased attention was given to soil properties and landform. At the same time in British Columbia the phytosociological appraisal of sites was broadened to take account of climate, topography and soil (Krajina, 1965). More recently the other provinces have moved in the same direction, under the coordination of a federal forest site classification working group. Progress has been fastest in British Columbia under the continued guidance of Krajina, and more recently Klinka, of the University of British Columbia. The biogeoclimatic ecosystem classification of British Columbia will be used as the example of Canadian methods.

British Columbia

British Columbia is a province of dramatic relief and a wide range of climate. The main part of the offshore islands and western slopes of the Coast Mountains fall into Köppen's cool mesothermal Cfb type, with the very cool Cfc type along the northern coast. In the extreme south, the climate around Vancouver is of the Mediterranean Csb type. To the east, the interior plateau and across to the Rocky Mountains form the western part of the vast Canadian boreal Df zone. Most of the country is covered in coniferous forest, that of the Cf subzones often being referred to as temperate rainforest.

The biogeoclimatic ecosystem classification (BEC) is three-dimensional, with climate, soil moisture regime and soil nutrient regime as the major variables (Pojar *et al.*, 1987). All three are recognized as synthetic factors, requiring classification and quantification. The climatic classification consists of a hierarchy of zones, subzones and further subdivisions. Instead of being formed by subdividing the range of selected climatic variables into a number of classes, the zones are derived from phytosociological studies and mapping of forest vegetation types, the climatic data being adduced later. Oceanicity/continentality is considered to be one of the most important variables, at least in the coastal region (Klinka *et al.*, 1989). Effective mapping depends on the recognition of zonal sites and climatic climax forest communities thereon. The importance of climatic gradients is readily apparent in influencing the composition of the forest on zonal sites.

Soil moisture regime (SMR) and soil nutrient regime (SNR) form the two axes of an ecogram known as the edatopic grid. There are eight classes of SMR and five classes of SNR. The classes of SMR are defined semi-quantitatively in terms of the ability of the soil to supply soil moisture during the summer. In the soil nutrient regimes of BEC more emphasis is given to nitrogen availability than to soil acidity, and humus form is considered to be a useful predictor of SNR.

The use of indicator plants in site diagnosis is a key feature of BEC. Indicator plants can be used to predict SMR and SNR and, to a limited degree, climatic zone. Numerical indicator values have not been assigned to individual species; instead, indicator species groups representative of different portions of the grid have been listed.

Phytosociology and the use of indicator species are important in BEC but are supported by the attention given to soil nitrogen, humus form and moisture supply. Relatively little direct attention is given to lithology, soil parent material or soil type, although it is of course recognized that variations in these are the major cause of the variations in soil quality expressed as SMR and SNR.

The BEC system has been in use by the Provincial Forest Service since 1975 and mapping of zones, subzones and lower subdivisions is well advanced. The classification has been accepted as highly relevant to a wide range of resource management matters.

Britain

Britain is a small country with a fairly narrow climatic range within Köppen's cool mesothermal Cfb and Cfc, but with a wide range of soils due to the variety of lithologies provided by the almost unbroken geological series from the Precambrian era to the present. Although the vast majority of the natural forest has been removed through agricultural development, the result is a diversity of arable, 'improved' pasture, unimproved or 'rough' grassland, heathland, moorland, bog, recent plantations and remnants of semi-natural forest. The conservation importance of some of the man-made or maintained vegetation, particularly the calcicolous grasslands, heathlands, bogs and some of the moorlands, is as high as that of the semi-natural communities.

A number of attempts have been made to produce bioclimatic maps and classifications; e.g. the soils of Britain have been completely mapped at a scale of 1:250 000. This has been a mapping of soil types and parent materials (soil series and associations) with an agricultural bias, but it provides a good picture of a

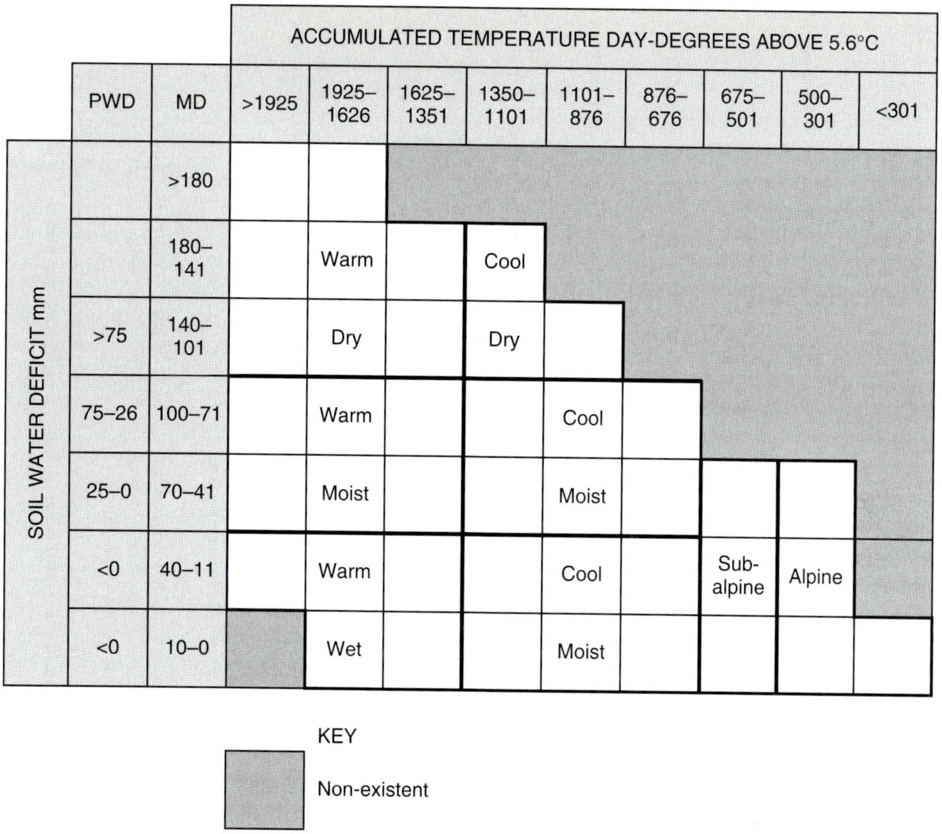

SOIL WATER DEFICIT mm		ACCUMULATED TEMPERATURE DAY-DEGREES ABOVE 5.6°C								
PWD	MD	>1925	1925–1626	1625–1351	1350–1101	1101–876	876–676	675–501	500–301	<301
	>180									
	180–141		Warm		Cool					
>75	140–101		Dry		Dry					
75–26	100–71		Warm			Cool				
25–0	70–41		Moist			Moist				
<0	40–11		Warm			Cool		Sub-alpine	Alpine	
<0	10–0		Wet			Moist				

KEY

Non-existent

Fig. 8.1 Climatic zones in Britain. PWD = potential soil water deficit (Birse, 1971), MD = moisture deficit (Bendelow and Hartnup, 1980).

wide range of soil properties. Since early this century there have been many attempts to classify and map the vegetation of parts of Britain and a few attempts either to classify or to map the whole country. Notable classifications include those of Tansley (1939) for the whole country, Poore and McVean (1957) for Scottish mountain vegetation, and McVean and Ratcliffe (1962) for the Scottish Highlands. There have been numerous ecological studies of particular types of vegetation, such as calcicolous grasslands, heaths, fens and saltmarshes, and of the range of vegetation occurring in particular areas of the country. However, it was not until the appearance of the National Vegetation Classification (NVC), being published in five volumes (Rodwell, 1991a,b 1992), that a comprehensive account of British plant communities has become available.

A climate grid can be formed from axes of accumulated temperature (day-degrees above 5.6°C) and potential soil water deficit (Birse, 1971) or moisture deficit (Bendelow and Hartnup, 1980). Eight zones (BEC usage) are defined in Figure 8.1. Subzones can be formed by finer divisions along the same axes. Oceanicity can be added as a third variable. It is measured by the annual range of mean monthly temperature, but is also related to length of growing season, atmospheric humidity, degree of winter cold and windiness. Each zone or subzone can fall in one or more of four classes of oceanicity.

An ecogram for British soils would have to have at least one class of calcareous soils as well as several classes of acid or neutral soils. Quantitative definition of the classes requires research directly relating site characteristics that can be recognized in the field (such as humus form or indicator plant species) to measured soil properties. Such research has not progressed far enough to provide full definitions of the classes. In the meantime it is possible to deduce tentative definitions using ecological indicator values and the species lists of plant communities of the NVC. Thompson *et al.*

Table 8.3 Site details of selected NVC sub-communities

Code	mR	mN	$mR + mN$	mF	Zones	Name (community and sub-community)	Comments[a]
W4c	3.22	3.08	6.30	8.07	Warm dry, Warm moist	*Betula pubescens–Molinia caerulea* woodland, *Dryopteris dilatata–Rubus fruticosus* s-c	Typically occurs on drained infertile peat
W5a	5.90	5.61	11.51	7.83	Warm dry	*Alnus glutinosa–Carex paniculata* woodland, *Phragmites australis* s-c	Swamp or fen woodland, often flooded with eutrophic waters
W7b	5.43	5.20	10.63	7.25	Warm moist, Cool moist	*Alnus glutinosa–Fraxinus excelsior–Lysimachia nemorum* woodland, *Carex remota–Cirsium palustre* s-c	Typical of soligenous, minerotrophic flushes
W10a	5.21	5.25	10.46	5.28	Warm dry, Warm moist	*Quercus robur–Pteridium aquilinum–Rubus fruticosus* woodland, typical s-c	Lowland 'bluebell woods' on brown earth soils
W11a	3.95	4.58	8.53	5.41	Warm moist, Warm wet	*Quercus petraea–Betula pubescens–Oxalis acetosella* woodland, *Dryopteris dilatata* s-c	Upland 'bluebell woods' on base-poor brown earth soils
W12a	6.51	6.08	12.59	5.24	Warm dry	*Fagus sylvatica–Mercurialis perennis* woodland, *Mercurialis perennis* s-c	Deeper rendzina soils and calcareous brown earths
W16a	3.39	3.76	7.15	5.40	Warm dry	*Quercus* spp.–*Betula* spp.–*Deschampsia flexuosa* woodland, *Quercus robur* s-c	Lowland oakwood on podzolized brown earths and podzols
W18a	2.33	2.18	4.51	5.18	Cool moist	*Pinus sylvestris–Hylocomium splendens* woodland, *Erica cinerea–Goodyera repens* s-c	Typically on freely drained podzols with mor humus
H9b	2.15	2.35	4.50	6.11	Warm dry to Cool moist	*Calluna vulgaris–Deschampsia flexuosa* heath, *Vaccinium myrtillus–Cladonia* spp, s-c	Maintained by burning and grazing. Podzols with mor humus or thin peat
H12b	2.15	2.02	4.17	5.83	Cool moist to Cool wet	*Calluna vulgaris–Vaccinium myrtillus* heath, *Vaccinium vitis-idaea–Cladonia impexa* s-c	As for H9b
M10a (i)	5.13	2.28	7.41	8.11	Cool moist, Cool wet	*Carex dioica–Pinguicula vulgaris* mire, *Carex demissa–Juncus bulbosus/kochii* s-c, *Eliocharis quinqueflora* variant	Soligenous mire of mineral soils and shallow peats
M19b	2.04	1.85	3.89	7.36	Cool wet	*Calluna vulgaris–Eriophorum vaginatum* blanket mire, *Empetrum nigrum nigrum* s-c	Deep ombrogenous blanket peat, usually at high elevation
M25a	3.33	2.39	5.72	7.95	Warm moist, Cool moist	*Molinia caerulea–Potentilla erecta* mire, *Erica tetralix* s-c	Shallow and deep peats on slopes with some lateral water movement

U4a	3.64	3.40	7.04	5.34	Warm moist to Cool wet	*Festuca ovina–Agrostis capillaris–Galium saxatile* grassland, typical s-c	Calcifugous grassland maintained by grazing and occasional lime and fertilizer. Poor brown earths
U4e	2.90	2.63	5.53	5.55	Cool moist to Cool wet	*Festuca ovina–Agrostis capillaris–Galium saxatile* grassland, *Vaccinium myrtillus–Deschampsia flexuosa* s-c	Calcifugous grassland on podzolized soils and rankers with mor humus or thin peat
CG2a (ii)	6.92	3.16	10.08	4.02	Warm dry	*Festuca ovina–Avenula pratensis*[b] grassland, *Cirsium acaule–Asperula cynanchica* s-c, typical variant	Calcicolous grassland maintained by grazing without fertilizing. Shallow rendzinas
CG3b	6.83	3.50	10.33	4.17	Warm dry	*Bromus erectus* grassland, *Centaurea nigra* s-c	Calcicolous grassland, lightly grazed or ungrazed. Rendzinas
MG5a	5.96	4.66	10.62	5.13	Warm dry, Warm moist	*Cynosurus cristatus–Centaurea nigra* grassland, *Lathyrus pratensis* s-c	Hay meadows, grazed in winter, organic manuring. Brown earths of loamy to clayey texture
MG6a	5.72	5.69	11.41	5.49	Warm dry, Warm moist	*Lolium perenne–Cynosurus cristatus* grassland, typical s-c	Permanent pasture, typically grazed by dairy cattle. Brown earths enriched with fertilizers and lime

[a] Based on Rodwell, 1991a,b, 1992).
[b] Synonym *Helictotrichon pratense*.

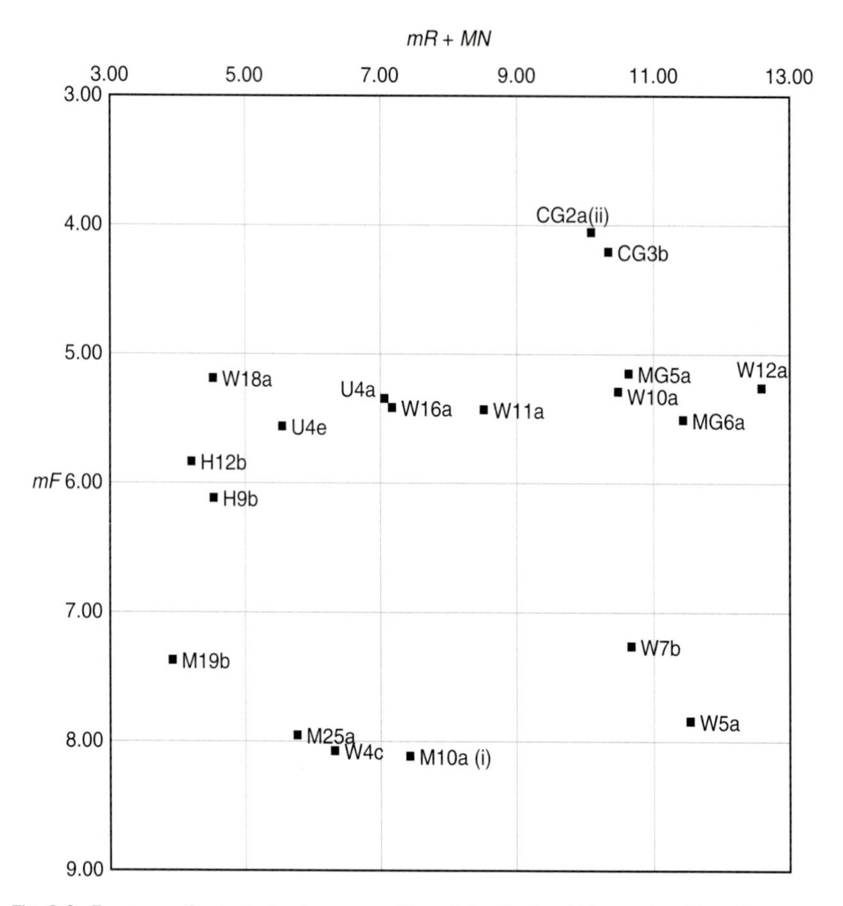

Fig. 8.2. Ecogram of selected sub-communities of the National Vegetation Classification.

(1993) have shown that Ellenberg's numbers are applicable in Britain.

Ellenberg's indicator values for soil moisture (*F*), soil acidity or reaction (*R*) and nitrogen availability (*N*) for each species are weighted by the frequency of the species in the community list, to obtain a site mean indicator value for each soil factor (*mF*, *mR* and *mN*). Frequency is scored 1–5 according to the proportion of quadrats in which the species is found, i.e. $1 = 0–20\%$, $2 = 20–40\%, \ldots, 5 = 80–100\%$. Results for a selection of plant communities from the NVC are given in Table 8.3. A soil quality grid formed from *mF* on one axis and $mR + mN$ on the other axis can be used to locate the communities (Figure 8.2).

The range of *mF* values, roughly 4.0 to 8.0, is not as wide as could be expected, particularly at the dry end. Nevertheless, each unit interval of the scale could serve as a class of soil moisture regime. The range of $mR + mN$ values is quite wide and it is suggested that intervals of 2.0 would form five classes of nutrient regime. The heathlands, ombrogenous mires and pine woodlands within the class 3.0 to 5.0 should be equivalent to the lowest SNR class (very poor) of BEC. The woodlands W12a and W5a, with large values of both *mR* and *mN*, might be equivalent to the fifth (very rich) class of SNR. The nature of a sixth, calcareous SNR class is suggested by the *mR* and *mN* values for the calcicolous grassland communities CG2a and CG3b. The highly calcareous nature of the rendzina soil (Rodwell, 1992) is reflected in the very large *mR* values, whereas the small *mN* values indicate that availability of nitrogen is lower than in the soils of the W12a and W5a woodlands. The calcareous SNR class could therefore tentatively be defined as having an *mR* greater than 6.5 together with an *mN* of less than 5.5.

Not surprisingly, heathlands represent consistently very poor or poor nutrient regimes; mires cover a wider range from very poor to at least medium. Woodlands and grasslands are found across all classes of nutrient regime. There are numerous grassland communities in the calcareous SNR class, but on similar lithologies woodlands tend to lie in the very rich class. Perhaps the woodland soils are less shallow and lime-rich or else a greater accumulation of surface organic matter has a moderating influence. The traditional hay-meadow community of MG5a on naturally rich soil treated annually with farmyard manure forms an interesting comparison with the very rich MG6 managed under a more intensive regime of grazing and enrichment with inorganic fertilizers. The poorest of the calcifugous grasslands (U4e) merges with the heathlands. The less poor U4a grassland is probably on the site of a former *Quercus–Betula* woodland (Rodwell, 1992).

Representing these various communities on a single ecogram is strictly inappropriate because they do not occur in the same climatic zone (Table 8.3). The three-dimensional classification with axes of climate, soil moisture regime and soil nutrient regime seems to provide an effective means of interpreting the site conditions under which these communities exist, particularly the relativities between communities of different physiognomy on broadly similar sites.

Conclusions

For the purposes of this chapter terrestrial ecosystems are defined as patches of land with uniform climate, soil and plant and animal communities, the combination of climate and soil being termed the site.

The use of climate (especially warmth and wetness), soil moisture regime and soil nutrient regime as the three main variables may provide a site classification that is both simple in concept and effective in predicting the nature of the plant community that the site can support. Such a classification can encompass all of the complexity of site variation caused by lithology, soil type and topography. It lends itself to quantitative definition of the classes on each axis. Just as ecological site quality influences the nature of the plant community, so the ecological amplitudes of all species can be measured and then used in reverse to predict site quality.

Although the plant community constitutes only a part of the ecosystem, it is often the most important part because it exerts a strong influence directly or indirectly on the other groups of organisms present. If it is accepted that site quality defined in terms of climate, soil moisture regime and soil nutrient regime controls the composition and functioning of the plant community, then any conservation policy or practice should be based first on such a classification of site quality. Also, mapping of 'ecological regions' requires a basis of physical and climatic factors.

References

Bendelow, V.C. and R. Hartnup (1980) *Climatic classification of England and Wales*. Soil Survey Technical Monograph No. 15. Harpenden: Rothamsted Experimental Station.

Birse, E.L. (1971) *Assessment of climatic conditions in Scotland, 3*: The bioclimatic sub-regions. Aberdeen: Macaulay Land Use Research Institute.

Cajander, A.K. (1926) The theory of forest types. *Acta Forestalia Fennica* **29**(3), 1–108.

Cajander, A.K. (1949) Forest types and their significance. *Acta Forestalia Fennica* **56**(5), 1–71.

Ellenberg, H. (1988) *Vegetation ecology of central Europe*, 4th edn. Cambridge: Cambridge University Press.

Ellenberg, H., H.E. Weber, R. Dull, V. Wirth, W. Werner and D. Paulissen (1992) *Zeigerwerte von Pflanzen in Mitteleuropa. Scripta Geobotanica*, Vol 18, 2nd edn. Gottingen: Goltze.

Griffiths, J.F. (1966) *Applied climatology*. London: Oxford University Press.

Klinka, K., V.J. Krajina, A. Ceska and A.M. Scagel (1989) *Indicator plants of coastal British Columbia*. Vancouver: UBC Press.

Krajina, V.J. (1965) Biogeoclimatic zones and classification of British Columbia. *Ecology of Western North America*, **1**, 1–17.

Kuusipalo, J. (1985) An ecological study of upland forest site classification in southern Finland. *Acta Forestalia Fennica*, **192**, 1–78.

McVean, D.N. and D.A. Ratcliffe (1962) *Plant communities of the Scottish Highlands*. London: HMSO.

Ministère de la Région Wallonne (1991a) *Le fichier écologique des essences, 1: Texte explicatif*. Namur, Belgium: Ministry of the Wallonia Region.

Ministère de la Région Wallonne (1991b) *Le fichier écologique des essences, 2: Le fichier écologique*. Namur, Belgium: Ministry of the Wallonia Region.

Moore, P.D. and D.J. Bellamy (1974) *Peatlands*. London: Elek Science.

Noirfalise, A. (1984) *Forêts et stations forestières en Belgique*. Gembloux, Belgium: Les Presses Agronomiques de Gembloux.

Pojar, J., K. Klinka and D. Meidinger (1987) Biogeoclimatic ecosystem classification in British Columbia. *Forest Ecology and Management* **22**, 119–54.

Poore, M.E.D. and D.N. McVean (1957) A new approach to Scottish mountain vegetation. *Journal of Ecology*, **45**, 401–39.

Rameau, J.C., D. Mansion and G. Dumé (1989) *Flore Forestière Française: guide écologique illustré; Vol 1: Plaines et Collines.* Paris: Institut pour le développement forestier, Ministère de l'Agriculture et de la Pêche.

Rameau, J.C., D. Mansion and G. Dumé 1993. *Flore Forestière Française: guide écologique illustré; Vol 2: Montagnes.* Paris: Institut pour le développement forestier, Ministère de l'Agriculture et de la Pêche.

Rodwell, J.S. (ed.) (1991a) *British plant communities. Volume 1. Woodlands and scrub.* Cambridge: Cambridge University Press.

Rodwell, J.S. (ed.) (1991b) *British plant communities. Volume 2. Mires and heaths.* Cambridge: Cambridge University Press.

Rodwell, J.S. (ed.) (1992) *British plant communities. Volume 3. Grasslands and montane communities.* Cambridge: Cambridge University Press.

Tansley, A.G. (1939) *The British Islands and their vegetation.* Cambridge: Cambridge University Press.

Thompson, K., K.G. Hodgson, J.P. Grime, I.H. Rorison, S.R. Band and R.E. Spencer (1993) Ellenberg numbers revisited. *Phytocoenologia* **23**, 277–89.

Trewartha, G.T. and L.H. Horn (1980) *An introduction to climate.* New York: McGraw-Hill.

CHAPTER 9

Taxonomy and systematics

NIGEL MAXTED

Introduction

An appreciation of the general principles of taxonomy and an understanding of the taxonomy of the group being studied is fundamental to the formulation of effective biodiversity conservation programmes. Taxonomy, through the organism naming system it provides, contributes the referencing system for the whole of biology. It provides the backbone onto which pieces of biological information, including conservation data, are attached and so communicated. Just as a human without a backbone would collapse, so biology could not function without taxonomy. The importance of this is underlined by the requirement for trained scientists with these skills as stated by Article 12 of the Convention on Biological Diversity:

> The Contracting Parties ... shall ... establish and maintain programmes for scientific and technical education and training in measures for identification, conservation and sustainable use of biological diversity and its components.

How can we expect to effectively conserve sea mammals or the wild relatives of maize, if we do not, for example, understand the species concepts applied or what are the distinguishing features that allow one species to be separated from another in the field? How can we produce Red Data Books or enact CITES (see Chapter 5) and other conservation legislation without a relatively stable and accepted nomenclature? An understanding of the taxonomy of a group of organisms helps the conservationist answer these and many other fundamental questions and provides access to the wealth of biological information available for a species once the correct name is known. It is also true that the inability of many conservationists to utilize the existing taxonomic and systematic data is undoubtedly hampering conservation activities and is resulting in poor utilization of biodiversity for the benefit of all humankind (UNCED, 1992).

Conservation resources are and will always be limited and use of existing taxonomic knowledge can enhance the conservationists' choice of conservation strategies and priorities. One of the primary problems facing conservationists is where to focus their conservation effort (see Chapter 3). One criterion has been the number of taxonomic groups, a criterion which requires a firm taxonomic base for the groups being studied; the diversity of taxa in the region must have been studied, taxonomic limits defined, accepted names provided and distributional patterns established. In this chapter, therefore, the scope of taxonomy is discussed, the methods of analysis of taxonomic data are described and the sources of taxonomic information are outlined.

What is taxonomy?

There are numerous definitions of what constitutes taxonomy and systematics; the two terms have often been used synonymously. Taxonomy, however, may be defined as the scientific study and description of biodiversity and how that diversity arose. Biological diversity, even in this period of rapid species extinction, is vast. The numbers of described and estimated species are provided in Table 9.1. It is only possible for the conservationist to work with and understand this vast level of diversity by being aware of existing taxonomic structures and applying taxonomic principles.

Table 9.1 Numbers of species in the groups of organisms

	Described species	Working number of estimated species
Viruses	5 000	500 000
Bacteria	4 000	400 000
Fungi	70 000	1 000 000
Protozoans	40 000	200 000
Algae	40 000	200 000
Plants	250 000	300 000
Nematodes	15 000	500 000
Molluscs	70 000	200 000
Crustaceans	40 000	150 000
Arachnids	75 000	750 000
Insects	950 000	8 000 000
Vertebrates	45 000	50 000
Totals	1 604 000	12 250 000

Source: Groombridge (1992).

Taxonomy as an experimental science

Like all sciences taxonomy has at its core experimentation and incremental growth of knowledge. The objective is to increase our knowledge and understanding of a group of organisms; how many species are there, how do we recognize these species, where are they to be found, are all questions of fundamental importance to the conservationist? In taxonomy the experimental study of patterns of diversity in biodiversity is commonly referred to as a *revision*. The term revision is used because following the experimental study our knowledge of the patterns of diversity within a particular taxon are 'revised'. The revision process involves the recording and analysis of features or characteristics both within and between taxa (populations, species, genera). The results of this analysis are then considered in conjunction with information from the literature and yield various products. The core product is the novel, 'revised' classification of the taxon and this is complemented by a range of secondary products, such as descriptions, keys, synonymized lists, taxon illustrations, critical notes, etc.

During the revision, features or characteristics are recorded and analysed from representative specimens for the taxa being investigated. These specimens are commonly found in *taxonomic collections*, which can have various forms: an herbarium is a collection of dried plants, an experimental garden is a collection of live plants grown under uniform conditions, a zoological garden is a collection of live animals, and a culture collection is a collection of live cultures. It is important to differentiate between live and dead collections; it is generally easier and less expensive to keep collections of dead material. However, it may then be impossible to record certain characteristics such as flower colour or behavioural characteristics.

The features or characteristics that are recorded and analysed during a revision are referred to as *characters*. Davis and Heywood (1973) define a character as any attribute (or descriptive phase) referring to form, structure or behaviour which the taxonomist separates from the whole organism for a particular purpose such as comparison or interpretation. They stress the abstract nature of characters; it is *character states* that taxonomists actually utilize, i.e. for the character human eye colour we might actually record the states green, blue and brown.

Characters may be divided into numerous categories, depending on both the source of the character and their intrinsic nature. Characters may be obtained from diverse sources, including: morphological (external, internal, microscopic, including chromosomal and developmental characters), physiological, chemical, behavioural, ecological and distributional sources. Traditionally, taxonomists have based their analysis on obvious characters, often using morphology and focusing on reproductive structures (flowers and genitals) because of their inherent conservatism, which in this context means their ability to remain relatively unchanged over long periods of evolutionary development. In general, the characters that are most useful in taxonomy, 'good' characters, are those that are fairly constant within a taxon (not genetically variable or susceptible to environmental modification but vary between taxa). These 'good' characters also tend to form correlations with other characters to produce a stable classification.

There are two important questions that must be addressed when selecting characters for use in taxonomic analysis: (a) whether or not each character selected should be given equal importance or weight in the analysis (for example, whether characters thought to have been important in evolutionary terms should be given increased importance), and (b) how many characters should be used in a particular taxonomic study. The most explicit form of weighting is that of character inclusion or exclusion from the analysis, but having made this point Sneath and Sokal (1973) consider any character weighting inappropriate. Each organism possesses an almost limitless number of characters; however, the taxonomist will always be limited temporally and economically. Thus the number of characters employed in a study requires skilful selection if a consistent classification is to be produced. Sneath and Sokal (1973) do not provide a fixed figure, rather they suggest the higher the number of characters used the better.

The analysis of taxonomic data

Traditionally, taxonomic data analysis was intuitive; the taxonomist mentally estimated resemblance or evolutionary ancestry on the basis of observable features. This was undoubtedly subjective and in recent years an attempt has been made to increase the objectivity of taxonomic data analysis by using computers to analyse the patterns in the recorded character sets. There are two predominant contemporary schools of taxonomic data analysis, phenetic and phylogenetic. This dichotomy is necessitated by the fundamental difference in these two methods of analysis and their approach to the problem of relating taxa. *Phenetic* techniques use the occurrence of character combinations as they are now perceived to construct taxonomic groups, while *phylogenetic* techniques add the dimension of time and ancestry to describing organisms and the characters selected are those that are thought to have been important in evolution. The subject of phenetic and phylogenetic analysis is introduced in detail by Stuessy (1990). Using both techniques, the presence of a consistent character combination is used to define a particular taxon and the greater the content of information used to construct a classification, the better the resultant classification.

An illustration of the practical sequence of taxonomic data analysis can be summarized as follows:

1 The study group is selected and delimited, e.g. skwarbs, a fictitious group of animals drawn in Figure 9.1.

Toothless Skwarb

Big Toothed Skwarb

Greater Spotted Skwarb

Lesser Spotted Skwarb

Bearded Skwarb

Small Skwarb

Fig. 9.1 Drawings of the six fictitious skwarb species.

Table 9.2 Recorded characters and character states for skwarb species

Character number	Character	Character states
1	Face shape	Rectangular (1), triangular (2)
2	Face base size	mm
3	Hair quality	Straight (1), curly (2)
4	Horns	Present (1), absent (2)
5	Horn orientation	Upward (1), downward (2)
6	Eye number	One (1), two (2)
7	Eye shape	Round (1), square (2)
8	Teeth	Present (1), absent (2)
9	Teeth size	mm
10	Spots	Present (1), absent (2)
11	Spot size	Large (1), small (2)
12	Beard	Present (1), absent (2)
13	Leg length	mm
14	Feet shape	Oval (1), triangular (2)

Table 9.3 Coded character data set for skwarb species

Species	Characters													
	1	2	3	4	5	6	7	8	9	10	11	12	13	14
Toothless	2	35	2	1	1	2	1	2	I[a]	2	I	2	5	1
Big Toothed	1	35	1	1	1	2	2	1	5	2	I	2	8	2
Greater Spotted	1	35	1	1	1	2	2	1	2.5	1	1	2	8	1
Lesser Spotted	2	33	2	1	1	2	1	1	2.5	1	2	2	5	1
Bearded	1	40	1	2	I	1	2	1	2.5	2	I	1	8	2
Small	1	17	1	1	2	2	2	1	1.5	2	I	2	5	1

[a] I = inapplicable.

2 Characters are recorded and 'good' characters are included in the character set, e.g. face shape or eye number.
3 All representative specimens or taxa (skwarbs) have their character states recorded for the character set, (see Table 9.2).
4 The recorded set of character states are converted to codes (see Table 9.3) and entered into the computer.
5 The resemblance or relationship between specimens or taxa (skwarbs) is calculated using multivariate statistics. For phenetic analysis this is divided into two phases:

(a) calculate a similarity or dissimilarity matrix between specimens or taxa (see Table 9.4);
(b) the triangular similarity matrix is then analysed using a possible range of techniques which generally fall into two categories, cluster and ordination methods. (It is recommended that more than one form of analysis is used to avoid any possible bias of the results introduced by any one algorithm.) Cluster analysis (see appropriate

Table 9.4 Similarity coefficient matrix for skwarb character data set

Species	Toothless (TL)	Greater Spotted (GS)	Lesser Spotted (LS)	Bearded (BD)	Small (SM)	Big Toothed (BT)
TL	1.0					
GS	678.0	1.0				
LS	136.3	208.3	1.0			
BD	134.7	212.7	4.3	1.0		
SM	144.7	216.7	8.3	19.3	1.0	
BT	141.3	519.3	144.3	118.7	212.7	1.0

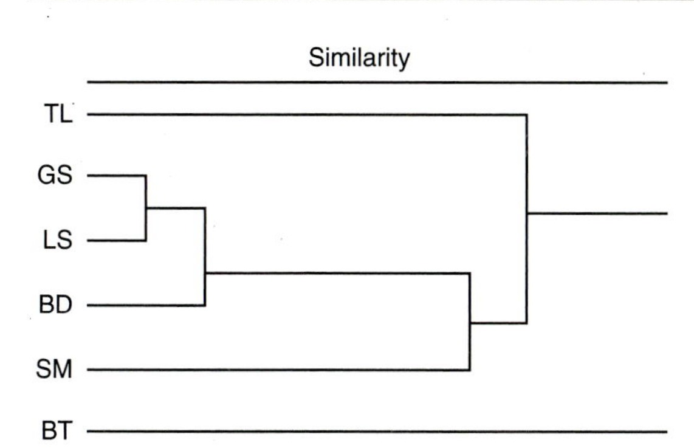

Fig. 9.2 Dendrogram produced by cluster analysis of skwarb species data set.

statistical texts such as Sokal and Rohlf, 1981) usually yields a hierarchical classification of specimens or taxa, which is depicted in the form of a dendrogram (see Figure 9.2). Ordinations usually present their results in the form of a scatter diagram, in which the most closely allied specimens or taxa are spatially juxtaposed (see Figure 9.3).

6 Production of a classification of the included taxa (skwarbs) based upon resemblances or evolutionary relationships as indicated by the analysis (see Table 9.5).
7 Once the classification has been produced and the specimens are attributed to accepted taxa, it is possible to make generalizations about each taxon based on the representative specimens. Therefore descriptions (Table 9.6), identification aids (Figures 9.6 and 9.7 later in this chapter, and Table 9.7), distribution lists, etc., can be generated for each taxon; see below.

The products of taxonomic data analysis

The primary product of any revision is the *classification*. This is usually presented in the form of a list

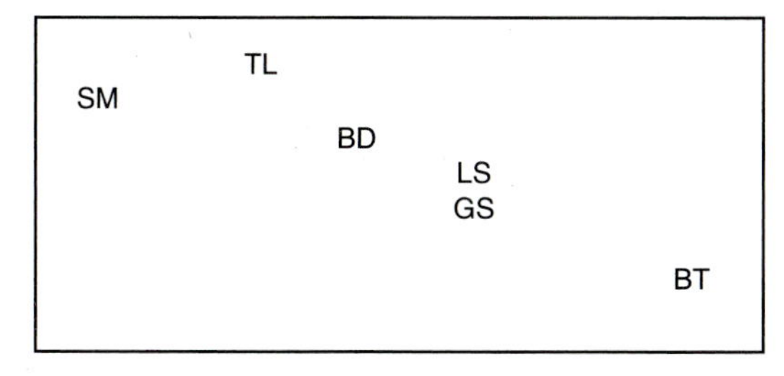

Fig. 9.3 Scatter diagram produced by ordination of skwarb species data set.

Table 9.5 Classification of skwarb species

Toothless
Greater Spotted
Lesser Spotted
Bearded
Small
Big Toothed

of accepted taxon names with their authors and publication details. The order of the taxa reflects their postulated taxonomic relationships. Closely allied taxa, in terms of overall resemblance or evolutionary descent, are juxtaposed, while more distant taxa will be placed more remotely. Once established, the classification will directly affect the production of each of the secondary products, such as taxon descriptions, identification aids, accepted nomenclature, etc. Many older classifications were *artificial*; they were based on a few morphological characters. Contemporary classifications are more commonly based on large numbers of characters from diverse sources. These *natural* classifications based on numerous characters are said to have a high information content and are predictive. This

Table 9.6 Description of skwarb species

Toothless Skwarb
Face triangular, base 35 mm. Hair curly. Horns present and pointing upwards. Two round eyes. Teeth absent. Spots absent. Beard absent. Legs 5 mm, feet oval.

Greater Spotted
Face rectangular, base 35 mm. Hair straight. Horns present and pointing upwards. Two square eyes. Teeth 2.5 mm long. Spots present and large. Beard absent. Legs 8 mm, feet oval.

Lesser Spotted
Face triangular, base 35 mm. Hair curly. Horns present and pointing upwards. Two round eyes. Teeth 2.5 mm long. Spots present and small. Beard absent. Legs 5 mm, feet oval.

Bearded
Face rectangular, base 35 mm. Hair straight. Horns absent. One square eye. Teeth 2.5 mm long. Spots absent. Beard present. Legs 8 mm, feet triangular.

Small
Face rectangular, base 17 mm. Hair straight. Horns present and pointing downwards. Two square eyes. Teeth 1.5 mm long. Spots absent. Beard absent. Legs 5 mm, feet oval.

Big Toothed
Face rectangular, base 35 mm. Hair straight. Horns present and pointing upwards. Two square eyes. Teeth 5 mm long. Spots absent. Beard absent. Legs 8 mm, feet triangular.

Table 9.7 Tabular key to skwarb species

Characters	Toothless	Big Toothed	Greater Spotted	Lesser Spotted	Bearded	Small
1 Face shape	T	R	R	T	R	R
2 Face base size (mm)	35	35	35	33	40	17
3 Hair quality	C	S	S	C	S	S
4 Horns	P	P	P	P	A	P
5 Horn orientation	U	U	U	U	I	D
6 Eye number	2	2	2	2	1	2
7 Eye shape	○	□	□	○	□	□
8 Teeth	A	P	P	P	P	P
9 Teeth size (mm)	I	5	2.5	2.5	2.5	1.5
10 Spots	A	A	P	P	A	A
11 Spot size	I	I	L	M	I	I
12 Beard	A	A	A	A	P	A
13 Leg length (mm)	5	8	8	5	8	5
14 Feet shape	V	T	V	V	T	V

Abbreviations: R rectangular, T triangular, S straight, C curly, P present, A absent, U upward, D downward, 1 one, 2 two, ○ round, □ square, L large, M small, V oval, I inapplicable.

means that if we know a species location in a classification we can predict certain characteristics for that species. We can predict that spiders, as members of the class Arachnida, will have four pairs of walking legs, a body divided into two regions, simple rather than compound eyes and no antennae or wings. A natural classification is considered to present a closer approximation to the intrinsic 'natural' classification that underlies the partitioning of biological diversity. A new classification is not automatically accepted by the scientific community. However, a classification is more likely to be accepted if it can be shown to be natural and has a high predictability or information content.

Within the classification each individual organism is placed in a specific category and given a Latin name which distinguishes it from other organisms. Each group of organisms given a name is referred to in general terms as a *taxon* (plural *taxa*); this term can be applied to any rank, e.g. species, genus, tribe, family. Classifications are built up from the raw materials of populations and species. Those species that share most characters are placed in a group called a genus, genera which share most characters are grouped into a tribe, etc. So the *taxonomic hierarchy* is composed of a series of increasingly inclusive groups based on decreasing similarity, each distinct level being given a name – species, genus, class, etc. The level of the hierarchy used for a taxon is referred to as its *rank*. The names used for various ranks in the hierarchy are slightly different for animals and plants, as can be seen in Table 9.8.

Table 9.8 A simplified taxonomic hierarchy for plants and animals

Botanical	*Zoological*
Kingdom	Kingdom
Subkingdom	Subkingdom
Division	Phylum
Subdivision	Subphylum
	Superclass
Class	Class
Subclass	Subclass
	Infraclass
{Superorder}	Superorder
Order	Order
{Suborder}	{Suborder}
	Infraorder
	Superfamily
Family	Family
Subfamily	Subfamily
	{Supertribe}
Tribe	Tribe
Subtribe	Subtribe
{Supergenus}	
Genus	*Genus*
Subgenus	*Subgenus*
Section	
Subsection	
Series	
Subseries	
Species	*Species*
Subspecies	*Subspecies*
Variety	

Notes: Italicized categories should be italicized in any text, names in { } are not commonly used and all names from the genus up begin with a capital letter.

Species

Species, unlike any other rank in the hierarchy, are generally and practically accepted to exist as real, distinct entities, unlike other taxonomic categories. There are many definitions used, but plant species are generally delimited in terms of morphological or biological concepts. *Morphological species* are species that share certain characteristics; however, the characteristics that separate species in one group may separate genera in another or varieties in yet another. In practice morphological species are often defined in the terms of their close relatives. If six species are defined by a certain level of morphological difference then a newly discovered seventh taxon with the same level of difference would also be regarded as a distinct, allied species. This is subjective and has often led to arguments between taxonomists concerning the most appropriate rank for a particular taxon. One proposed solution to this problem is to adopt the *biological species* concept. This is more easily defined; if two specimens can successfully interbreed then they belong to the same species. However, this raises the question of definition of successful; does it mean that fertilization can take place, an embryo begins to develop, the next generation is produced or the next generation is fertile? In practice the application of this definition would, however, necessitate the crossing of all specimens of each species in a species complex to establish if they can successfully interbreed, which is impractical. Concepts of what defines a particular rank of the hierarchy above the species are even more vague and open to dispute.

An understanding of which species concept is applied is of fundamental importance for conservation, especially for plants where some gene flow between species is common. The majority of plant species have been described and classified using the morphological concept. If the conservationists' aim is to conserve maximum genetic variation, then a classification based on biological species concepts is more appropriate, because the aim would be to sample species between which there is no gene flow, thus ensuring conservation of maximum genetic variation. However, there are still few biologically-based classifications available. They tend to be restricted to well-known crops and their allies (e.g. wheat, potato, maize, rice, etc.), where the genetic relationships among the taxa have been extensively studied and the genetic make-up of the species complex is relatively well understood.

Nomenclature

Having established which species and other taxa exist, the taxonomist must provide each taxon with a name. It is important that the conservationist uses the correct, accepted name for a taxon, so that when they communicate the results of their research others have the same concept of the taxon for which the results apply. The procedure of applying the correct scientific name to an organism is *nomenclature*. A detailed introduction to biological nomenclature is provided by Jeffrey (1977). Carl Linnaeus (1707–78) is known as the founder of taxonomic nomenclature. His life's work was to name and describe biological diversity and he attempted to produce a catalogue of all animals and plants. His work was presented in numerous publications (*Systema Naturae* of 1735 for animals, plants and minerals, and *Species Plantarum* of 1753 for plants). The first edition of *Species Plantarum* (1753) for plants and the tenth edition of *Systema Naturae* (1758) for animals have subsequently been taken as the starting points for nomenclature. Names included in these publications have *priority* (nomenclatural precedence) over earlier or more recently published names. However, in an attempt to simplify the application of this principle, groups of biologists started to draw up codes for zoological and botanical taxa that contained supplementary rules to aid the naming of organisms.

Linnaeus also established the convention, later to become a nomenclatural rule, that each species should be described in Latin and provided with a *binomial*. For example, in the Latin phrase *Avena sativa*, *Avena* is the genus name and *sativa* the species name (or specific epithet). An *accepted name* is the nomenclaturally correct name that is generally used for a taxon, the name accepted and used by the appropriate expert for that group. The taxon expert will decide which taxa they recognize and then attach the appropriate names. The name used will be either the Linnaean name, if Linnaeus recognized that taxon, or the first name subsequently validly published. This procedure is referred to as the principle of *priority*. However, if a name is widely used and a name change would cause great inconvenience to other biologists then an invalid name can be correctly used or conserved; e.g. the correct name for the legume family is Fabaceae, but Leguminosae is widely used and so has been conserved. The Latin name used for a taxon does not have to have any biological meaning but often the author of a name will use a name that does imply some meaning, e.g. *sativa* or *vulgaris* = common, *viridus* = green, *hirsuta* = hairy, *palustris* = marshy, *peruviana* = from Peru. Taxon names for different ranks in the same classification are often drawn from the same stem, for example the field or common vetch of the legume family, *Vicia sativa*, belongs to the legume series *Vicia*, section *Vicia*, subgenus *Vicia* and genus *Vicia* of the tribe Vicieae.

Anyone who finds a distinct taxon that has not previously been described can publish it as a new species. To validly publish a new name the various nomenclatural codes demand that the author provide certain details (precise details vary depending on the code):

1 The new name (binomial), the combination of genus and species name, cannot have been used previously.
2 A diagnostic (short and important) description must be given in Latin, often with an illustration.
3 A type (representative) specimen(s) or drawing must be designated.
4 The details must be published in an accepted journal.

Synonyms

One of the most important secondary products of revision is a list of *synonyms*. These are nomenclaturally illegal names which refer to a taxon with a different accepted name. In a recent study of the temperate forage legume vetches (*Vicia*), Maxted (1995) listed 149 synonyms for the field or common vetch, which was described originally by Linnaeus 250 years ago as *Vicia sativa*. So several names can refer to the same taxon; however, only one is accepted and correct. A knowledge of synonymy is appropriate for the conservationist; if they know a particular name is a synonym of another name, they can still obtain access to the biological information associated with the taxon via the accepted name.

Description and diagnosis

Having observed a range of specimens during the revision and formed a concept of species, genera, tribes, etc., each specimen can be assigned to a particular taxon. The range of character states for the specimens that represent each taxon can then be generalized to describe the taxon. A logically laid-out statement of the characteristics of a taxon is referred to as a *description*. Descriptions for the six skwarb species are provided in Table 9.6. This is distinguished from a *diagnosis*, which is a short description covering only those characters (diagnostic characters) which are necessary to distinguish a particular taxon from related taxa. Diagnostic characters are those used in identification aids to distinguish related taxa.

Distribution

Just as the range of character states for the specimens that represent each taxon can be generalized to produce a description of the taxon, so ecological and geographical location (ecogeographic) details of representative specimens can be synthesized to produce generalized data concerning the taxon's ecology and geographical distribution. The identification of a species' ecological niches and geographical localities is of vital importance when planning conservation. The level of ecogeographic information may vary in detail from a simple list of countries and ecosystems to a detailed enumeration of the localities and habitats in which specimens have been found.

Geographical data are commonly displayed in the form of *distribution maps*. Taxon distribution is indicated in two basic ways: by shading or enclosing areas with a single line that contains sampled populations, as shown in Figure 9.4, or by using dots to indicate geographical position of the populations sampled, as shown in Figure 9.5. Stace (1989) points out that enclosing line maps are ambiguous, because they provide no indication of the frequency of the taxon within the region. A single outlying specimen might erroneously suggest that the taxon is continuously present throughout an entire region. The occurrence of a species is often sparse at the periphery of its range and there is rarely a distinct cut-off line. Indicating presence in this manner also means that any variation due to local ecological, climate and edaphic factors (see Chapter 8) within the individual provinces or countries cannot be shown. Therefore to represent detailed distribution patterns there is a general trend towards the use of dot distribution maps.

Identification

To effectively conserve species the conservationist must be able to distinguish and identify species. It is important for the conservationist to obtain a correct identification, so that he or she can amass information on the species and communicate the information to

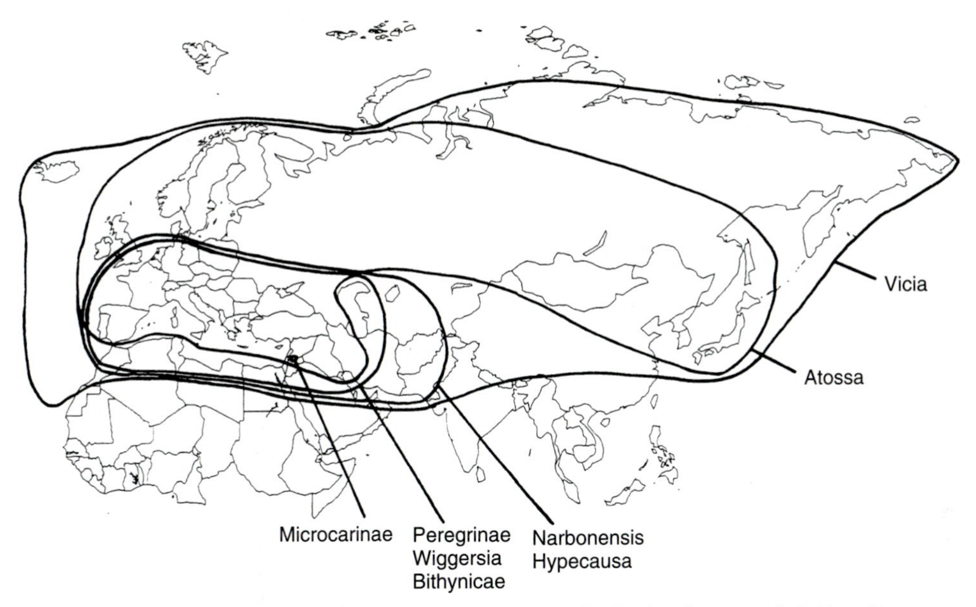

Fig. 9.4 Enclosed line map for vetch (*Vicia*) data indicating general distributional ranges of eight sections.

others without the confusion that would result from publishing data that were not correctly attributed to that species. Once a specimen is correctly identified the conservationist can gain access to the existing wealth of information available in the literature for that species, whereas if a specimen is either unidentified or wrongly identified any data that may be associated with that specimen are likely to

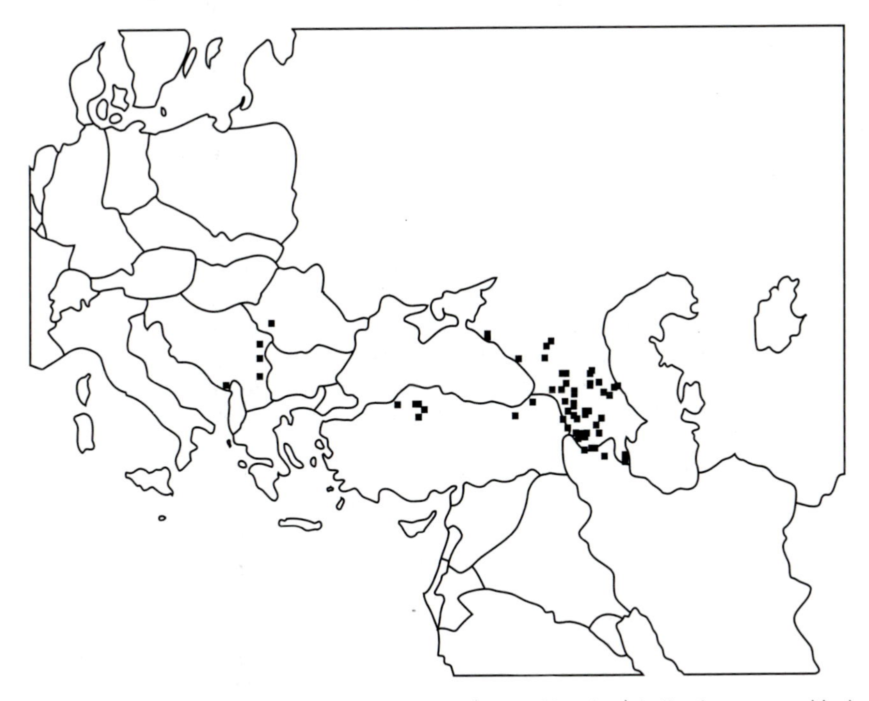

Fig. 9.5 Dot distribution map for vetch species (*Vicia abbreviata*) indicating geographical position of the populations sampled.

be misleading and utilization of the specimen or population is likely to be of limited value.

The process of specimen *identification* or *determination* involves two steps: firstly, the decision as to which taxon (e.g. genus, species or subspecies) the specimen represents, and secondly, the decision as to what is the 'accepted' name to use for it, if more than one name has been used for that taxon. The first step is achieved by use of some form of key or identification aid, discussed below. The second step involves finding out from the latest taxonomic treatment of the group what the expert on that group considers to be the accepted name. The latter step is discussed in the last section of this chapter.

Specimens are commonly named to species, but if lower taxonomic entities have been described they could be named to subspecies or variety. The correct identification of the specimen is achieved by comparing its characteristics to sets of diagnostic characteristics possessed by species. If the specimen's characteristics fall within the range of a species' diagnostic characteristics, then the specimen is a representative of that species, the range of the diagnostic characteristics for the species having been previously determined by a detailed study of a broad range of specimens representing the species.

There are basically two forms of identification, matching and elimination. *Matching* involves the comparison of the specimen to taxon description or some form of exemplar, such as a named herbarium sheet. For example, when trying to identify a skwarb species, the characteristics of the specimen would be compared to each of the descriptions provided in Table 9.6 and the description that provided the best match would indicate the identification. This would be relatively easy for the six skwarb species, but to match a specimen to one of a large number of possible taxa could prove impossibly time-consuming. Some means is required to narrow down the potential possibilities. Identification by *elimination* involves the user in comparing a specimen to a set of mutually exclusive short descriptions and making a decision as to which fits the specimen better, repeating the process for another set of descriptions until only one taxon remains, giving the identification. Returning to the skwarb example, one might start by taking the character 'spot presence or absence'; if the specimen you are attempting to identify does not have spots then immediately the greater spotted and lesser spotted skwarbs are eliminated as potential species. Often in practice, identification will begin by elimination, and proceed by matching when the range of possible taxa has been narrowed down to manageable proportions.

Keys

The traditional tool used by biologists for identifying specimens is a *key*. Keys are commonly based on gross morphological features, the characteristics that are readily observable in the field, laboratory, museum or herbarium. Traditional keys are commonly referred to as being:

- *dichotomous*, meaning that the user is presented with two brief, diagnostic descriptions at each stage in the key, only one of which matches the specimen being identified;
- *single-access*, meaning there is only one point of entry, via the choice between the first pair of brief, alternative descriptions;
- *sequential*, meaning that the user is faced with a sequence of alternative character states; once the first choice is made the user is faced with a second and third choice of alternative diagnostic information until by making the final choice the name of the specimen is obtained; and
- *diagnostic*, meaning that the few characters used in the key distinguish the taxa included, but are not intended to provide a full description of the taxon.

The nature of the key design means that the process of identification follows a specific and structured sequence until the identification is finally made. At each step in the key two alternative descriptions (couplet) are provided; these descriptions should be constant for the particular species and mutually exclusive without any overlap, so that only one description in each couplet could fit the specimen and there is no ambiguity as to which path the identification should follow. There are two basic styles of single-access key: the parallel (or bracketed) and the yoked (or indented), but neither has an obvious advantage over the other. A parallel key to the skwarb data set is provided in Figure 9.6, and the same data set is shown in the yoked style in Figure 9.7. Once an identification is achieved using any form of key, the specimen should always be compared to a detailed description of the taxon to confirm the identification and to find out whether mistakes have been made during the keying-out process.

Multi-access keys were developed to overcome some of the problems associated with single-access keys. A multi-access key does not force the user to go through the character set in a specific preordained sequence. It

1. Face triangular, hair curly, eye shape round 2
   Face rectangular, hair straight, eye shape square 3
2. Spots absent, teeth absent . . . . Toothless Skwarb
   Spots present, teeth present . . . . Lesser Spotted Skwarb
3. Feet triangular 4
   Feet oval 5
4. Horns present, two eyes, teeth 5 mm long, beard absent. . Big Toothed Skwarb
   Horns absent, one eye, teeth 2.5 mm long, beard present . . . . . Bearded Skwarb
5. Face base size 35 mm, horns pointing up, teeth 2.5 mm long, spots present, leg 8 mm long . . . . Greater Spotted Skwarb
   Face base size 17mm, horns pointing down, teeth 1.5 mm long, spots absent, leg 5 mm long . . . . Small Skwarb

Fig. 9.6 Parallel or bracketed key to six skwarb species.

allows the user to ignore particular characters and still obtain an identification. For example, if the specimen lacks seed or is in the larval form, the identification can be based on vegetative and flower or larval characters alone. Most commonly computers are now used for multi-access identification. However, various forms of printed tabular keys are available; an example of a tabular key for the skwarb data is provided in Table 9.7, with species in the columns and characters in the rows. The table is filled with coded character states for each taxon. To identify a new specimen, the user records its characteristics using the same codes and row structure as the tabular key and then compares the specimen data with each column of the table. A match between the user's score card and a particular taxon provides the identification.

Illustration

One of the unfortunate problems for non-specialists in learning to use traditional keys is the amount of technical terminology involved. For example, where is the trochantin on a beetle or what shape is a capitate style? One means of avoiding technical terms is to use *illustrations* (line drawings, photographs, paintings, etc.) of the key features of the species, hence avoiding the use of complex terminology. The relative efficacy of using line drawings, paintings or photographs to aid identification is a matter of subjective assessment and individual preference. However, one problem associated with using photographs for identification is that they can only show what is observed at that time in a two-dimensional image, whereas with a drawing or painting the illustrator can enhance the observed two-dimensional image to include features that may be less obvious on an individual specimen or in that particular plane of view. For this reason it is less likely that a specimen could be accurately identified by comparison with photographs and as a general rule illustrations should be used in conjunction with other aids.

1\. Face triangular, hair curly, eye shape round 2
   - 2\. Spots absent, teeth absent . . . . Toothless Skwarb
   - 2\. Spots present, teeth present . . . . Lesser Spotted Skwarb

1\. Face rectangular, hair straight, eye shape square 3
   - 3\. Feet triangular 4
     - 4\. Horns present, two eyes, teeth 5 mm long, beard absent . . . . Big Toothed Skwarb
     - 4\. Horns absent, one eye, teeth 2.5 mm long, beard present . . . . Bearded Skwarb
   - 3\. Feet oval. 5
     - 5\. Face base size 35 mm, horns pointing up, teeth 2.5 mm long, spots present, leg 8 mm long . . . . Greater Spotted Skwarb
     - 5\. Face base size 17 mm, horns pointing down, teeth 1.5 mm long, spots absent, leg 5 mm long . . . . Small Skwarb

Fig. 9.7 Yoked or indented key to six skwarb species.

Data processing

Computers are now often and will increasingly be used in identification. Several traditional dichotomous key-generating programs have been produced (see Pankhurst, 1991), but since the 1960s computers have had an increasing impact by providing *interactive identification*. The multi-access key can easily be implemented in a computer program, as the standard data structures used in high-level programming languages naturally accommodate the concept of a taxonomic data matrix, each taxon a row and each character a column, say. The choice of characters to use is decided by the user, but the program may assist by ranking the characters and presenting them in different ways – perhaps in a new sequence. Discarded rows and columns may be restored, however, at the user's request. Thus, the active part of the matrix diminishes or expands until the identification is obtained, i.e., when only one taxon remains.

Sources of taxonomic information

Having introduced the procedure and products of taxonomy which are of most importance to the conservationist, where do we obtain this information? The primary sources of taxonomic information are specialist publications (revisions, monographs, checklists, taxonomic journals, floras and faunas), educational materials (field guides, lectures, multimedia programmes) and taxonomic experts. Specialist publications are likely to be found in the libraries associated with various kinds of taxonomic collections, e.g. museums, herbaria, botanic gardens, germplasm collections and zoos, while general libraries will provide field guides and other media guides. The conservationist may gain access to taxonomic experts by contacting the taxonomic collections or educational establishments with which they are usually associated. There are published lists of taxonomic literature, such as the *Kew Record of Taxonomic Literature*, which is published by the Royal Botanic Gardens, Kew, annually. Traditionally a conservationist studying a group would have undertaken a literature survey, but a contemporary review will include all media, microfiche, diskettes, multimedia CD-ROM and on-line services (BIDS, BIOSIS, CABI, etc.), as well as traditional printed text.

Taxonomy involves large quantities of data capture, storage and retrieval. In recent years to enhance the efficiency of data handling, computer database technology is being increasingly used to organize taxonomic data. Since the 1980s many multi-institutional database projects have begun to collate biodiversity information for either specific taxonomic groups or geographical regions. For example, ILDIS (International Legume Database and Information Service) have established a botanical diversity database for the 17 000 legume species (Zarucchi *et al.*, 1993). This project is managed as a cooperative, involving approximately 20 research groups from five continents, and the information system is available internationally. Similar projects have been established for several other groups. Bisby (1994) lists the following examples:

- *Bacteria*: DSM, List of valid bacterial names (Deutsche Sammlung von Mikroorganismen und Zellkulturen, Braunschweig, Germany); BIOSIS TRF, BIOSIS Bacterial Taxonomic Reference File (BIOSIS, Philadelphia, PA, USA)
- *Protists*: ETI, Linnaeus Protists (Expert Center for Taxonomic Identification, Amsterdam, The Netherlands)
- *Insects*: ANI, Arthropod Name Index (CAB International, Wallingford, UK)
- *Molluscs*: ETI, Linnaeus Molluscs (Expert Center for Taxonomic Identification, Amsterdam, The Netherlands)
- *Fish*: W.N. Eschmeyer, Taxonomic Database for Fishes (California Academy of Sciences, San Francisco, CA, USA)
- *Fungi*: IMI, Species Fungorum Database (International Mycology Institute)
- *Plants*: ILDIS, International Legume Database and Information Service (ILDIS Coordinating Centre, Southampton, UK); CITES, CITES Cactaceae Checklist (Royal Botanic Gardens, Kew, UK)

The long-term goal must be to combine these and other taxonomic databases to make a single biodiversity database, which would obviously prove an invaluable tool to the conservation community.

The examples provided above are largely centred on specific taxonomic groups, but there are also several geographically restricted projects, such as ERIN (Environmental Resources Information Network). ERIN is an Australian initiative to provide a geographically related environmental information system that will aid environmental impact assessment and monitoring of species, vegetation types and heritage sites (ERIN, 1991).

Conclusions

So at what stage does the conservationist need to utilize taxonomic information? Taxonomic understanding of the relationships between taxa and use of the range of taxonomic products is vital throughout the application of any effective biodiversity conservation programme. To be able to conserve, the conservationist must have a basic understanding of how the taxonomy works; he or she must have a basic appreciation of taxonomic collections, revisions, characters, phenetic and phylogenetic analysis, classification, rank and the taxonomic hierarchy, species concepts, nomenclature, synonymy, description and identification and where this information can be obtained. For it is these diverse data that provide us with the basic biodiversity data necessary to formulate effective conservation programmes and priorities. This understanding is hampered by two factors, the fact that numerous conservation workers from areas rich in biodiversity have received only rudimentary training in the use of taxonomic data and the apparent reluctance of taxonomists to ensure that their wealth of data is made accessible to the non-taxonomic community. When habitat, species and genetic diversity are diminishing with such speed there is a priority requirement for a closer working relationship between the two communities if conservation activities in the future are not to be unnecessarily hampered.

References

Bisby, F.A. (1994) Global master species databases and biodiversity. *Biology International*, **29**, 33–40.

Davis, P.H. and V.H. Heywood (1973) *Principles of Angiosperm taxonomy*. New York: Krieger.

ERIN (1991) *ERIN newsletter*. ERIN, Canberra, Australia.

Groombridge, B. (ed.) (1992) *Global Biodiversity: Status of the Earth's living resources*. WCMC (World Conservation Monitoring Centre). London: Chapman & Hall.

Jeffrey, C. (1977) *Biological nomenclature* (2nd edn). London: Edward Arnold.

Maxted, N. (1995) *Ecogeographic study of* Vicia *subgenus* Vicia. Rome: International Plant Genetic Resources Institute.

Pankhurst, R.J. (1991) *Practical taxonomic computing*. Cambridge: Cambridge University Press.

Sneath, R.R. and P.H.A. Sokal, (1973) *Numerical taxonomy*. San Francisco: W.H. Freeman.

Sokal, R.R. and F.J. Rohlf (1981) *Biometry. The principles and practice of statistics in biological research*. San Francisco: W.H. Freeman.

Stace, C.A. (1989) *Plant taxonomy and biosystematics* (2nd edn). London: Edward Arnold.

Stuessy, T.F. (1990) *Plant taxonomy: The systematic evaluation of comparative data*. New York and Oxford: Columbia University Press.

UNCED (1992) *International convention on biodiversity*. Geneva: UNCED.

Zarucchi, J.L., P.J. Winfield, R.M. Polhill, S. Hollis, F.A. Bisby and R. Allkin (1993) The ILDIS Project on the world's legume species diversity. In F.A. Bisby, G.F. Russell and R.J. Pankhurst (eds) *Designs for a global plant species information system*. Oxford: Oxford University Press, pp. 131–44.

SECTION C Conservation biology in practice

The eight chapters in this section provide examples of case studies in conservation of biological diversity at different levels of biological organization, from genetic through habitats to communities and ecosystems. There are also examples of different methods of conservation from protected areas through *ex situ* conservation and from ecological restoration to modelling. All contributors write from their own experience and balance selected case studies with theory. In all cases, the contributors commence with introductory material then go on to describe the more specialized aspects.

Alan Gray (Institute of Terrestrial Ecology, UK) provides an assessment of the importance of genetic considerations in conservation biology and also provides an up-to-date assessment of the implications of low genetic diversity in small populations. The overlap between conservation genetics and the conservation of populations is taken up by Mike Hutchings and Alan Stewart (University of Sussex). They provide an introduction to population dynamics and demography as well as discussing recovery programmes for populations. Keith Kirby (English Nature) describes the concepts of habitat and niche and goes on to lend his experience in the UK to assessment of ecological surveys at different spatial scales. Conservation of remnant biological communities in modified landscapes can require intensive management and new ways of overcoming this problem need to be found. The combined expertise of Sue McIntyre (CSIRO, Australia) and Geoff Barrett and H.A. Ford (University of New England, Australia) is used to assess conservation at the community and ecosystem level of organization. They provide a classification of communities and ecosystems as modified by human impact and emphasize the importance of identifying historical processes which have brought about fragmented and modified ecosystems. This theme is then picked up by Bob Pressey when he looks at a series of case studies involving protected areas. He also assesses the effectiveness of protected areas and suggests some new directions. Another approach to conservation is *ex situ* conservation. David Worley is well experienced in the work of zoological gardens and also in conservation education. He assesses the limitations and difficulties of *ex situ* conservation but also draws attention to the need for zoological and botanical gardens. Addressing the damage to biological communities is considered with the combined and complementary expertise of Richard Pywell (Institute of Terrestrial Ecology, UK) and P.D. Putwain (Liverpool University). They describe the theory of ecological restoration (noting the importance of climate and physical factors) and provide examples of case studies in some detail. Setting objectives and monitoring progress is an important part of ecological restoration programmes. There may be risks attached to those programmes and that is one of the themes in the find chapter by Hugh Possingham (Imperial College, University of London and University of Adelaide). His chapter could arguably come first because it describes modelling and decision making as applied to conservation biology. There are risks and uncertainties to be addressed and in times of limited resources these need to be addressed early in any conservation exercise.

This section is not just about applied ecology, it is about the practice of conservation biology. The chapters in this section are therefore written within the perspective of conservation biology and pick up recurring themes from Sections A and B. For example, common to all contributions in this section is the importance of involving local people. All too often conservation projects have not succeeded because of lack of community involvement which in turn has sometimes been a product of poor consultation and communication.

The ecological basis of conservation has not always been communicated very well and there are some poor examples of both misinterpretation and misapplication of ecology. There are two examples in this section: attempts to use island biogeographical theory and the popularization of 'wildlife corridors'. For many years there has been a huge effort to try and apply island biogeographical theories to the selection, design and management of protected areas. Regretfully the theories of island biogeography have often been misinterpreted. The fascination with 'wildlife corridors' has led to a popular belief that they do work and could help to reduce effects of habitat fragmentation. By way of contrast there is good ecological science to be found in the studies on the population dynamics of populations in habitat matrices and also in studies of effects of habitat fragmentation on population dynamics.

Habitat fragmentation, spatial scale and spatial heterogeneity are dominating conservation thinking, from problems of conservation of small remnants of biological communities (where intensive management is required) to conservation strategies for entire geographical regions where innovative thinking is required. For reasons of differences in local ecology and for reasons of differences in the scale of human impacts, the conservation strategies which may emerge from innovative thinking are not always applicable to all regions. There are, for example, notable differences between conservation strategies adopted in the UK (high population density) and those adopted in New Zealand and in parts of Australia where population density is much lower. The implications of differences in population density and extent of landscape change become very clear in Chapters 12, 13 and 14. There have been some recent and bold attempts to look at conservation on a large scale involving whole countries or large parts of continents. An interesting review of such bold attempts was given by Charles Mann and Mark Plummer in the journal *Science* (Vol. 260, 25 June 1993, pp. 1868–71). Large-scale conservation must be given real and serious consideration and there are appropriate tools such as Geographical Information Systems to help. However, there are many lessons to be learnt from conservation biology; conservation cannot be achieved by ecology and other sciences alone.

CHAPTER 10 The genetic basis of conservation biology

ALAN J. GRAY

Introduction

The importance of genetic considerations in conservation biology remains controversial. One view believes them to be paramount, holding the maintenance of high levels of genetic diversity to be a major goal of conservation and the loss of genetic diversity, usually in small populations, to be a prime cause of species' vulnerability to extinction. Those with the opposite view point to the many examples of abundant and successful plants and animals which have little genetic variation, and emphasize that extinction is essentially a demographic process and is likely to happen in small populations purely by chance.

The aim of this chapter is to explore this debate by tracing its origins and by outlining enough of the genetic background for the reader to understand what the arguments are – and, if especially interested, to explore them further. The introductory genetics are contained mainly in the first two sections. These, respectively, attempt to answer the following questions:

1 What is genetic diversity, how do we measure it, and how is it distributed within and among populations of different species?
2 What processes, internal and external, affect the maintenance of, and bring about changes in, genetic diversity in natural populations?

The relationship between genetics and conservation biology is explored in the third main section, which considers the origins of the debate about genetic diversity and why it is thought to be important in conservation biology. The final section provides some concluding remarks. First, however, we need to discuss some basic genetics.

Genetic diversity – its measurement and distribution

What is genetic diversity?

A widely acceptable definition of 'genetic diversity' is hard to find. That in *The World Conservation Strategy* prepared by IUCN reads 'the range of genetic material found in the world's organisms' (IUCN *et al.*, 1980) – a definition which is very nearly equivalent to that of the term 'biodiversity'. Geneticists prefer a rather narrower meaning, applying the term in a wide sense to individual species and in a narrower sense to individual organisms (at its narrowest, of course, it could even refer to specific sequences of DNA – see the glossary for an explanation). Thus, we can speak of one species, or even groups of species, having more (or less) genetic diversity than another, or of an individual plant or animal being genetically more diverse than another. Mostly the comparison is made at the population level. Indeed, the concept of diversity (genetic or otherwise) only makes sense through the process of comparison. The key question is how species/populations/individuals compare in terms of their amounts and distribution patterns of measurable genetic variation. Perhaps we can adopt the phrase 'the amount of measurable genetic variation' as a useful working definition of genetic diversity for now, and move on to look at what sorts of variation can be measured.

We have learnt most of what is known about the genetic diversity of organisms from measurement of variation in proteins, and specifically isoenzymes. The first reports of species-level isoenzyme variation, in 1966 by Lewontin and Hubby working on the fruit fly, *Drosophila*, and by Harris working on human genetics, heralded an explosion of studies measuring the extent of such variation in a wide range of organisms. Different forms of an enzyme can be detected if they differ in electrical charge or shape, and consequently have different mobilities when subjected to an electric field across an inert porous matrix, usually a gel of starch or acrylamide. Using this process (called electrophoresis), the different forms of the enzyme can be visualized as bands on the gel by staining with a dye precipitated by an appropriate enzyme-catalysed reaction. Such bands may represent alternative products, or alleles, of a single gene (allozymes) or products of different genes (isozymes).

Set alongside modern molecular methods of detecting genetic variation, protein electrophoresis probably seems a little old-fashioned and has some clear disadvantages. Perhaps the major one is that it detects only a proportion, perhaps less than a third, of the underlying genetic variation. Unless variation in the coding sequence in the DNA, or in the sequence of amino acids in the protein, translates into variation in the electrophoretic mobility of the molecule, such variation will remain undetected. The use of several different gels, varying in pH or ionic concentration, and of heat denaturation, has revealed a great deal of hidden variation, usually as additional alleles at a gene locus. Apart from the problem of such cryptic variation, it is unclear whether the variation revealed by allozymes is typical of the variation in the genome of the organism as a whole (or even of variation in other proteins including more substrate-specific enzymes not detectable by routine electrophoretic analysis). We may be getting a fuzzy, even distorted, picture of genetic diversity.

Despite these difficulties, the fact remains that protein electrophoresis has given us our most complete picture to date of the comparative genetic diversity of different plants and animals. This is because it is a technique which is relatively inexpensive and easy to use, and has therefore been extensively used. Virtually any tissue (leaves, muscle, liver, blood) can be ground up in a buffer solution to provide a crude extract of water-soluble proteins, and many samples (usually 20 to 50) can be run on one gel. More importantly, the genes coding for allozyme variation usually display codominance. In other words, individuals that are heterozygous – having, in diploid organisms, two different alleles at a locus – can be recognized by the presence on the gel of bands produced by each allele. (Depending on the structure of the protein, heterozygotes may produce additional bands of intermediate mobility – see the example of glutamate oxaloacetic transaminase (GOT) in Figure 10.2 below.) For these reasons it is often possible to assess variation in a large number of individuals in a comparatively short time. For many species this has enabled the calculation of key genetic parameters, such as **P**, the proportion of loci that are polymorphic; **A**, the average number of alleles at a polymorphic locus; and **H**, the proportion of heterozygous loci in the average individual. Derivations of these basic parameters, especially **P** and **H**, calculated at the level of populations, species or taxonomic groups, provide us with the statistical means of assessing genetic diversity. For example, it is possible to estimate how similar, or different, populations are by using various measures of 'genetic distance' (Nei, 1987 gives a review of methods), or to detect the genetic 'structure' of populations by calculating how their levels of heterozygosity depart from those predicted by the preconceived model of random mating. As we shall see, measures of heterozygosity are particularly important in conservation biology, especially in the context of endangered or rare species' conservation.

Almost thirty years of allozyme studies have demonstrated enormous variability among higher organisms in levels of genetic polymorphism and heterozygosity, so much so that it is somewhat misleading to talk about averages. However, if we do, it can be said that overall, at loci detectable by protein electrophoresis, between a quarter and a third of all genes are polymorphic and around 10% of loci are heterozygous in an average individual. Most variable genetically are the invertebrates, followed by plants, and then by vertebrates, which have the lowest average values of both **P** and **H**. (Prokaryotes such as *Escherichia coli* have two or three times the amount of variation.) Of more interest than this broad picture are specific comparisons between groups of plants or animals with different life-history characteristics, breeding systems, distribution patterns, and so on. Such comparisons are germane to discussions about conservation genetics and will be developed later in this chapter. Meanwhile, good reviews of protein diversity which give detailed analyses of **P** and **H** over sufficient loci for a large number of species are Hamrick and Godt (1989) for plants, Ward *et al.* (1992) for animals and Nevo (1983) for both. (See also Avise, 1994.)

Turning to other ways of measuring genetic diversity, one can either focus down to the DNA level, or draw back from individual genes and their products to measure variation in traits under the control of many genes – so-called quantitative trait loci, or qtls. There is much to be gained from doing both, although both are more complex, and expensive, than allozyme survey. Paradoxically, both involve some loss of resolution, either ecological or genetic, as the genome or the phenotype, respectively, are brought into sharper focus. There is room here to do little more than list some methods and comment on their value to conservation biology.

First, the expanding field of molecular genetics. Detection of diversity at the DNA level is now possible using a large, and growing, number of techniques, several of which have revolutionized parts of evolutionary biology and genetics. Perhaps the most germane to conservation biology are the various forms of restriction site analysis. These employ enzymes derived from bacteria which are capable of cutting double-stranded DNA at the sites of specific sequences of base-pairs. Each enzyme or 'restriction endonuclease', of which several hundred have been isolated, cleaves DNA at a particular sequence of 5–6 base-pairs. The resulting fragments of DNA can, like protein variants, be separated by electrophoresis – usually on an agarose or acrylamide gel. In one method they may then be transferred as single strands to a nylon or nitrocellulose membrane by a process known as Southern blotting (after Southern, 1975), where fragments of interest can be detected using an appropriate DNA 'probe'. The probe, of previously isolated DNA, can be radioactively labelled and, following hybridization to the homologous sequence(s), their position on the gel visualized by autoradiography. DNA fragments may also be visualized chemically, particularly where the gel contains extremely pure DNA, from, say, mitochondria. These methods reveal Restriction Fragment Length Polymorphisms, or RFLPs, and can be used to detect variation in different parts of the genome.

Here lies one of the major advances of this technology: not only is small-scale variation at the DNA level detectable but also, depending on the probe, variation can be detected in single nuclear or cytoplasmic genes, in non-coding DNA, and in mitochondrial (mtDNA) or chloroplast (cpDNA) DNAs. This allows access to genetic diversity in DNA which may be evolving at different rates and have different modes of transmission (most mtDNA and most cpDNA (but not in gymnosperms) is inherited maternally). Southern blots can be used also to assay highly variable ('hypervariable') regions of DNA from which all individuals (except monozygotic twins) can be distinguished – the 'DNA fingerprints' of Jeffreys (1987).

To the above restriction site analyses can be added an expanding list of methods for studying diversity at the DNA level: DNA–DNA hybridization, DNA sequencing and techniques based on amplifying regions of DNA using the Polymerase Chain Reaction (PCR), such as the analysis of simple sequence length polymorphisms known as microsatellites and of polymorphisms in anonymous DNA detected by short lengths of random nucleotide sequences (primers of around 10 base-pairs) – the polymorphisms known as Random Amplified Polymorphic DNA, or RAPDs. Discussion of these and other 'molecular tools' (Avise, 1994) is beyond the scope of this book, save to say that they are adding, in varying but sometimes enormous amounts, to our knowledge of genetic variation at the species, population and individual levels, and that some have provided insights into genetic variation in rare and endangered species – and thus will be mentioned later. Even the study of quantitative trait loci is being revolutionized by their use as markers of RFLPs, microsatellites and RAPDs.

For most people interested in natural populations, whether as ecologists, plant- or animal-breeders, or conservation biologists, it is variation at the phenotypic level in traits often controlled by many genes with individually small effects which first draws their attention. Variation in size, plumage, flowering, leg length, beak shape, frost resistance, fin length, seed production, leaf shape, relative growth rate, lifetime fecundity – these are the stuff of conservation biology. Although the genetic control of such traits has not always been studied individually, their distribution in natural populations is often very well known. They have provided the basis for differentiating species into subspecies, local races and varieties, and a whole range of subspecific taxa. Yet, despite the vast literature, and the predominance of continuously varying traits over discrete variation in natural populations, they are rarely discussed in the context of conservation genetics. This is presumably because the bewildering array of character diversity does not lend itself to ready generalization, and case-studies of the genetic analysis of quantitative traits, particularly 'ecologically relevant' or 'fitness' traits, are accumulating relatively slowly. Almost wherever it has been looked for, environmentally correlated heritable variation for continuously varying traits has been found, at scales from microhabitat to geographical and in patterns from clinal to abrupt. We cannot make the assumption that such diversity is distributed in a similar way to allozyme diversity, although there is some limited evidence for broad comparisons. But probably we can assume that the ability of the genotype to respond to environmental variation means that, when there are no other data available, it is reasonable to expect that, for many species, individual samples from the widest array of different environments will capture the most genetic diversity.

How is genetic diversity distributed?

The reviews of protein diversity referred to earlier have highlighted significant differences in the variation at these loci displayed by different groups of organisms. These differences occur not only between animals and plants but also between groups of species with different ecological and life-history features. For example, gymnosperms and monocotyledonous plant species generally have more genetic diversity than dicotyledonous species; cosmopolitan and widespread species of both plants and animals generally have more genetic diversity than narrow-ranging and endemic species; those mammals which live above ground have more diversity than those which live predominantly underground; outcrossing plants have more diversity than self-pollinating plants, and so on. These broad generalizations, based on empirical studies of large numbers of species, populations and loci, provide a general 'expectation' of the genetic diversity which a species might have.

Perhaps more important from a conservation biology point of view are expectations about the way in which the genetic diversity within a species might be distributed among its populations. How much of the total species diversity is the average population likely to contain? How many populations must we sample to be sure of including, say, 90% of the diversity? These are the sorts of questions which the manager will need to know to design his or her scheme for conserving a species' genetic diversity.

The allozyme literature does provide general answers to these questions, but it is necessary first to look more closely at the genetic parameters used to measure diversity. This is because a different parameter can give a different picture. Figure 10.1 compares 28 inbreeding and 23 outbreeding diploid flowering plant species for which the parameters **P** and **H** have been calculated (based on data from an early review by Gottlieb, 1981). If **H** is used as the measure of diversity, outbreeding species can be seen to be considerably more diverse than inbreeders. However, many inbreeding species have a high proportion of polymorphic loci and as a group they are more similar to outbreeders when **P** is the parameter used. In other words, inbreeding species may contain high variation, but have few heterozygous individuals. To allow for this, the parameter 'gene diversity' can be calculated as one minus the sum of squares of allelic frequencies at a locus ($\mathbf{H}_T = 1 - \Sigma p_i^2$, where p_i is the mean frequency of the ith allele). The expression Σp_i^2 is used as a basis for calculating an index of species diversity (Spellerberg, 1991). It turns out that using this and derived gene diversity parameters, predominantly outcrossed plant species still have significantly higher levels of overall genetic diversity than self-pollinating species. Especially interesting is the fact that the genetic diversity of the two groups of species is distributed very differently among their populations. Hamrick and Godt's (1989) review of nearly 450 plant species showed that, on average, selfing species had more than 50% of their total diversity distributed among populations, whereas outcrossed, wind-pollinated species had less than 10%.

This sort of contrasting distribution pattern is illustrated in Figure 10.2, which shows the frequencies of the three alleles at the common GOT enzyme locus (glutamate oxaloacetic transaminase) in two grass species in roughly the same geographical area of south-west Britain. Both species have all three alleles but in nitgrass (*Gastridium ventricosum*), a self-pollinating annual, each of the ten populations is fixed for only one (in eight out of ten it is the same allele), whereas in bristle bent-grass (*Agrostis curtisii*), a self-incompatible perennial, all three alleles are present, albeit in different frequencies, in all 30 populations. Furthermore, all other loci in *A. curtisii* display a similar pattern to GOT, as do most, but not all, loci in *G. ventricosum*. The important point to make here is that a very different sampling strategy would be required to capture most of the genetic diversity in the two species. If for some reason we were interested in conserving the diversity at the GOT locus, a single population of *A. curtisii* could be sampled. In contrast, in *G. ventricosum*, even knowing the allelic distribution in advance, a minimum of three populations would be needed. This suggests that, with no advance knowledge of the way in which genetic diversity is shared among populations, different

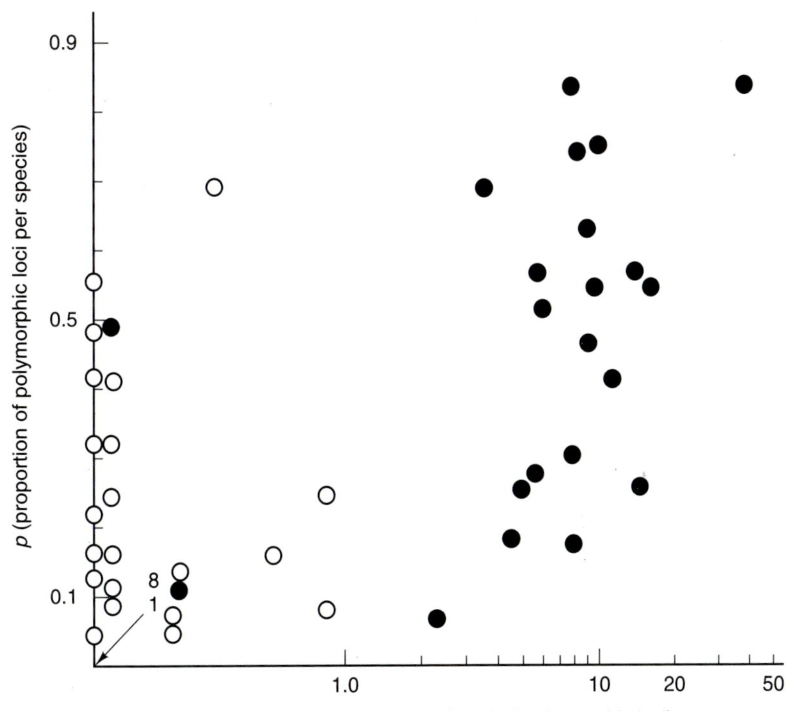

Fig. 10.1 Variation in *P* (the proportion of polymorphic loci per species) and *H* (the mean heterozygosity of all polymorphic loci) for several inbreeding (○) and outbreeding (●) diploid plant species. (Drawn from data in Gottlieb, 1981.)

strategies will be necessary for conserving the diversity in different species. For example, in the absence of any other evidence, a conservationist would be advised to devote more effort to preventing the loss of, say, one of only six remaining populations of an inbreeder than he might, other things being equal, were it an outbreeder.

As this simple example illustrates, electrophoretic surveys, in addition to giving us a general expectation of the amount of genetic diversity to be found in different species based on their taxonomic status, geographical distribution, reproductive biology, and so on, may also suggest how that diversity might be distributed within and among populations.

Such surveys have also, often unexpectedly, turned up examples of species with low genetic diversity, and some with apparently no diversity at all. Ten of the best known of these (one or two of which will be discussed in detail later) are given in Table 10.1; others are listed in Avise (1994). The low genetic diversity of such species has been one of the main triggers for the concern and debate in conservation genetics. The question which arises is: does their low genetic diversity make these species particularly vulnerable to extinction? Concern is heightened by the fact that in all cases there are close relatives, even congeners or subspecies, with what appear to be 'normal' levels of diversity (e.g. southern elephant seals, Serengeti lions, white-tailed deer). Indeed, the lack of allozyme variation is sometimes in surprising contrast to visible morphological variation – the variants in shell pattern in the Florida tree snail, the coat-colour polymorphism in fallow deer.

A more immediate question than that of their vulnerability is how these and other species are likely to have become so depauperate genetically. This is the subject of the next section.

Genetic processes – the dynamics of conservation genetics

The genetic diversity measured in any particular population should not be thought of as a constant

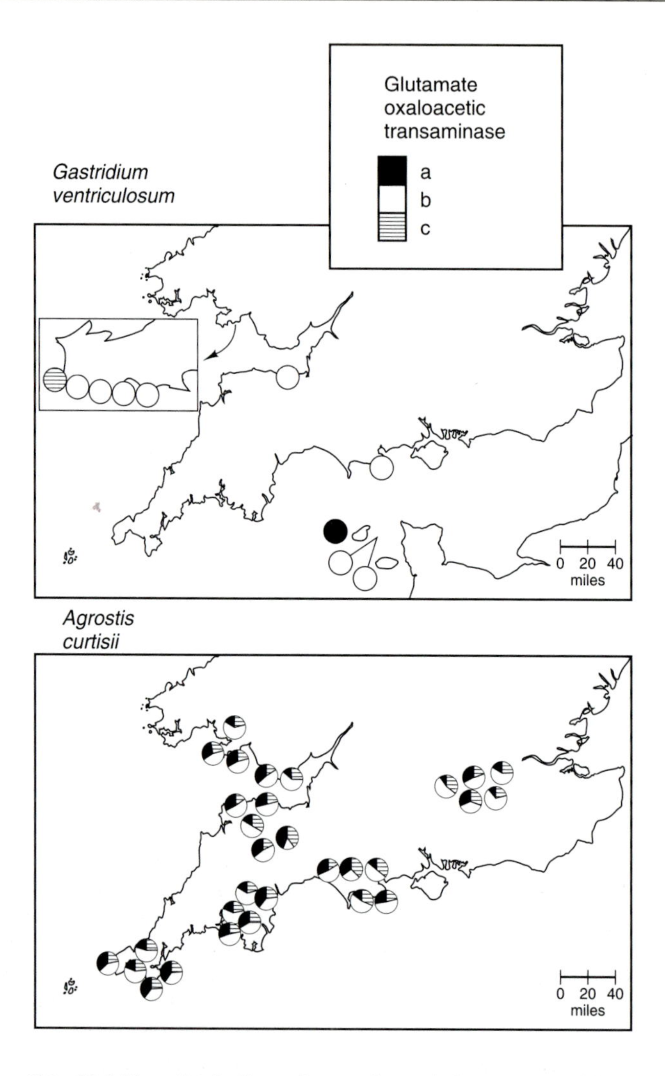

Fig. 10.2 The distribution of genetic variation at the GOT 3 (glutamate oxaloacetic trans-aminase) locus in two grass species of contrasting breeding systems and life histories in southern Britain. There are three alleles at this locus and in *Gastridium* each population ($n = 10$) is fixed for one of these whereas in *Agrostis* all three alleles are found (at different frequencies) in all 30 populations. Data for *Gastridium* are taken with permission from R. John (PhD Thesis, University of Swansea); data for *Agrostis* are from the author's unpublished work.

property of that population. It is there because of the dynamic processes, internal and external, which have produced it and are continuing to change it. These processes range from the behaviour of the chromosomes at meiosis to the behaviour of individuals in choosing a mate, and are influenced by factors ranging from historical changes in population size to rates of between-population migration. In fact, if conservation genetics could be said to have a single objective, it might well be that of conserving the range of those processes which act to shape diversity in a set of species populations – in other words, conserving the ability of a species to change and adapt on ecological and evolutionary time scales. With this objective in mind, what are the main processes acting to reduce diversity in populations, and under what

Table 10.1 Examples of species with extremely low (or apparently no) genetic diversity

Animal or plant	Remarks
Animals	
Cheetah (*Acinonyx jubatus*)	No variation at 47 allozyme loci in 55 individuals from southern Africa, 155 proteins from 6 animals gave *H* of 0.013 (O'Brien *et al.*, 1983)
Asiatic lion (*Panthera leo persica*)	No variation in 46 allozyme loci in 28 animals in Gir forest (Wildt *et al.*, 1987)
Northern elephant seal (*Microunga angustirostris*)	No variation in 24 allozyme loci in 159 seals from 5 populations (Bonnell and Selander, 1974); no variation in 55 allozyme loci, and only 2 mtDNA haplotypes (Hoelzel *et al.*, 1993)
Fallow deer (*Dama dama*)	No variation at 30 loci in 794 animals from 37 sites in Britain (Pemberton and Smith, 1985)
Bison (European Bison, *Bison bonasus*; and North American bison, *B. bison*)	*H* of 0.012 in 69 allozyme loci in 35 animals from Poland (*B. bonasus*) (Hartl and Pucek, 1994); only one out of 24 loci polymorphic in a South Dakota herd of *B. bison* (McClenaghan *et al.*, 1990)
Florida tree snail (*Liguus fasciatus*)	Only one polymorphic allozyme locus out of 34 from 60 individuals, with $H = 0.002$ (Hillis *et al.*, 1991)
Plants	
Howellia aquaticus	No variation in 18 allozyme loci in several populations from USA (Lesica *et al.*, 1988)
Bensoniella oregana	No variation in 24 allozyme loci in this North American endemic (Soltis *et al.*, 1992)
Spartina anglica	Very low levels of variation at 21 allozyme loci in England, with most populations genetically uniform (Raybould *et al.*, 1991)
Pedicularis furbishiae	No variation in 22 allozyme loci in populations in northern Maine, USA (Waller *et al.*, 1987)

circumstances may they have important consequences for conservation biology? These are discussed in this section under two headings which deal respectively with the genetic consequences of inbreeding and of small population size. We should note that these processes are strongly linked.

What are the genetic consequences of inbreeding?

Inbreeding can be defined as mating between relatives. That it can have deleterious consequences for the fitness of offspring is well known from human societies and also from animal breeding, where protocols to avoid its effects are an essential part of captive breeding programmes for rare vertebrates (see Chapter 15). Plants, too, are affected by inbreeding, but to varying degrees, depending on their breeding systems and past breeding history.

An obvious consequence of inbreeding is a loss of heterozygosity. If we consider the fate of a heterozygous gene locus **Aa** under the severest form of inbreeding, repeated selfing (a common occurrence in plants), it can be seen that in a diploid species, from a 1 to 1 ratio of heterozygotes (**Aa**) and homozygotes (**AA, aa**) after the first selfing, the proportion of **Aa** genotypes, given equal descent, declines to 1 in 253 after only seven generations of selfing. In addition, half of the homozygotes in this (highly unlikely) example would be **aa** recessives. The exposure and increase of recessive alleles is one of the potential causes of the loss of fitness which may occur under inbreeding and which is given the general term 'inbreeding depression'. For instance, it is relatively easy to demonstrate that recessive alleles with harmful and even lethal effects can occur at quite high frequencies in habitually outbreeding species. An example which springs to mind is that of outbreeding grasses, where a round of artificial selfing can produce the strangest set of weakling individuals, including, in sea meadow-grass (*Puccinellia maritima*) and cocksfoot (*Dactylis glomerata*) and probably other species, extremely short-lived albino seedlings incapable of manufacturing chlorophyll. If these had been the **aa** recessives of the rather fanciful example used above, the outcome would demonstrate one of the other potential genetic effects of inbreeding – the purging from populations under selection of alleles with deleterious or lethal effects. Therefore, under some circumstances inbreeding might actually improve the average fitness of a population, and has even been prescribed and used effectively in some captive breeding programmes (e.g. in Speke's gazelle).

Inbreeding depression was observed by Charles Darwin (1876), who, among other things, compared the performance of plants produced by self-fertilization with those produced by cross-fertilization. In plants, the reduction in performance, or fitness, due to inbreeding depression can be expressed at all stages of the lifecycle from seed abortion rates, germination, and juvenile survivorship through to fecundity in breeding individuals. Thus, in an experiment with the rose clover (*Trifolium hirtum*), seed from selfed and outcrossed plants had equal germination rates but the seedlings of selfed plants were less vigorous and had reduced survivorship (Molina-Freaner and Jain, 1993). As adults, there was no difference in reproductive performance between plants from selfed or outcrossed parents, yet the difference in fitness acting largely at the seedling stage was sufficient to reduce the relative overall fitness of selfed plants (compared to open-pollinated plants) by 36% for a single selfed generation and 77% for those produced by two generations of selfing. By contrast, in another rare species, the lakeside daisy (*Hymenoxys acaulis*), a single generation of selfing produced significant reductions in the amount of seed set in selfed versus outcross lines (Demauro, 1993). In fact, this sort of finding, a reduction in flower production or seed set, is rather common in the empirical comparisons of selfed and outcrossed progeny in plants. It may indicate that much inbreeding depression occurs from the effects of many genes of individually small effect, as opposed to harmful or lethal recessives. These sorts of genes will not be easily purged by inbreeding and are likely to be present even in plants, and indeed invertebrates, which habitually reproduce by selfing. That they are having an effect on fitness may be detectable only by measuring the performance of outcrossed individuals, using different inbred lines. In many cases these will display hybrid vigour, or heterosis, linked to increased heterozygosity. In others, where local populations may have diverged genetically and evolved internal coadapted combinations of genes, a reduction of fitness in the hybrids may be observed – a phenomenon known as 'outbreeding depression'.

The effect of inbreeding is therefore likely to depend on a large number of factors including variation in the genetic basis of inbreeding depression. In *Drosophila* populations, approximately half the inbreeding depression results from rare recessive lethal or semi-lethal alleles, individuals in large outbred populations being normally heterozygous for one or two recessive lethals (at about 5000 loci), and half results from many slightly detrimental mutations that are mildly recessive (see Lande, 1988). For most species we simply do not know what the genetic basis of inbreeding depression is, and would find it difficult to predict the outcome of any particular bout of inbreeding, say in an attempt to purge a captive population of its genetic load of harmful recessives.

An example in the conservation genetics literature which nicely illustrates this point and attempts to compare the effects of inbreeding in different populations of the same species is the work of Lacy and others on deer mice of the genus *Peromyscus* in North America (Lacy, 1992). As shown in Figure 10.3, the response to experimental cumulative inbreeding as measured by changes in litter viability was very different in different populations of mice – most showed inbreeding depression of varying severity, but two displayed increased offspring survival with inbreeding. Importantly, but disappointingly, there was no apparent relationship between the presumed history of inbreeding in these populations (as indicated by their size, genetic diversity and insularity) and the effect of the experimental inbreeding. If inbreeding depression in this species was caused mainly by deleterious recessive alleles (the 'dominance' model), one might expect small, isolated populations with long histories of inbreeding to have been purged already of

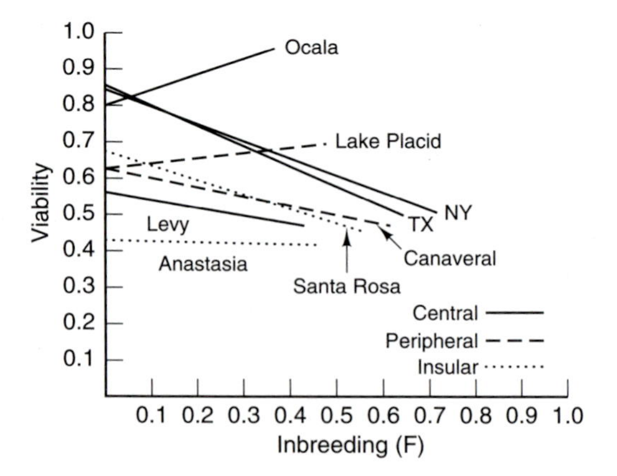

Fig. 10.3 The effect on litter viability (*y* axis) of cumulative inbreeding (*x* axis) in eight populations of the deer mouse *Peromyscus*. Mice were from central, peripheral or insular populations with presumadly different histories of inbreeding. (After Lacy, 1992.)

their genetic load, and thus be tolerant of further inbreeding. Although such populations were less diverse genetically, they showed in some cases greater inbreeding depression than mice from the large heterogeneous central populations. Nor can we ascribe easily the variation in inbreeding depression to the fact that it may result from some form of general heterozygote advantage throughout the genome (the 'over-dominance' model). Both previously inbred and previously outbred populations displayed variation in the effects of inbreeding.

In summary, the genetic consequences of inbreeding are not, in particular populations or species, easy to predict. They are most likely to produce inbreeding depression, a reduction in fitness relative to outbred individuals, in normally outbreeding species which suddenly face a bout of inbreeding (caused perhaps by reduced population size), but even species of plants which are regularly selfing display inbreeding depression in controlled experiments where the progeny of selfed and outcrossed individuals are compared. Indeed, a third of all flowering plant species, and many animals, regularly inbreed in natural populations. We do not have a generic model of the effects of inbreeding which we can take off the shelf and apply to all species or populations. This is not only because of variation in the breeding system or breeding history of different populations but also because of uncertainty in particular cases about the genetic basis of inbreeding effects.

What are the genetic consequences of small population size?

The first point to make is that a likely consequence of small population size is an increase in the amount of mating between relatives, and hence inbreeding; therefore, small population size may have the genetic consequences ascribed to those of inbreeding discussed above. Indeed, the debate in conservation biology about what constitutes a Minimum Viable Population size (MVP – see Chapter 11) embraces the concept that the population of concern should be large enough to prevent, or avoid the effects of, inbreeding depression (and to retain genetic diversity).

Secondly, a discussion of population size should be prefaced by some remarks about 'effective population size'. Rarely, if ever, does every individual in a natural population, of size N, have an equal chance of contributing genes to the next generation. Among the host of possible reasons for this are unequal sex ratios, differences in individual fertility, non-random mating (say, as a result of social grouping in animals or limited gene flow in plants), variation in age structure, and fluctuations in the number of breeding individuals. From these and other causes, the size of the breeding population is unlikely to be that of the total population, N. A parameter often estimated in population genetics is the genetically effective population size, N_e. This is the size of an idealized population in which all individuals have an equal contribution to the gamete pool (the genes for the next generation), and have the same variation in allele frequencies and inbreeding rate as the observed population. N_e has been calculated from allozyme data in several animal (but hardly any plant) populations, and is generally much smaller than the actual, or census, number. Empirically estimated ratios of effective population number to actual number (N_e/N) in animal populations show enormous variation, those listed by Crawford (1984) varying from 0.01 to 0.95 (i.e. between 1% and 95% of the census number). This suggests that we should exercise extreme caution in inferring the genetically effective size of a population from the actual number of individuals in it.

Returning briefly to the effect of population size on inbreeding, it can be shown that the level of inbreeding, expressed as the inbreeding coefficient F, is dependent on the effective population size N_e and changes over time as $\Delta F = 1/(2N_e)$ per generation. Thus, populations with small effective size become inbred more rapidly.

A second genetic consequence of small population size is the random fluctuations, and even loss, of alleles during the sampling process which occurs between generations. This is the process known as 'genetic drift'. The probability of an allele being lost, or alternatively fixed, purely by chance increases as the effective population size decreases (and as the allele frequency decreases or increases, respectively). Alleles at low frequency are more likely to be lost from a population of a given small size than alleles at high frequency. Where the starting frequency of each allele at a locus is equal, as in our selfing heterozygote example earlier (**Aa**), the probability of loss or fixation of an allele in the first generation (i.e. getting **AA** or **aa**) is 0.5 when $N = 1$, and decreases as N increases.

Genetic drift has two effects on the genetic diversity of populations. First, it reduces diversity through decreasing heterozygosity (as the low frequency drifting allele becomes rarer) and eventual loss or fixation of alleles, and secondly, it increases the amount of genetic differentiation between populations (since by chance different alleles may be lost from different isolated, small populations). Because it impacts on small populations, genetic drift is likely to be important during processes such as the colonizing of new sites by a few individuals, a 'founder' event, or the temporary reduction in size which a population may experience by accident, isolation, or following high predation or disease, a 'bottleneck'. A currently large population may of course have experienced a founder event or a bottleneck in its past and, as we shall see below, knowledge of the history and population dynamics of a species is important in trying to understand the pattern of genetic diversity and structure it displays.

Not all temporary reductions in population size will produce genetic bottlenecks. The genetic effects of a bottleneck depend on several factors, including how small the population becomes, how long the bottleneck persists, and how quickly the population recovers or increases in size following the bottleneck. There is a good deal of theoretical and experimental work to do in this area, particularly in regard to the effects of bottlenecks on quantitative traits. Nevertheless, the occurrence of a bottleneck in recent, or sometimes evolutionary, times is commonly argued to be the major cause of low genetic diversity in species populations. The animals listed as examples in Table 10.1 have in all cases had known or suspected population bottlenecks which could explain their extremely low genetic diversity. Two bottlenecks may have occurred in cheetahs, one prior to the division into two subspecies (*c.* 10 000 years ago) and a more recent one within the South African population. The Asiatic lion of the Gir Forest was reduced to less than 20 individuals in the early part of this century; the northern elephant seal was hunted almost to extinction in the nineteenth century, with today's large population having descended from fewer than 30 animals; European fallow deer are thought to have experienced a bottleneck during the glaciation of Europe and a period of captivity by early man; and so on. Random drift in small, isolated populations has been invoked to explain the fact that the small, scattered populations of the west European brown bear (*Ursus arctos*), are monomorphic and fixed for different mtDNA haplotypes in different parts of Europe. There are many other examples.

Many plant species, too, including those in Table 10.1, have experienced extremely low population numbers during which genetic bottlenecks may have occurred. *Howellia aquatilis* and *Peduncularis furbishiae* are both thought to have been reduced to small, isolated populations during Pleistocene glaciation. *Spartina anglica*, by contrast, experienced a genetic bottleneck during its speciation around 100 years ago when it was generated by allopolyploidy following hybridization between *S. maritima*, a European species, and *S. alterniflora*, a North American species accidentally introduced into southern England by shipping. *S. anglica* is believed to be genetically uniform because it was created by a rare, even single, hybridization event and chromosome doubling, because its parents are relatively uniform, it spreads extensively by clonal growth, and because preferential pairing at meiosis prevents recombination between parental genomes (Gray *et al.*, 1991).

Despite its low genetic diversity, *S. anglica* is a successful and invasive species, having colonized extensive areas of intertidal mudflats in temperate zone saltmarshes around the world. Saltmarsh dominated by *S. anglica* occupies about a quarter – more than 10 000 ha – of the saltmarsh area in Great Britain and, remarkably, from only a handful of plants sent to China in 1963 there are now more than 40 000 ha of *Spartina* marsh along the Chinese coast. Similarly, several plant species with low genetic diversity as a result of a few founders being introduced into another country have become successful colonizers, and even weeds. Examples include *Avena barbata* in California, *Bromus tectorum* in North America, and *Chondrilla junceum* in Australia (see Barrett and Kohn, 1991). Therefore, at least on some occasions, species with low genetic diversity, far from being endangered, appear to be expanding rapidly in numbers. This brings us to the subject-matter of the next section – the origins of conservation genetics and the issues which it has generated.

Conservation genetics – origins and issues

Although problems such as the potential loss of genetic diversity in small populations had been issues for some time before, the major stimulus to the new science of conservation genetics was provided in 1981 by Frankel and Soulé's classic book *Conservation and Evolution*. This book set the agenda for the modern debate. It brought together concerns about the loss of genetic resources for crop breeding and improvement voiced by Otto Frankel in the early 1970s, with questions about the wider implications of reduced heterozygosity in small populations on species survival and evolution, which Michael Soulé and others had been addressing in earlier studies. The book gave prominence to the '50/500 rule of thumb' developed by Franklin (1980) and Soulé (1980) in an earlier influential text on conservation biology edited by Soulé and Wilcox (1980). These numbers represent the approximate genetic effective size at which a tolerable level of inbreeding will occur and fitness will not be seriously reduced ($N_e = 50$), and the minimum genetically effective size required to prevent the gradual erosion of genetic variation and to allow for future adaptive and evolutionary change ($N_e = 500$). The universal validity of the 50/500 rule has been challenged frequently, particularly as a general prescription for managing natural populations (see Chapter 11). Nevertheless it embodies the major concern and thrust of Frankel and Soulé's argument, which is that human activity has led (and continues to lead), via habitat loss and fragmentation or the effects of pollution, to the reduction of many species' populations to sizes at which they face a threat from genetic effects on both fitness and evolutionary potential. They describe the genetics of nature conservation as 'the genetics of scarcity'. From the outset, conservation genetics has been focused on small populations and rare species, and on the threat presented by low genetic diversity.

In advancing their argument, Frankel and Soulé (1981) criticize the view, which they label as 'phyletic optimism', that there is sufficient genetic variation in most species to enable them to adapt to changing environments, and that natural selection is a universal and powerful force capable of overriding random genetic drift and facilitating rapid and precise adaptation. In support of the optimists' view, typified by Berry (1971), there are several examples of rapid genetic response to environmental challenges, such as the evolution of heavy metal tolerance in several plant species on contaminated soils from metal mines (which has implications for ecological restoration: see Chapter 16, page 207) and the spread of melanic (black) forms of a number of moths and other invertebrate species in response to industrial pollution. However, Frankel and Soulé maintain that this picture of high genetic diversity and adaptability does not fit all species and is totally inappropriate for large, slowly reproducing organisms. Large organisms do not live in very large populations, cannot recover quickly from catastrophic reductions in number because of low reproductive rates, are often inefficiently dispersed and unable to colonize new patches, have generally less genetic diversity than small organisms, and are prone to lose genetic variation in very small populations by genetic drift and inbreeding. Indeed, for most large organisms, notably vertebrates, there is probably not enough space in existing nature reserves for future speciation, or even for continuing gradual evolution by adaptive genetic change. Genetically effective population sizes are often below the 500 threshold.

Challenges to the views of Frankel and Soulé (whose arguments and evidence are a good deal more subtle and complex than it is possible to do justice to here) have been of two sorts. We can crudely label these (1) demography is more important than genetics, and (2) low genetic diversity does not matter. The first of these was initially advanced by Lande (1988), who argued that demographic considerations (those factors which affect population growth and age structure) are more immediately important than population genetics in determining minumum viable population sizes – and hence the likelihood of local extinction. Lande warns us against the assumption that low levels of electrophoretically detectable variation in enzymes necessarily mean that a population lacks heritable variation in quantitative traits or is suffering from inbreeding depression. Furthermore, he points out that there is little evidence that heterozygosity *per se* actually increases fitness, as opposed to simply avoiding inbreeding depression or allowing adaptation to environmental change. Much more important in small populations are demographic factors such as social structure (a critical family size, the need for group defence, problems in finding a mate), stochasticity (random fluctuations created demographically or by environmental effects), and the patterns of local extinction and colonization in species which exist as subdivided populations.

The idea that genetic diversity does not matter, or more specifically that a lack of genetic diversity does not constitute a threat to species survival, has grown from the realization that there are several widespread and successful species which are genetically depauperate. It can also be pointed out that in those cases where the cause of low diversity was known or believed to be a population or genetic bottleneck (e.g. fallow deer, northern elephant seal, *Spartina* grass), populations have recovered remarkably well, often rapidly, from very low numbers. As we saw in the first two sections, low genetic diversity in a population may signal several different phenomena, depending on the species, its breeding system and its history. In many cases it will be there as a result of a bottleneck rather than as a cause of a bottleneck. In others it will reflect a high level of inbreeding, but without any perceptible effects on fitness. In yet others it will not reflect the levels of genetic variation in other parts of the genome, say in quantitative trait loci.

Taken together, and applied to the problems of conserving rare and endangered species, these factors demonstrate the difficulty of untangling genetic from demographic causes of population decline and extinction. Indeed, genetic and demographic processes are likely to interact synergistically. When populations become small enough for demographic factors to further reduce their size (e.g. mates are difficult to find or, as in the lakeside daisy, there are not enough mating types for a self-incompatible breeding system to operate), then loss of fitness due to genetic drift and inbreeding can lead to a further reduction in numbers, creating more demographic problems, and so on. This has been dubbed the 'extinction vortex' (Gilpin and Soulé, 1986). A key question, of course, is to determine at what size the extinction vortex begins to operate (i.e. the Minimum Viable Population size). It may be a size for which N_e is less than 50, or it may not, but it is likely to vary greatly between species.

The intimacy of demographic and genetic factors and their heightened interaction in small populations mean that it may never be possible to say of a particular extinction event that it was caused by a lack of genetic diversity. The relative importance attached to genetic factors in biological conservation may therefore depend on the experience and background of the observer and the intellectual baggage he or she brings to the debate. It can be argued that, despite population genetic theory and experience with plant and animal breeding, it has yet to be shown that inbreeding depression has been the cause of decline in abundance in any natural population, and that no wild population of any species has ever been shown to have gone extinct because of loss of heterozygosity (see Chapter 11).

Caro and Laurenson (1994) marshal these arguments in relation to the case of the cheetah. As we have seen, this animal is extremely homozygous (Table 10.1), and further evidence of lack of genetic variation has been provided by the successful acceptance of skin grafts between pairs of widely separated, 'unrelated' individuals (suggesting they have very little variation at a major histocompatibility complex). This has been said by several authors (mostly not the original researchers) to be the reason why cheetahs suffer high juvenile mortality, both in the wild and in captivity, and why they appear to be highly susceptible to disease. In fact, research on Serengeti cheetahs demonstrated that the major cause of cub mortality was predation, by lions and spotted hyenas, followed by environmental hazards (fire and exposure), and abandonment. Only two deaths out of 35 in the lair could be attributed to possible genetic defects, and in all other respects females showed 'normal' reproductive performance. In captivity, the main problem turns out to be extremely low conception rates. Here, again, the evidence points overwhelmingly to non-genetic causes, such as poor husbandry, difficulty in detecting oestrus, and unsuitable social conditions for mating. Finally, recent studies have demonstrated variation in the susceptibility of cheetahs to a range of diseases, indicating that their immune systems can recognize and respond to a range of pathogens. Interestingly, recent studies have also unearthed moderate levels of genetic variation in other parts of the cheetah's genome (the more rapidly evolving parts such as mtDNA and minisatellite nuclear loci).

From this evidence, Caro and Laurenson conclude that, while genetic considerations are clearly important in managing captive populations, they may have limited relevance in most wild populations – mainly because 'they impact populations on a slower time scale than environmental or demographic problems'. A perspective on this point is provided by the example of the Virunga mountain gorilla population, which numbers about 300 animals. Harcourt (1991) has calculated that even if N_e is only 50, more than 100 years will pass before heterozygosity is reduced by 10%. In that time the local human population, at current rates of population increase, will have doubled, and doubled again. Under these circumstances it is difficult to maintain that the gorilla population is threatened by loss of genetic diversity.

Although it is generally accepted that human disturbance presents a greater, and more rapidly acting, threat to natural populations than loss of genetic diversity, it is nonetheless true that, in the human management of such populations, whether by captive breeding or in the location and design of nature reserves, genetic considerations are of prime importance. Assays of genetic diversity, even at a limited number of isoenzyme loci, can provide powerful insight into the breeding system and history of the species. They can also form the basis of nature reserve design, species recovery programmes involving manipulation of populations or reintroductions, and of course, any controlled breeding programme

(Chapter 15). A knowledge of genetic diversity, as we saw in the case of nitgrass and bristle bent-grass, can help in specific instances with the choice of which populations to defend or conserve (it being highly unlikely that the option of conserving everything will be possible).

The application of population genetics principles and practice to the management and monitoring of natural populations is in its infancy and a very good account of developments in plant conservation genetics is given by Ellstrand and Elam (1993). The warning signs that populations may be vulnerable (changes in size, isolation and fitness) are identified and some management options discussed. Prescriptions for the management of populations of wild animals are somewhat scattered through the literature, but Avise (1994) provides an excellent introduction and lists examples.

Conclusions

A widely accepted goal of nature conservation is the maintenance of biodiversity, and this includes genetic diversity. Genetic diversity is believed to be important for species survival and evolution. (It is also regarded as important for the future human exploitation of genetic resources, usually for food, medicine or fibres, and, on ethical grounds, as a legacy to future generations.)

Although new molecular techniques are adding enormously to our knowledge, most measures of genetic diversity have been made by sampling the genetic variation at a restricted number of gene loci – overwhelmingly those which code for variation in soluble proteins. Nevertheless, they reveal that the forces which shape genetic diversity in populations include intrinsic factors such as the breeding system and extrinsic factors such as the history, size and isolation of populations.

Such forces are particularly sensitive to population size, and to rapid reductions in size, often to very low numbers, which may create a genetic bottleneck. The main concern of conservation genetics has therefore been focused on small populations and rare species – 'the genetics of scarcity'. This is because the loss of genetic diversity from inbreeding and genetic drift in small populations is seen to be a major cause of species vulnerability to extinction.

There is no universally applicable population size below which a serious reduction in genetic diversity will occur, or above which a population is safe from genetic erosion. For any particular species, the genetically effective population size (the size of an idealized population in which all individuals contribute equally to the gene pool for the next generation) is likely to be different (usually smaller), and may be very different, from the actual number of individuals. As a first approximation, a genetically effective population size of 50 has been advanced as that necessary to prevent serious inbreeding effects, and one of 500 as that which will enable future adaptive and evolutionary change.

It may never be possible to directly attribute the extinction of a particular population or species to a lack of genetic diversity. In small populations, demographic and genetic processes will often interact synergistically to further reduce numbers – the extinction vortex. The extent to which low diversity affects survivorship and fitness varies greatly between different species. Several successful and widespread species appear to be genetically depauperate. So are many rare and endangered species. There are divergent views on the effect of inbreeding and loss of genetic diversity on population persistence.

Population genetics principles can be applied to the management of natural populations and, together with ecological and behavioural considerations, should guide recovery programmes for endangered species. They can also guide the design of nature reserves. (The maintenance, and occasionally the enhancement, of genetic diversity in captive populations is benefiting from new molecular methods for assessing relatedness and constructing pedigrees.) A major aim of conservation genetics is to conserve the range of processes which allow evolutionary changes to proceed in natural populations, i.e. to conserve dynamic processes, not to fix levels of genetic diversity.

Finally, we should not forget that conservation genetics in an infant science. General principles are slow to evolve, and, despite well-established population genetics theory, will take time to emerge satisfactorily from the growing caseload of empirical studies of specific plants and animals. If we accept that the greatest threat to wild populations is the burgeoning human population, and that unravelling and understanding the ecological and genetic processes which lead to extinction mostly involves detailed, long-term research, we must face the reality that conservation biologists, to say the least, will tend to have other things on their minds. Recognising this, Frankel and Soulé (1981) declared that 'Conservationists cannot afford the luxury and excitement of adversary science' (see page 4 of this volume for full quote). Such realities have added urgency, spice, and not a little emotion to the debate about the importance of genetic considerations in conservation biology. And that is no bad thing.

References

Avise, J.C. (1994) *Molecular markers, natural history and evolution.* New York: Chapman & Hall.

Barrett, S.C.H. and J.R. Kohn (1991) Genetic and evolutionary consequences of small population sizes in plants: implications for conservation. In D.A. Falk and K.E. Holsinger (eds) *Genetics and conservation of rare plants*, Oxford: Oxford University Press.

Berry, R.J. (1971) Conservation aspects of the genetical constitution of populations. In E. Duffey and A.S. Watt (eds) *The scientific management of plant and animal communities for conservation.* Oxford: Blackwell Scientific, pp. 177–206.

Bonnell, M.L. and R.K. Selander (1974) Elephant seals: genetic variation and near extinction. *Science*, **184**, 908–9.

Caro, T.M. and M.K. Laurenson (1994) Ecological and genetic factors in conservation: a cautionary tale. *Science*, **263**, 485–6.

Crawford, T.J. (1984) What is a population? In B. Shorrocks (ed.) *Evolutionary ecology.* Oxford: Blackwell Scientific, pp. 135–74

Darwin, C.R. (1876) *The effects of cross and self-fertilization in the vegetable kingdom.* London: John Murray.

Demauro, M.M. (1993) Relationship of breeding system to rarity in the lakeside daisy (*Hymenoxys acaulis* var. *glabra*). *Conservation Biology*, **7**, 542–50.

Ellstrand, N.C. and D.R. Elam (1993) Population genetic consequences of small population size: implications for plant conservation. *Annual Review of Ecology and Systematics*, **24**, 217–42.

Frankel, O.H. and M.E. Soulé (1981) *Conservation and evolution.* Cambridge: Cambridge University Press.

Franklin, I.A. (1980) Evolutionary change in small populations. In M.E. Soulé and B.A. Wilcox (eds) *Conservation biology: an evolutionary–ecological perspective.* Sunderland, MA: Sinauer Associates, pp. 135–50.

Gilpin, M.E. and M.E. Soulé (1986) Minimum viable populations: processes of species extinction. In M.E. Soulé (ed.) *Conservation biology: the science of scarcity and diversity.* Sunderland, MA: Sinauer Associates, pp. 19–34.

Gottlieb, L.D. (1981) Electrophoretic evidence and plant populations. *Progress in Phytochemistry*, **7**, 2–46

Gray, A.J., D.F. Marshall and A.F. Raybould (1991) A century of evolution in *Spartina anglica*. *Advances in Ecological Research*, **21**, 1–62.

Hamrick, J.L. and M.J.W. Godt (1989) Allozyme diversity in plants. In A.H.D. Brown, M.T. Clegg, A.L. Kahler and B.S. Wier (eds) *Plant population genetics, breeding and genetic resources.* Sunderland, MA: Sinauer Associates.

Harcourt, S. (1991) Endangered species. *Nature (London)*, **354**, 10.

Hartl, G.B. and Z. Pucek (1994) Genetic depletion in the European bison (*Bison bonasus*) and the significance of electrophoretic heterozygosity for conservation. *Conservation Biology*, **8**, 167–74.

Hillis. D.M., M.T. Dixon and A.L. Jones (1991) Minimal genetic variation in a morphologically diverse species (Florida tree snail. *Liguus fasciatus*). *Journal of Heredity*, **82**, 282–6.

Hoelzel. A.R., J. Halley, C. Campagna, T. Ambon, B. LeBoeuf, S.J. O'Brien, K. Ralls and G.A. Dover (1993) Elephant seal genetic variation and the use of simulation models to investigate historical population bottlenecks. *Journal of Heredity*, **84**, 443–9.

IUCN, UNEP, WWF (1980) *The World Conservation Strategy. Living resource conservation for sustainable development.* Gland: IUCN.

Jeffreys, A.J. (1987) Highly variable minisatellite and DNA fingerprints. *Biochemical Society Transactions*, **15**, 309–17.

Lacy, R.C. (1992) The effects of inbreeding on isolated populations: are minimum viable population sizes predictable? In P.L. Fiedler and S.K. Jain (eds) *Conservation Biology.* New York: Chapman & Hall, pp. 277–96.

Lande, R.C. (1988) Genetics and demography in biological conservation. *Science*, **241**, 1455–60.

Lesica, P., R.F. Leary, F.W. Allendorf and D.E. Bilderback (1988) Lack of genic diversity within and among populations of an endangered plant. *Howellia aquatilis*. *Conservation Biology*, **2**, 275–82.

Lewontin, R.C. and J.L. Hubby (1966) A molecular approach to the study of genic heterozygosity in natural populations. II. Amount of variation and degree of heterozygosity in natural populations of *Drosophila pseudoobscura*. *Genetics*, **54**, 595–609.

McClenaghan. L.R., J. Berger and H.D. Truesdale (1990) Founding lineages and genic variability in plains bison (*Bison bison*) from Badlands National Park, S. Dakota. *Conservation Biology*, **4**, 285–9.

Molina-Freaner, F. and S.K. Jain (1993) Inbreeding effects in a gynodioecious population of the colonizing species *Trifolium hirtum* All. *Evolution*, **47**, 1472–9.

Nei, M. (1987) *Molecular evolutionary genetics*, New York: Columbia University Press.

Nevo, E. (1983) Adaptive significance of protein variation. In G.S. Oxford and D. Rollinson (eds) *Protein polymorphism: adaptive and taxonomic significance.* London: Academic Press, pp. 239–82.

O'Brien. S.J., D.E. Wildt, D. Goldman, C.R. Merril and M. Bush (1983) The cheetah is depauperate in genetic variation. *Science*, **221**, 459–62.

Pemberton, J.M. and R.H. Smith (1985) Lack of biochemical polymorphism in British fallow deer. *Heredity*, **55**, 199–207.

Raybould. A.F., A.J. Gray, M.J. Lawrence and D.F. Marshall (1991) The evolution of *Spartina anglica* C.E. Hubbard (Gramineae): origin and genetic variation. *Biological Journal of the Linnean Society*, **43**, 111–26.

Soltis, P.S., D.E. Soltis, T.L. Tucker and F.A. Lang (1992) Allozyme variability is absent in the narrow endemic *Bensionella oregona* (Saxifragaceae). *Conservation Biology*, **6**, 131–4.

Soulé, M.E. (1980) Thresholds for survival: maintaining fitness and evolutionary potential. In M.E. Soulé and B.A. Wilcox (eds) *Conservation biology: an evolutionary–ecological perspective*. Sunderland, MA: Sinauer Associates, pp. 111–24.

Soulé M.E. and B.A. Wilcox (1980) *Conservation biology: an evolutionary–ecological perspective*, Sunderland, MA: Sinauer Associates.

Southern, E.M. (1975) Detection of specific sequences among DNA fragments separated by gel electrophoresis. *Journal of Molecular Biology*, **98**, 503–17.

Spellerberg, I.F. (1991) *Monitoring ecological change*. Cambridge: Cambridge University Press.

Waller, D.M., D.M. O'Malley and S.C. Gawler (1987) Genetic variation in the extreme endemic *Pedicularis furbishiae* (Scrophulariaceae). *Conservation Biology*, **1**, 335–40.

Ward. R.D., D.O.F. Skibinski and M. Woodwark (1992) Protein heterozygosity, protein structure, and taxonomic differentiation. *Evolutionary Biology*, **26**, 73–159.

Wildt, D.E., M. Bush, K.L. Goodrowe, C. Packer, A.E. Pusey, L.J. Brown, P. Joslin and S.J. O'Brien (1987) Reproductive and genetic consequences of founding isolated lion populations. *Nature (London)*, **329**, 328–31.

CHAPTER 11 Conservation of populations

ALAN J.A. STEWART and
MICHAEL J. HUTCHINGS

Introduction

Conservation biologists can target their efforts towards ecosystems, habitats, communities, species, populations, lower taxonomic units or even specific alleles. Many conservation programmes are seen as ultimately directed towards conservation of species, although the practical work is invariably aimed at *populations*. In any case, conservation of populations often approximates closely to conservation of species; for example, 73% of the 2290 native North American plant taxa regarded as priorities for conservation are believed to exist in five or fewer populations.

Charged with safeguarding a declining or endangered species, the challenge for the population biologist is at least to arrest, if not reverse, the adverse population trend. Several types of ecological information are needed before informed action can be taken: (1) an understanding of what factors control the population density of the species at any one place; coupled with (2) how and why population size changes over time; (3) the identification and assessment of potential and actual threats to the population; and (4) knowledge of how particular management interventions are likely to affect the population positively.

Historically, management regimes have often been developed using a subjective mixture of wisdom, intuition and reaction to circumstances. This approach has often been successful in spite of the paucity of scientific information about most rare species. However, subjective approaches to management are risky, as errors of judgement may cause accidental extinction of the 'protected' species. Even when the outcome is not so dire, it is still difficult to be certain why the programme failed, so that no positive lesson is learnt and similar mistakes may be repeated. It is now widely agreed that conservation action has to be based on sound scientific data and rooted in the principles of population ecology.

This chapter is concerned specifically with the conservation of populations. Examples are drawn from both animals and plants, although the literature shows rather different emphases and less overlap than might be expected or indeed desired. Recent reviews providing substantial coverage of the field can be found in Caughley (1994), Falk and Holsinger (1991), Fiedler and Jain (1992), Schemske *et al.* (1994), Simberloff (1988) and Western and Pearl (1989).

Defining the population

Population status can only be defined in the context of a fixed geographical area. While some species form discrete colonies with very restricted movement between them, the majority do not and so the definition of a population is often based as much on the geopolitical unit most convenient to the investigator as that which makes most biological sense. Scale of measurement is critically important, to the extent that it is possible for a species to be common at one scale but rare at another. This sometimes results in significant resources being directed towards conservation of a species that is rare or endangered at a relatively parochial level, but neither at a larger, perhaps international, scale.

Population processes

Once the boundaries of a population have been defined, it can be characterized at three basic levels.

The primary descriptor of a population is its density, expressed as numbers per unit of area, volume or habitat. A large literature exists on the various methods for determining or estimating either absolute or relative population densities; these methods are outside the scope of this review and are amply covered elsewhere (Greig-Smith, 1983; Krebs, 1989). The second level of description highlights the four population parameters that affect density directly: birth (sometimes referred to as natality), immigration, death (mortality) and emigration. The first two have a net positive effect on density, the second two have a negative effect and the combined effects can be described by the basic equation of 'population flux' (often referred to by the acronym BIDE):

$$N_{t+1} = N_t + B + I - D - E$$

where N_t = the population density at time t. A third level of population description involves its structure, including characteristics such as age, size and spatial distribution and genetic structure. These features may have important influences on the components of BIDE.

Ecologists have traditionally concentrated on the birth and death components of BIDE as the most important processes controlling population dynamics. The dispersal processes of immigration and emigration have rarely been estimated, perhaps mainly because of the technical difficulties involved in marking and tracking highly mobile organisms or their mobile propagules. For this reason, ecologists have often chosen to work with populations on real or habitat 'islands', where the assumption that immigration and emigration are negligible is most likely to be tenable. Where population boundaries are less obvious, it has usually been assumed that the effects of the two processes on population numbers are equal, cancelling each other out, and that they can therefore be ignored. This assumption is likely to be invalid and, even if it were numerically true, it is highly unlikely that the secondary characteristics of the immigrants and emigrants would be the same.

Monitoring population dynamics and demography

Two approaches can be adopted to determine the demographic structure of a population:

1. One method is to classify individuals into states of development. For plants, the categories would include seeds, seedlings, juveniles (i.e. pre-reproductive), vegetative and reproductive adults, senescent and dormant plants. The proportion of individuals in different age states can then be used to indicate the condition or 'health' of the population. This approach has yielded valuable results for the rare marsh gentian (*Gentiana pneumonanthe*). In a survey of populations in The Netherlands, Oostermeijer *et al.* (1994) were able to distinguish 'invasive' populations, in which seedlings and juveniles far outnumbered adults, from 'regressive' populations, in which seedlings and juveniles were absent. Invasive populations occurred where there was much bare soil and a low percentage cover of litter, bryophytes, herbs and shrubs, while populations were regressive where there was a lack of bare ground and a high percentage cover of both litter and shrubs. Thus, without vegetation disturbance a population will enter the regressive state as vegetation cover increases. As *G. pneumonanthe* does not have a persistent seed bank, the population goes extinct when the long-lived adult plants die and are not replaced. Management is therefore aimed at producing a more open vegetation, in which at least a proportion of the population produces seeds; in the absence of such management, populations are likely to go extinct in 30–50 years.
2. A more comprehensive approach is to follow the fate of individuals within a population by repeated censusing of their presence and performance. The lifespans of individual organisms, and the flux in births and deaths (i.e. the demographic details of the population's behaviour) can be determined from such data, and predictions can be made about the rate of growth (or decline) of the population that is likely to ensue.

Demographic analyses based on census data of the early spider orchid (*Ophrys sphegodes*) have been used to test the efficacy of different management regimes. The range of this species within the British Isles has contracted by over 80% in the last fifty years (Hutchings, 1987) and the numbers of plants in many populations have also declined. It is now an endangered species in Britain. The species grows in chalk and limestone grassland – communities created by sheep grazing on nutrient-impoverished base-rich soils. Habitat destruction or degradation is a major cause of decline and the effects of climate change may also be involved. Long-term monitoring of a colony of the species in south-east England since 1975 has

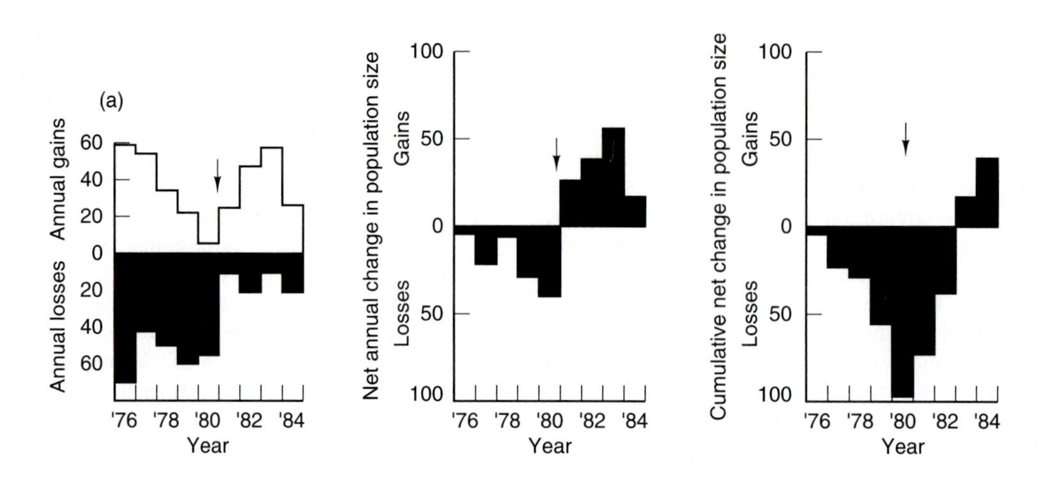

Fig. 11.1 Flux analysis over ten years in a population of the Early Spider Orchid *Ophrys sphegodes* at Castle Hill National Nature Reserve, England. (a) Annual recruitment (above) and annual mortality (below). (b) Annual net change in population size. (c) Cumulative net change in population size. Arrows mark a change in grassland management regime from cattle grazing (up to 1980) to sheep grazing (after 1980); note the immediate change from a declining population to an increasing one, but also the delay before the cumulative population change moves into credit. From Hutchings (1991).

covered a period of site management by cattle grazing, followed by a change to sheep grazing. During the period when cattle grazed the site, there was a deficit of recruitment compared with mortality, resulting in a progressive population decline. This was thought to be caused by physical damage to the underground tissues in the thin rendzina soil caused by trampling by the heavy cattle. Replacement of the cattle by sheep brought about a dramatic change in the demographic status of the population. At least initially, recruitment did not alter noticeably, but mortality fell strikingly, causing the population to achieve a net gain in numbers each year. Within two years of the change in management, the population had made up the deficit in numbers suffered during the four years of the cattle grazing regime, and moved into numerical credit over the whole period of the study. Analysis of flux in the population revealed the beneficial effect of the management change before it could be detected from a simple annual population count (Hutchings, 1991; Waite and Hutchings, 1991) (Figure 11.1). More recently collected data demonstrate a very clear increase in the size of the population.

Clearly, census data which allow the investigator to analyse the fates of individual organisms through time, are more valuable than survey counts of individuals or inventories, which simply record what is present and not the way in which it changes through time. The value of census data increases with the number of censuses carried out, but sets of long-term ecological data are regrettably scarce, especially on rare species, despite the widespread recognition of their importance. In practice, collection of census data is often easier on plants than on animals, mainly because plants are relatively immobile. Plants can be tagged or their positions can be accurately recorded as coordinates from fixed locations on the ground, enabling the same plant to be revisited many times to follow its life-history. Large animals can also be tagged, although specialized and expensive techniques may be needed. For smaller animals such as invertebrates, however, it is not usually possible to follow individual organisms, and changes in population size may be all that can be determined.

What is rarity?

Several terms, including rare, threatened and endangered, have been used to describe the level of threat to species survival. Rarity *per se* does not necessarily mean that a species is in danger of extinction. Some species are naturally rare, possessing attributes that confer a greater ability to persist in small populations. From the vulnerability point of view, changes in population size are probably more important than absolute densities.

The most widely accepted framework for classifying types of rarity originates with Rabinowitz (1981). She categorized species status based on three characteristics: (1) size of geographical range (large versus small), (2) habitat specificity (wide versus narrow), and (3) local population size (high versus low). Seven of the eight possible combinations of these categories (i.e. all except the 'large–wide–high' category, illustrated by many common species) confer different types of rarity. This classification emphasizes the fact that species do not necessarily need to have small local populations or to be highly habitat specific in order to be rare. It views rarity as multidimensional and recognizes that species can show different degrees of rarity at global (range size), regional (habitat), or local scales. Many top predators have large ranges and are not particularly habitat specific, but nevertheless occur at low densities; the barn owl (*Tyto alba*) would be an example. Other birds of prey such as the osprey (*Pandion haliaetus*) are widespread but have narrow habitat requirements; it is also a top predator but feeds exclusively on fish.

Causes of rarity

Although many species are naturally rare by virtue of their ecology, there is now widespread agreement that many of the common causes of species rarity, and ultimately extinction, are anthropomorphic in origin. The ecologies of some species will mean that they are better buffered than others against the effects of man-induced changes such as habitat fragmentation. One challenge for conservationists is to identify the traits that predispose species to rarity, in the expectation that these traits are likely also to confer greater sensitivity to habitat changes such as fragmentation. If vulnerability can be predicted from a knowledge of a species' ecology, scarce resources can be better focused on conservation of the species in greatest danger of extinction.

The search for patterns in the ecology of rare species has generated a number of correlates of extinction proneness, although this remains an under-worked and inexact field. A weakness of many such studies is that they tend to be focused on comparing a small number of perhaps unrepresentative species, where a wider ranging and less taxonomically selective approach could be more revealing. Amongst the more comprehensive and unbiased surveys, a recurrent theme is that many rare species are characterized by poor dispersal abilities. Quinn *et al.* (1994) found that rarity, spatial aggregation and poor dispersal were all intercorrelated amongst 139 scarce British vascular plants. Similarly, it is widely suggested that many threatened invertebrates show either physical or behavioural impediments to dispersal. Many of the butterflies that have undergone rapid recent declines in Britain tend to have closed and sedentary populations (Thomas, 1984), where individuals are reluctant to cross even quite short distances of unfavourable habitat to colonize newly created favourable habitat close by (Thomas, 1991). Warren and Key (1991) suggest that many of the rare and threatened insects associated with dead and decaying wood habitats have similarly low dispersal abilities.

Studies that demonstrate associations between such ecological traits and rarity suffer from the weakness that it may not be clear whether the trait is the cause or the consequence of rarity. As Kunin and Gaston (1993) point out, a trait such as low self-compatibility or poor dispersal rate may be a *response* to rarity rather than a cause of it. Indeed, Dempster (1991) studied specimens of selected scarce butterfly species collected over a range of years when populations were in decline and suggested that morphometric changes, such as a reduction in the wing:body size ratio, reflected the evolution of local races that had adapted to habitat fragmentation through reduced dispersal ability.

Why do populations decline, sometimes to extinction?

It is axiomatic that, in order to have a better chance of saving a population or indeed a species from extinction, we need to have a clear understanding of the reasons for its original decline. A very wide range of factors may result in a decline in numbers or a contraction of range, but a useful distinction can be made between factors that operate externally on the population and those that result from the demographic or genetic properties of the individuals within the population. Many authors have chosen to distinguish between *deterministic processes* that reflect directional causal relationships and *stochastic processes* that are due to random chance events. Declining populations are usually subject to more than one negative influence and indeed it is often a combination of factors, operating either concurrently or consecutively, that finally drives a population to extinction.

External influences

External factors can be divided between those that are deterministic, in the sense that the processes are both predictable and directional, and those that are stochastic by virtue of both their timing and severity being the result of chance.

Habitat change

At a global level, habitat change has probably been the most frequent known cause of population decline and species extinction. Clearly, this category ranges from catastrophic changes resulting from wholesale habitat destruction through to more subtle qualitative changes that render the habitat less suitable for the species in question. Jenkins (1992) estimates that 36% of animal extinctions with known causes were due to habitat destruction. Ehrlich and Wilson (1991) estimate that habitat destruction is responsible for the extinction of over 100 species per day. Changes in habitat management, sometimes through ignorance of the species' requirements, can have profound consequences. In plagioclimax communities, held in suspended succession often by external influences such as grazing or periodic fire, removal of management will allow succession to proceed to a different community that may be inhospitable for many of the original species. Changes consequent on habitat fragmentation will be dealt with later. Finally, climatic changes may induce changes in habitats that have profound effects on particular species.

Other organisms

Interactions with other organisms, often those which have been artificially introduced into a community, have been the cause of some of the better documented cases of extinction. These introduced species have frequently been animal predators that follow man; black rats (*Rattus rattus*) have been strongly implicated in the extinctions of many island bird species, while a single cat was responsible for killing all the Stephen Island wrens (*Xenicus lyalli*) in the same year as the species was first discovered.

Regrettably, the history of the biological control of animal pests and weeds is rich in examples where introduced control agents have had adverse effects on other non-target species. One of the more notorious examples concerns pulmonate land snail species in the genus *Partula*, endemic to the island of Moorea in French Polynesia, which have been the subject of extensive evolutionary studies. The giant African snail (*Achatina fulica*), widely introduced for food, had become a serious pest of many crops and garden plants. A predatory snail *Euglandina rosea* was introduced to control *A. fulica*, but unexpectedly switched to preying on the seven endemic species of *Partula*, driving them to extinction in the wild (Clarke *et al.*, 1984). A captive breeding programme was initiated for six surviving species of *Partula*, which are now the subject of a controlled reintroduction experiment.

The precipitous decline of other species has not always been discovered in time. The coconut moth (*Levuana viridis*), paradoxically once a serious pest threatening the Fijian coconut economy, was driven to extinction within one to three years of the introduction of the parasitic fly *Bessa remota*. The control agent then switched to other prey and has since been strongly implicated in the extinction of several other tortricid moth species endemic to Fiji. Amongst plants, US biological control programmes against certain thistles in the genus *Cirsium* using introduced herbivores have been so successful that the plants are now endangered. Competition with introduced species may be equally damaging for native species. Although causality is often hard to prove conclusively, introduced species have been implicated in the decline and in some cases extinction of native species of birds in Hawaii and New Zealand. Finally, the role of infectious diseases, whether introduced or endemic, is a hitherto underestimated factor in the survival of natural populations.

Direct human influences

Many human impacts on natural populations operate proximally, for example through habitat destruction or the introduction of alien competitors or predators. However, a number of man's activities operate directly on natural populations. Several species have been brought to the brink of extinction through commercial exploitation (certain cetaceans are obvious examples), whilst others have actually succumbed completely to direct persecution, perhaps the best known being the great auk (*Pinguinus impennis*). Birds of prey are often threatened by human persecution (shooting, poisoning, trapping), in response to perceived threats to commercial or sporting interests. Even if not threatened with overall extinction, geographic ranges may be severely restricted by such persecution.

Recreational hunting has often posed severe threats to animal populations in the past, although there is a

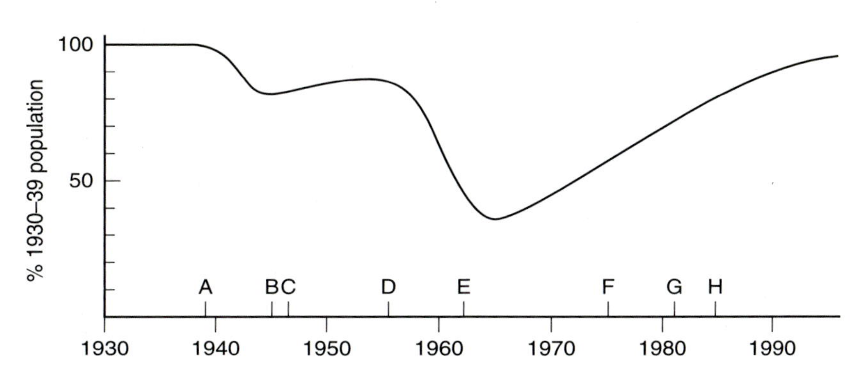

Fig. 11.2 Decline and recovery of the Peregrine (*Falco peregrinus*) population in Britain following introduction and subsequent withdrawal of organochlorine pesticides. Landmark events include: introduction (A) and cessation (B) of wartime control of peregrines, introduction of DDT (C) and cyclodiene insecticides (aldrin, dieldrin, etc.) (D), introduction of restrictions on cyclodiene seed dressing for spring-sown (E) and autumn-sown (F) cereals, banning of all uses of dieldrin (G) and DDT (H). From Ratcliffe (1980) with additional information from Cadbury *et al.* (1988).

growing realization that the future of the activity itself depends critically upon sustainable levels of exploitation. Unregulated, often illegal, hunting is a less easily quantifiable but probably more severe threat to population survival for many endangered species. However, adverse human impacts are not restricted to deliberate actions; population sizes and ranges may be restricted by accidental killing. Non-target species may even be accidentally destroyed during commercial exploitation of natural stocks; for example, common dolphins (*Delphinus delphis*) and seabirds are regular casualties in fishing nets. Other human impacts with sublethal consequences may also contribute to population declines or restrictions in range. Disturbance, often through recreational activities, may adversely affect breeding performance and individual survival, for example through disruption of feeding.

Environmental contaminants

The effects of pollution are often difficult to trace. The exposure of a causal link between the use of persistent organochlorine pesticides (such as DDT) in agriculture and catastrophic population declines in various birds of prey in Britain and North America in the 1960s is widely referred to. The progressive concentration of pesticide residues through the food chain caused severe eggshell thinning and reproductive failure in top predators such as the peregrine falcon (*Falco peregrinus*) (Ratcliffe, 1980). A strict ban on the sale of these chemicals has since allowed populations of this and many other raptor species to recover to pre-DDT levels (Newton, 1979) (Figure 11.2).

Environmental stochasticity

All populations are subject to the effects of unpredictable changes in environmental conditions. Although some authors have chosen to embrace all external influences in this category (including populations of other organisms, such as predators or competitors), such *environmental stochasticity* generally refers to chance fluctuations in the weather. Such variation has a direct effect on the survival of ectothermic organisms (plants, invertebrates and 'cold-blooded vertebrates'), which have limited tolerance to extremes of temperature. Endothermic vertebrates, on the other hand, are affected indirectly through their food supply. Some authors distinguish another category of environmental factor: *natural catastrophes*, such as fire, flood or drought. The distinction is largely artificial, however, since the latter two can be regarded simply as extreme weather events with very low probabilities of recurrence. The important point about environmental stochasticity is that its effects are largely independent of population size; an unusually cold winter will hit populations of small insectivorous birds hard irrespective of initial population size or density.

Intrinsic factors

Demographic stochasticity

In the absence of external influences, a population's fate is determined by a balance between the collective fecundity and mortality of the individuals within it. In a large population with a stable age structure and sex ratio, future population size is predictable from life table and fecundity data. When populations fall to critically low levels, rates of population change become progressively dependent on the chance fate of the small number of surviving individuals. Imbalances in the age structure or sex ratio of the remaining members of the population and chance variation in their reproductive success will produce unpredictable oscillations between population growth and decline. This stochasticity increases as population size declines, thus increasing the probability that the population will go extinct purely by chance. It is important to realize that a small population can slide to extinction in this random way, even when all the individuals are perfectly healthy and there are no adverse external influences. Put another way, an uneven age structure or sex ratio can precipitate a decline to extinction, even in a population with a balance between birth and death rate that would otherwise ensure rapid growth.

In animals, low numbers may also disrupt the social functioning of a population. Some species are dependent on types of aggregative behaviour for defence (schooling in fish), mating (lekking in some bird species), migration (mass assemblies in butterflies) or other more obscure functions. Critically small population size may jeopardize the functional value of these behaviours. Small population size in conjunction with low population density may also significantly reduce the chance of individuals locating a mate, thus reducing their reproductive success (the 'Allee effect').

Genetic stochasticity

The third set of chance factors that can hasten the decline of populations towards extinction concerns the various genetic constraints that come into operation when population size drops below a certain critical level. Small populations may be subject to one or more of these genetic influences depending on circumstances. In general, they centre around the premiss that genetically uniform populations are poorly buffered against chance environmental fluctuations, because they have lost the genetic variability to respond through selection to newly created conditions.

1. Loss of heterozygosity Heterozygosity is a measure of the genetic variance in a population. Formally, mean heterozygosity across a range of genetic loci can be estimated as:

$$H' = (1/L)h_j$$

where L is the number of loci considered and h_j is the proportion of individuals that are heterozygous at locus j, which in turn is estimated as:

$$h_j = 1 - {p_{ij}}^2$$

where p_{ij} is the frequency of allele i at locus j.

It is reasonable to assume that small populations have lower levels of mean heterozygosity. This may or may not coincide with reduced levels of genetic *variation*, manifested in a reduced *number of alleles* present in the population. The important point is that heterozygosity is directly related to individual fitness, because increasing homozygosity results in increased exposure of recessive lethal alleles. Levels of heterozygosity for a species need to be set in the context of related taxa, since there are major differences in mean heterozygosity between widely different taxonomic groupings. Invertebrate populations generally display heterozygosity levels twice those in vertebrate populations, whilst many mammals appear to be completely homozygous at several major loci.

2. Inbreeding depression Matings between closely related individuals become more likely as population size falls. This reduces heterozygosity and increases the probability of exposing lethal and semilethal recessive alleles. The resultant loss of fecundity and increase in mortality is known as inbreeding depression. It is rarely manifested unless the small population size is maintained for several generations.

3. Genetic drift As population size falls, there is an increase in the probability that an individual allele will be lost from the population purely by chance, for example because the individual(s) bearing the allele do not produce any viable progeny. Progressive loss of alleles in this way is referred to as 'genetic drift' which directly affects mean heterozygosity. In the absence of either immigration or mutation, heterozygosity will decrease at a rate that is inversely proportional to population size:

$$H_{t+1} = H_t(1 - 1/(2N))$$

This means that heterozygosity will have dropped to approximately one-third of its original level after $2N$ generations. Consequently, the loss of genetic variance will be faster when the population is small. Small losses of variation are not critical and may be offset in part by mutation. Loss of alleles may also occur through a sudden catastrophic decline in population size, even if the population recovers to its former level again. This type of 'bottleneck' may induce a type of 'founder effect', in which the genetic composition of the extant population is dependent on that of the small number of surviving individuals or 'founders'.

4. Outbreeding depression In certain circumstances, outcrossing between members of genetically divergent populations, which would normally enhance fitness through hybrid vigour, can actually result in a decline in fitness. Such outbreeding depression may result from incompatibilities between locally adapted gene complexes. This may pose problems under two sets of circumstances: captive breeding programmes using individuals taken from widely separated and genetically divergent populations, and the augmentation of natural populations through reintroduction of captive-bred animals which have diverged genetically as a result of directional selection during captivity. The disastrous consequences of such mixing of gene pools was shown in the mountain ibex (*Capra ibex*). A population of the Tatran subspecies (*C. ibex ibex*) was augmented with individuals from other subspecies (*C. ibex nubiana* and *C. ibex aegagrus*), resulting in disruption of the normal breeding cycle and complete extinction of the population (Templeton, 1986).

5. Minimum Viable Populations (MVP) Knowledge of these genetic pitfalls remains purely academic unless we can answer the very practical question: is there a critical minimum size below which the population risks imminent extinction? This question has exercised conservation biologists perhaps more than any other since the seminal works collated by Soulé (1987). It is clearly a critically important question for zoos attempting to maintain either a single-site or dispersed population of an endangered species, but it is also increasingly pertinent for reserve managers and those charged with safeguarding species distributed across fragmented natural populations. For the latter, the practical question often becomes: what is the Minimum Viable Area (MVA) of habitat needed to support the population?

Several important points need to be borne in mind when considering Minimum Viable Population (MVP) estimates. Firstly, an MVP can only be defined in the context of a probability of population survival over a defined time period (i.e. the population size, x, required to ensure a y% chance of surviving a minimum of z years). Secondly, these estimates will undoubtedly vary between species and perhaps between populations, so that there is no unique number that can be applied universally. Thirdly, the concept of an MVP refers to the risks of extinction from genetic causes; to this we must add the risks imposed by environmental stochasticity which are often harder to estimate. Finally, it is important to realize that the figures that have been widely quoted and adopted by conservation managers are probably little more than arbitrary guidelines, perhaps only accurate to one or two orders of magnitude.

Two figures have become tentatively accepted as working MVP estimates. A population size of 50 is considered the minimum to avoid inbreeding depression and is based on the experience of animal breeders, who generally accept a 1% increase in the inbreeding coefficient F which increases by $1/(2N)$ each generation. A population of 500 is deemed sufficient to avoid the risk of genetic drift, a figure derived empirically using evidence from experiments on fruit flies (*Drosophila* spp.).

MVP estimates are now widely used in endangered species management. As there are very few empirical tests of the underlying assumptions, the estimation of MVPs remains an inexact science. The scarcity of useful case studies, particularly experimental ones, largely reflects the length of time over which populations need to be studied; many species need to be rescued over much shorter time scales. In a study of 122 bighorn sheep populations in mountain habitats in southwestern USA, all populations containing less than 50 animals went extinct within 50 years, whilst those containing at least 100 individuals persisted throughout the 70-year study period (Berger, 1990). This suggests that an MVP of 50–100 is at least in the right order of magnitude.

6. Effective population size The 50/500 guidelines make a number of assumptions about the functioning of the population; it is assumed to show an equal sex ratio, random mating, non-overlapping generations and a Poisson-distributed variance in offspring number. Violation of any of these assumptions means that the *effective population size*, N_e, will be less than the actual population size, N, as censused in the field. For example, a 3:1 sex ratio will mean that $N_e = 0.75N$. This means that the adverse effects of inbreeding depression and genetic drift may manifest themselves much earlier than would be expected purely on the basis of population census data. For example, Wilcox (1986) suggests that the actual population of 200 grizzly bears (*Ursus arctos horribilis*), in Yellowstone National Park, USA, may in fact have an effective population size as low as 38, a level at which inbreeding depression and long-term

loss of genetic variation are likely to be serious problems. Considerable controversy remains as to the level of N_e/N to assume for practical management purposes. Recommendations vary from 0.4 down to as low as 0.05 or less. Whatever the value, it is important to realize that the 50/500 guideline applies to the *effective* population size N_e rather than the censused population size N.

Summary of influences

Faced with the preceding list of adverse influences on populations, which factors are likely to be most important and under what circumstances? There is widespread agreement that deterministic change in a population's environment is the most important mechanism, at least in initiating a population decline. Pre-eminent amongst such changes must be habitat loss, degradation and fragmentation. Few species will be able to adjust to such changes on an ecological time scale. The result will inevitably be a decline in population size. Of the stochastic processes, random catastrophic events when they occur are likely to have more profound effects on populations than environmental stochasticity, since populations will have evolved a degree of resilience to the less extreme vagaries of their environment. Demographic stochasticity probably poses a lesser threat, partly because it requires the population to have been reduced to low numbers first. In fact, there is considerable interaction between the latter two effects, such that their combined effect increases as population size drops.

A secondary debate has centred around the relative importance of genetic considerations. This argument has somewhat polarized between two schools of thought. One treats genetic limitations as being of paramount importance, while the other argues that adverse genetic effects only become important when the size of the population has dropped to levels where it is unlikely to be recoverable on ecological grounds anyway. To some extent, the different philosophies may reflect fundamental differences in the conservation status of the organisms under consideration. Genetic considerations have tended to preoccupy animal conservationists involved in captive breeding programmes, where the number of surviving individuals is often perilously low and where inbreeding depression and loss of genetic variability pose very real threats. Under such circumstances, sampling programmes for *ex situ* conservation must be devised to conserve as much genetic variation as possible, and breeding programmes must be implemented to adapt a species to a high level of inbreeding.

As is often the case, real populations may behave in a rather more complex manner depending on circumstances. It is likely that the relative importance of the different factors will change as the population declines to extinction and that one mechanism causing decline may prepare the way for another to continue the process. The following sequence of events might form a typical scenario. A large, healthy and demographically balanced population is subjected to some sort of adverse deterministic effect, such as destruction of a significant proportion of its habitat. This inevitably results in a reduced population size, perhaps accompanied by fragmentation. The species has an upper population limit fixed by the extent of remnant habitat; it can only remain stable in size or decline. Once the extant sub-populations are sufficiently small and widely separated, stochastic processes will come into play. Chance extreme environmental events, such as a long, hot and dry summer perhaps followed shortly by a particularly hard winter, sooner or later will conspire to reduce the population drastically. A reduced population density may ease competition for some resources, but demographic imbalances prevent this being translated into a population recovery. In addition, the population may have lost so much genetic variability that it cannot respond through selection to newly created conditions. The small size of what is left of the population inevitably provokes an increase in the frequency of matings between close relatives, leading to increased homozygosity, full expression of recessive lethals and the classic symptoms of inbreeding depression – reduced fecundity and increased mortality – resulting in further population decline. The population is then locked into a feedback loop of stochastic pressures that force it repeatedly downwards. Once events have progressed this far, if not well before, the population is almost certainly doomed.

This chain of events has been dubbed the 'extinction vortex' (Gilpin and Soulé, 1986; Caughley, 1994). Its general applicability is difficult to judge, as very few species have been monitored through to extinction. A notable exception is the heath hen (*Tympanuchus cupido cupido*), a member of the grouse family that was once common in scrub-woodland habitats in the northeastern United States. Widespread habitat destruction and hunting forced a drastic contraction of its geographic range, until the population was confined to a single small area and reduced to 50 birds. Establishment of a reserve to safeguard the

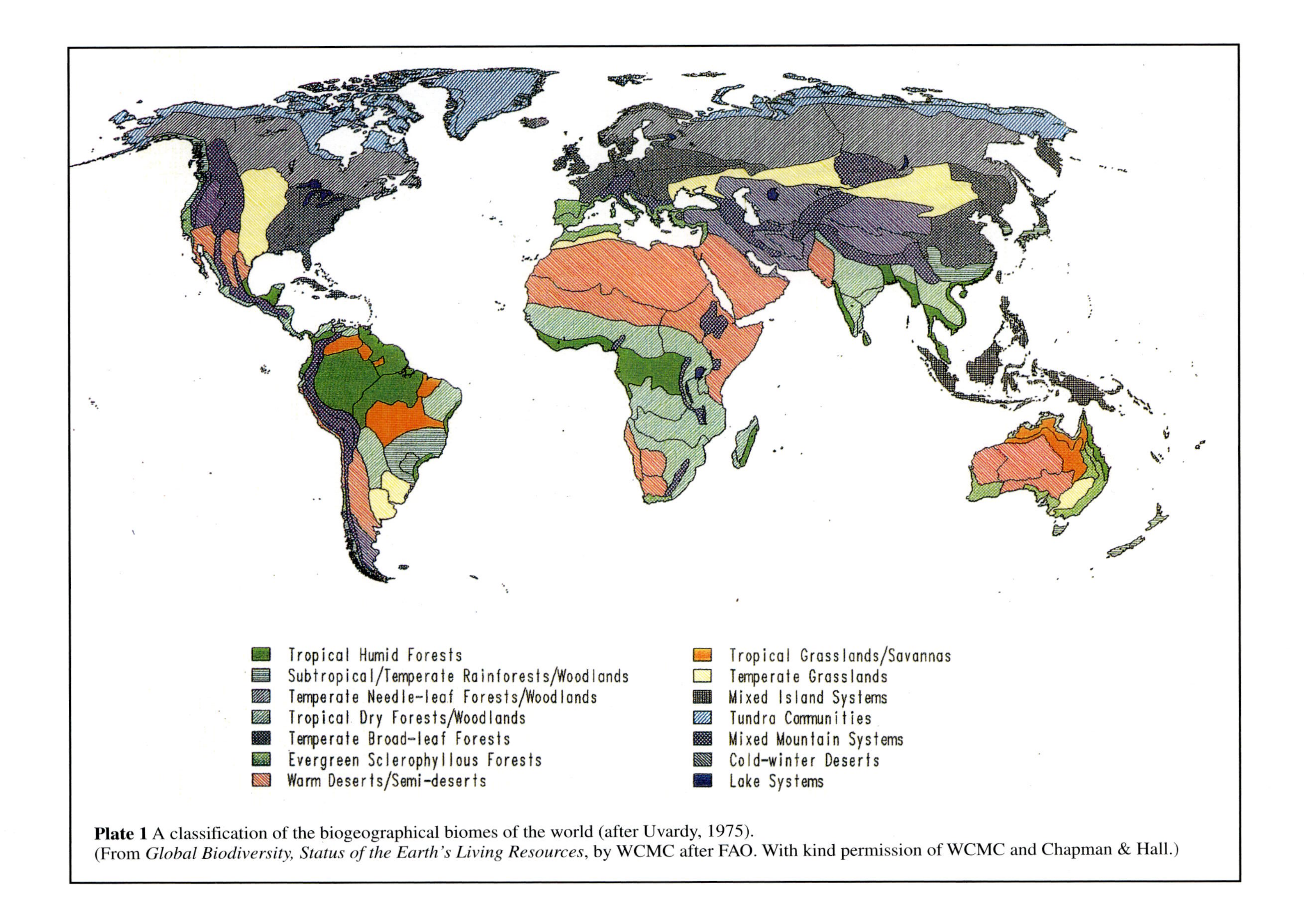

Plate 1 A classification of the biogeographical biomes of the world (after Uvardy, 1975).
(From *Global Biodiversity, Status of the Earth's Living Resources*, by WCMC after FAO. With kind permission of WCMC and Chapman & Hall.)

Plate 2 Twyford Down, Hampshire, UK. This is an area of chalk grassland communities where sites had previously been designated for scientific and archaeological interest, now destroyed with the cutting. A tunnel was rejected. (Photograph: Ian Spellerberg.)

Plate 3 Temperate rain forest, Olympic National Park, Washington, USA, a World Heritage Site. (Photograph: Michael O'Connell.)

Plate 4 Urban development, Southeast Florida, USA. Explosive growth in populations increases the need for conservation laws to provide standards for human activity.
(With kind permission of The Nature Conservancy, Inc.)

Plate 5 Sturt National Park, Western New South Wales, Australia. (Photograph: Peter Clarke.)

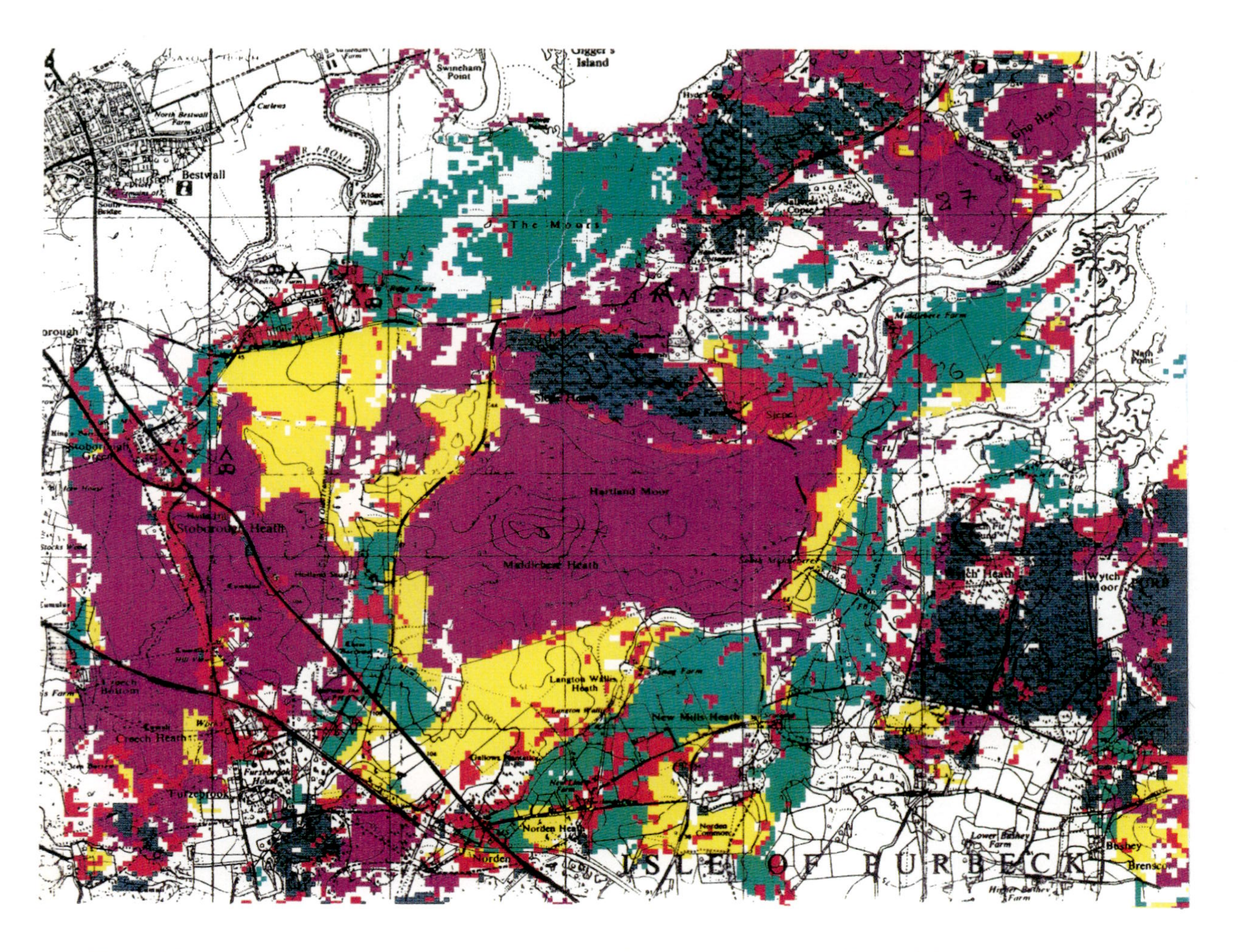

The information shown on this map is derived from the ITE Land Cover map of Great Britain.

All results are provisional.

Topographic overlay copyright of the Ordnance Survey.

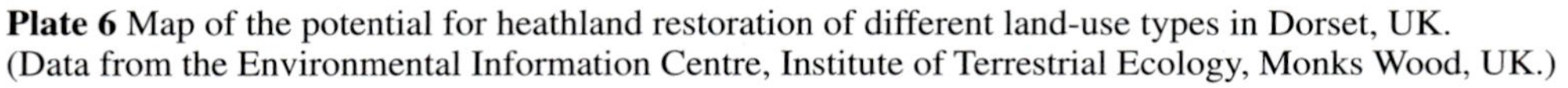

Plate 6 Map of the potential for heathland restoration of different land-use types in Dorset, UK. (Data from the Environmental Information Centre, Institute of Terrestrial Ecology, Monks Wood, UK.)

surviving population prompted some hope as numbers rose to 2000, but a series of disasters eventually brought about its demise. First, a fire destroyed a considerable area of the habitat and many birds. This was followed by a particularly harsh winter and an episode of severe predation by hawks. By this stage, the population was small enough to manifest the signs of inbreeding depression in the form of reduced reproductive output, which was followed by a disease outbreak. Over the space of ten years, the population was reduced to 13 birds, only two of which were females, and within a further five years the species had gone extinct. It is likely that once the population had fallen to 50 individuals, even despite the subsequent temporary recovery, the combined effects of inbreeding depression and demographic imbalance (primarily in the sex ratio) meant that the population was unlikely to survive. Halliday (1978) suggests that, despite an intrinsically high reproductive potential, the species was dependent on large aggregations for breeding and this may have resulted in a breakdown of social structure once the population had fallen below a critical minimum.

The interaction of several proximal factors can also be discerned in the decline and eventual extinction of the butterfly known as the large blue (*Maculinea arion*) in Britain (Thomas, 1980). This species once occurred at approximately 90 open grassland sites across central and southern England. Half of these were destroyed through conversion of the habitat to agriculture or forestry, while the remainder became unsuitable as a result of changes in vegetation structure. A widespread reduction in sheep grazing, together with an outbreak of the myxoma virus which devastated the rabbit (*Oryctolagus cuniculus*) population, resulted in a general increase in vegetation height on many sites. The butterfly is intimately linked with the ant *Myrmica sabuleti*, which carries larvae of the butterfly into its nest where they pupate. The ant in particular is dependent on the hot microclimate of a short and open sward produced by heavy grazing; once the sites became unsuitable for the ant, the butterfly quickly went extinct as well.

Prediction models

A number of predictive models (MVP analysis being the most popular) are now widely used in devising conservation strategies for endangered species. However, there is a danger of too great a reliance being placed on such models, particularly if their limitations are not fully understood. Models for endangered species are rarely based on sufficient data. Studies are very rarely replicated (often because research is not initiated until there is only a single surviving population), which means that the resultant models are not robust. Spatially separated populations may experience very different influences and the relative importance of these may vary temporally. Furthermore, as stated earlier, there are very few cases where a population has been fully monitored through to extinction. Nevertheless, predictions which are accurate even to within an order of magnitude can still be useful.

Perhaps the most comprehensive modelling approach involves the process of Population Viability Analysis (PVA). At its most simplistic, this is an attempt to estimate the probability that a population will survive for a given number of generations. PVA includes MVP analysis but takes it a stage further. It is essentially 'a long-term iterative process of modelling and research . . . to develop a better understanding of the behavior of the system' (Boyce, 1992). In practice, this means several activities, connected in a positive feedback loop: (1) development of an initial model that estimates population survival time, based on current conservation practice; (2) the evaluation of various management options, perhaps applied to separate populations, whether this involves adjusting the population demography or genetic composition, the habitat, key resources, natural enemies or competitors; and (3) monitoring the results of such 'experiments' to provide data which are then fed back into improving the original model.

Considerable attention has been devoted to developing PVA models for populations of high-profile species such as the northern spotted owl (*Strix occidentalis caurina*) and the grizzly bear (*Ursus arctos horribilis*). However, the empirically derived parameters that provide the input to these models are based on considerable research effort and it is widely acknowledged that only a very small fraction of endangered species will receive this level of attention. The inexorable pace of current extinctions is too fast for this approach to be applied to more than a handful of species. However, opinions differ as to whether more generalized 'rules-of-thumb' that can be applied across a range of species are likely to be helpful or dangerous. Measured against this is the recognition that efforts directed towards the conservation of 'keystone species' may facilitate the conservation of a wide spectrum of species within the whole community.

A further concern is that hitherto most models have been based on processes applicable to critically small populations; Caughley (1994) calls this the 'small-population paradigm'. Furthermore, they deal with the *consequences* of small population size (demographic and genetic stochasticity), rather than its *cause*. Caughley calls for greater emphasis on the 'declining-population paradigm' that addresses the question of what external influences actually cause populations to decline in size and range in the first place, how to detect such trends and ultimately how to halt or reverse them.

The spatial perspective

Populations rarely exist in complete functional isolation from other conspecific populations. Consequently, it should not be assumed that full knowledge of the internal dynamics of an individual population is sufficient to predict its fate. Populations operate within a geographical framework and depend to a greater or lesser extent on exchange of individuals or alleles between them. Conservation measures aimed at individual populations cannot therefore ignore the spatial context of the population within the species' range.

Fragmentation

At least in the developed world, there are intense pressures on natural habitats from various influences, including urbanization, industry, agriculture and forestry, resulting in natural and semi-natural habitats becoming increasingly confined to economically marginal land in progressively smaller parcels. Wilcox and Murphy (1985) unequivocally state the consequences: 'habitat fragmentation is the most serious threat to biological diversity and is the primary cause of the present extinction crisis'. Fragmentation, which has become a feature of many semi-natural habitats, has three consequences: a progressive diminution both of the total area of habitat and of the mean area of the remaining patches and an increase in the mean distance between patches. A classic example concerns the lowland heaths of Dorset in southern England, where urban development and conversion to agriculture has left only 15% of the extensive area of heathland that existed 200 years ago; the surviving fragments are both small and comparatively isolated (Webb and Haskins, 1980).

The consequences of the first two effects can only be adverse. Declining habitat area reduces the total carrying capacity, exerting a downward pressure on population size, and removes potential sources of immigrants. In extreme circumstances, populations may be reduced to levels at which the intrinsic factors outlined above become important. Fragmentation of the East African savannah has resulted in populations of many large mammals falling to critically low levels, while there is plentiful evidence from island biogeographic studies of depauperate biodiversity and increased extinction risk on small habitat islands or real islands. The probability of recolonization after chance extinction of a local population will decrease as patches become progressively isolated and as the width of alien habitat over which immigrants have to migrate increases. However, isolation of patches may not be unconditionally bad; spatial separation may slow the spread of contagious disease, fire or introduced predators.

Metapopulations

Regional populations of many species are naturally subdivided into a number of local populations occupying discrete habitat patches. These local, often small, populations are subject to periodic chance extinction, but the habitat patches which they occupied may be subsequently recolonized by individuals dispersing from other populations. As long as the rate of recolonization matches or exceeds the rate of local extinction, the regional population will persist. Such a 'population of populations', functionally linked and sustained by dispersing individuals, is termed a 'metapopulation'.

Metapopulation models were originally applied to species that occupy habitats that are naturally patchy or transient. However, the potential relevance of this type of population model to species occupying fragmented habitats was soon recognized by conservation biologists, who have since enthusiastically applied metapopulation theory to a number of endangered species. However, the justification for applying it across a range of species remains largely untested. Indeed, there is growing evidence that many if not most species do not readily fit a classical metapopulation pattern (Harrison, 1991).

The long-term stability of a metapopulation depends critically upon a number of important factors. Metapopulation persistence increases with the number and size of local populations and the rate of dispersal between them. Fragmentation produces

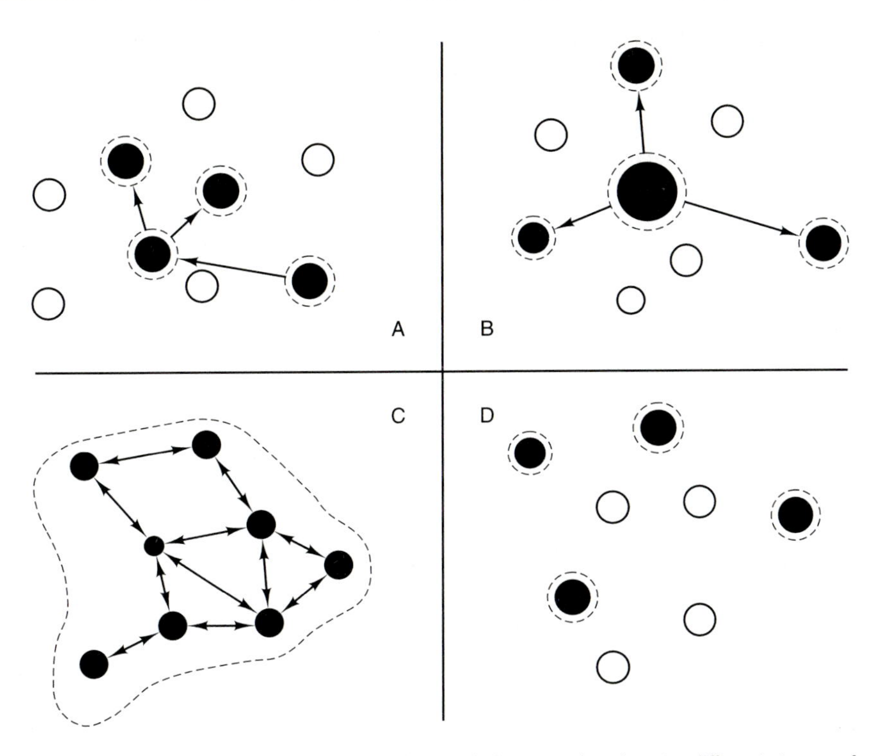

Fig. 11.3 Spatial arrangements of local populations conforming to different types of metapopulations. Circles represent habitat patches, which are either occupied (filled) or vacant (unfilled). Population boundaries indicated by dashed lines. Arrows indicate local migration (+/− colonization). A, Classic metapopulation. B, 'Mainland-island' metapopulation. C, 'Patchy' population. D, 'Non-equilibrium' metapopulation. (After Harrison, 1991).

smaller local populations which become progressively isolated from each other; the former increases the risk of local extinction, whilst the latter reduces the chance of subsequent recolonization. The implication is that metapopulations may suddenly plummet to extinction, either when the remnant fragments become too few or too small or if dispersal between them is interfered with.

The persistence of metapopulations is also dependent on the extinction probabilities for individual local populations being temporally independent. Only then will local populations survive to provide dispersing individuals which will recolonize the vacant habitat patches. Clearly, if the extinction of local populations becomes synchronized, the persistence time of the metapopulation will be drastically reduced. There is growing evidence that the population dynamics of many species are driven by environmental factors operating at a regional level, which are thus likely to affect all local populations simultaneously. Pollard (1991) and Hanski and Woiwod (1993) provide strong evidence that spatially separated populations of many species of butterfly and moth respectively fluctuate synchronously in response to meteorological conditions.

Harrison (1991, 1994) regards the classical metapopulation as just one of several possible configurations for a group of local populations (Figure 11.3). In the 'mainland–island' situation, long-term metapopulation persistence is assured by one or a few very large ('mainland') populations. Smaller nearby 'island' populations are, however, vulnerable to periodic extinction; these can be recolonized by dispersers from the large mainland population, but have no reciprocal effect on it because they are too small. The remaining populations of the rare bay checkerspot butterfly (*Euphydryas editha bayensis*) in California (Harrison *et al.*, 1988) apparently conform to this pattern.

Other species exhibit 'patchy populations' where patch fragmentation occurs within the area covered by the population. Such species tend to be good at dispersing between the habitat patches, so that the local populations operate demographically as a single unit and are therefore unlikely to undergo cycles of extinction and recolonization. There has been some disagreement between bird ecologists as to whether birds in woodlands fragmented across lowland agricultural landscapes form metapopulations. High dispersal ability may mean that most species treat woodland parcels as favourable patches between which they move freely. In this case, habitat fragmentation is at a relatively fine scale compared to the trivial dispersal range of the birds.

At the other end of the spectrum, 'non-equilibrium metapopulations' comprise local populations that are effectively isolated from one another; recolonization of vacant patches, if it occurs at all, is too slow to offset the rate of extinction. In the absence of any recolonization, each local population must be regarded as an isolated unit; the regional population gradually declines as individual local populations go extinct and are not replaced. This scenario must apply to many rare species that have become confined to isolated habitat patches that are too remote from each other to allow any realistic chance of natural recolonization. It is likely to apply most rigidly to species that are doubly disadvantaged by extreme specificity to scarce and localized habitats and limited powers of dispersal. Many invertebrates fit this description. Thomas (1984) emphasizes the exacting habitat requirements of many scarce butterflies in Britain and how they tend to form small closed populations, implying limited dispersal between them.

Harrison (1994) concludes that there are remarkably few species that conform well to the classic metapopulation model. Those that do apparently share two features: modest, rather than high, dispersal rates and a predilection for early successional habitats. The latter ensures local population extinction as the habitat patch matures; regional persistence can only be assured if new early succession patches are continually being created within the dispersal range of the species. In British deciduous woodlands, several species of fritillary butterfly exploit rather narrow time windows (perhaps only 3–4 years) in the early stages of the post-coppicing succession. Woodland management for these species must therefore ensure that coppicing is done rotationally, with sequential stages arranged contiguously to facilitate movement between adjacent blocks. The heath fritillary (*Mellicta athalia*), for example, is apparently unable to colonize developing patches of suitably aged woodland if they are separated from the source population by as little as 100 m of unfavourable habitat (Warren *et al.*, 1984).

Ranges

A number of population parameters will vary across a species' range. In general, population densities will be greatest at the centre of the range and decline towards the edge. In this context, the centre may be the area where optimal environmental conditions prevail, rather than the geographical centre. The variegated nature of geology, soils and habitats may mean that there is not one but several high population density foci; density profiles across ranges may be uni- or multi-modal. Similarly, density changes towards the range edge may be gradual or stepped.

We can assume that changes in density reflect changes in the balance between the basic demographic processes. Under optimal conditions towards the centre of the species' range, the birth rate will exceed the death rate and the population size will increase; population stability can only be maintained by a net emigration of the surplus individuals. By contrast, somewhere towards the edge of the range, the birth rate will equal the death rate; beyond this point, local populations can only be maintained by individuals emigrating from the core area. Pulliam (1988) described these two situations as 'source' and 'sink' populations. It is clear that the population dynamics of a species must vary considerably between the core and edge of its range.

Implications for conservation

The foregoing largely theoretical treatment of the spatial structure of populations has several important practical implications for conservation. Firstly, the information that we need to conserve a population extends far beyond its mere numerical size. Pulliam (1988) emphasizes that the demographic status of a population as a source or a sink is unrelated to its density. We cannot therefore rely on large populations being sources and vice versa; consequently, for the purpose of making conservation decisions, it is unsafe to assume that a small population is dispensable, since it may be an important source that is helping to maintain another, perhaps larger, population which is acting as a sink. Thus, there is no short cut that avoids measuring all the components of BIDE. Furthermore, research done in sink habitats may yield atypical, and potentially misleading, information about aspects of the species' basic ecology, such as habitat requirements or regulating factors. Population density in a sink habitat may reflect the proximity and nature of nearby sources more than the conditions actually at the site itself.

Secondly, large-scale habitat destruction, especially in the core of a species' range, may leave peripheral populations very vulnerable to natural extinction. Such isolates may be bolstered or even wholly maintained by emigrants from the core. These populations may be confined to marginal habitats where the suboptimal conditions restrain reproductive success rates and make conservation efforts all the harder.

The final point arises from the growing evidence of a positive correlation between overall range size and local abundance. In short, adverse local factors may affect the species far beyond the local population. Translated into metapopulation terms, this predicts a positive relationship between the number or proportion of habitat patches occupied and the mean population density within those patches. As Lawton *et al.* (1994) point out, if these two variables are indeed tightly linked, human-induced changes to one can be expected to affect the other. More specifically, destruction of local sites (i.e. reducing the number of occupied patches) will depress mean densities on the remaining ones. Similarly, adverse influences on population size (for example, through persecution or habitat change) will prompt a reduction in the number of sites occupied. In this context, the density profile of the species across its range becomes important, since a species with a former unimodal distribution will shrink to a single core site. Conversely, a species with a multimodal density pattern will contract to a number of sites centred around the former density peaks, this fragmentation occurring even in the absence of any habitat destruction.

Corridors

The rapidly contracting total area and increasing fragmentation of many habitats, together with a realization of the importance of dispersal and movement between fragments, has led conservationists to consider the notion of preserving or creating wildlife 'corridors' – strips of semi-natural habitat connecting wildlife sanctuaries, along which plants and particularly animals can disperse. The scale and ambitiousness of these proposals varies considerably. At one level, hedges and rough grassland such as occur along road verges and disused railway lines are seen as potential dispersal routes for species moving between larger patches of semi-natural habitat within predominantly agricultural landscapes. At a considerably larger scale, 'global change corridors' have been proposed to help offset the effects of climatic change by allowing species to move along a north–south axis in response to expected changes in temperature.

The ideas behind wildlife corridors have considerable intuitive appeal, but much controversy remains as to whether they are likely to be successful in practice (Hobbs, 1992). On the positive side, corridors should facilitate immigration into isolated habitat patches, to (1) maintain or increase species richness in those patches, (2) augment population size and encourage recolonization, thus reducing the risk of extinction (the 'rescue effect') and (3) reduce the genetic problems associated with small population size by introducing new genetic material. Measured against this, corridors may help to spread 'contagious disasters', such as fire, disease and introduced predators, pests or weeds which may disrupt the existing native plant and animal communities. Also, longer corridors will have to be wide enough to support breeding populations and not just dispersing individuals. For large mammals with extensive territories, the necessary width may be prohibitive. Given that all corridors will suffer from strong edge effects, widths must be great enough to preserve a central zone of pristine habitat. Some conservationists have argued that, at least for larger animals, dispersal of individuals between populations to avoid the adverse genetic and demographic effects of small population size may be more effectively and economically achieved by artificial translocation than by elaborate corridor systems. Despite considerable debate (and proposals for potentially costly schemes), there is remarkably little evidence to support the effectiveness of wildlife corridors, probably because the necessary studies are logistically very difficult to do. In short, corridors are potentially hugely expensive to create and maintain and the resources may be better allocated to maintaining or enlarging existing reserves.

Role of reserves

Nature reserves, national parks or sites with other conservation designations will always retain a central role in the conservation of animal and plant populations. In highly developed landscapes, they may represent the only refuges for some rare and endangered species. Issues concerned with reserve 'design' have focused on the question: what size and configuration of reserves is needed to conserve this species? A subsidiary question concerns whether a species is best preserved by designating a single large or several small reserves (coined the SLOSS debate). If two or more reserves are being considered, several 'design' issues emerge beyond simply the combined area of habitat preserved. Aspects of the spatial distribution of the reserves, including the distances between them, their degree of connectance and whether they are clumped or arranged linearly, will all affect their collective effectiveness in conserving the species. Although some very general principles may be derived from such considerations, what really matters is how the design of the reserve network matches the spatial dispersion of the species.

For any one reserve, the question of shape is critically important. The more the shape of the boundary departs from a perfect circle, the greater the ratio of edge to area becomes. This effect is compounded by the elongation of the shape and the tortuosity of the boundary. Since all habitat patches will be subject to edge effects up to a fixed distance from the boundary, the ratio of pristine habitat to buffer zone will decline as the shape elongates. Edge effects are not necessarily wholly negative, however, since some species positively select the ecotone habitats found towards the boundary.

Whilst reserve acquisition is a powerful tool for the conservation of many species, it is not universally appropriate. Firstly, the fine-scale distribution of some species is temporally fluid, with the exact location of local concentrations changing from year to year. Secondly, some species naturally occur at low densities and are highly dispersed. Many birds of prey, for example, have very substantial home ranges for which reserve designation is not a realistic option. Finally, the total area of land under some sort of conservation designation will inevitably be comparatively small; large areas of the 'wider countryside' will remain unprotected.

Recovery measures

Conservation *ex situ* and reintroductions

As a population declines, a critical point is reached below which any realistic hope of avoiding extinction in the wild must be abandoned. Some considerable time before this point of no return is passed, attention should turn to whether a programme of captive breeding (often referred to as *ex situ* conservation) would be appropriate and feasible. Whether the artificially maintained population is held in a zoo, aquarium, seed bank or botanic garden, the strategic principles of management remain substantially the same, although most experience to date has been gained with animals, and vertebrates in particular. The role of zoos has changed substantially in recent years from one of exhibiting animals primarily for public amusement to the maintenance of captive populations of endangered species for eventual reintroduction back into the wild. Most frequently, the objective is to enhance remaining wild populations numerically and/or genetically with captive-bred stock. *In extremis*, where the species is extinct in the wild, a captive population may be held until habitats are restored sufficiently to reinstate a population under semi-natural conditions.

Bringing all surviving individuals of a population into captivity is clearly a last resort. Once the global population of Californian condors (*Gymnogyps californianus*) had fallen to three individual birds or that of the black-footed ferret (*Mustela nigripes*) to twelve, conservationists had no realistic chance of retaining either population *in situ*. A generally accepted guideline is that captive breeding should be considered once the global population of a species has fallen below 1000 individuals. Soulé *et al.* (1986) estimate that some 2000 land vertebrate taxa will require captive maintenance in the next 200 years. *Ex situ* conservation strategies therefore become critically important in preserving global biodiversity, since many of these species may persist only as captive populations. Several high profile species have already been saved from extinction through captive breeding, including the golden lion tamarin (*Leontopithecus rosalia*), Père David's deer (*Elaphurus davidianus*), Arabian oryx (*Oryx leucoryx*), Przewalski's horse (*Equus przewalski*), European bison (*Bison bonasus*), black-footed ferret, Californian condor, Hawaiian goose (*Branta sandvicencis*) and Spix's macaw (*Cyanopsitta spixii*).

Captive populations are generally small and therefore subject to similar demographic and genetic constraints as equally sized wild populations, although measures can normally be taken to minimize adverse external influences. Management of such populations relies on the same principles of population biology, although arguably with the benefit of greater control, for example over matings. The principal objective is to maintain maximal genetic variation; preservation of 90% of the quantitative genetic variation (measured as heterozygosity) over 200 years is a widely accepted target (Soulé *et al.*, 1986). However, by the time *ex situ* strategies are implemented, genetic variation in the surviving natural population is often already severely reduced. Consequently, most captive populations, particularly when based upon few founders, are prone to inbreeding depression. One strategy for combating inbreeding depression is to inoculate the captive population with one or more individuals taken from a genetically different background. However, this runs the risk of outbreeding depression, through the disruption of locally adapted gene complexes, the symptoms of which (reduced fecundity, poor survival, etc.) may be hard to distinguish from those of inbreeding depression. Outbreeding depression may also occur when captive-bred animals are introduced back into wild populations.

All captive populations will be subjected to a degree of unintentional selection in their new artificial environment. Captive and wild populations will progressively diverge genetically with time, so the number of generations between capture of wild stock and reintroduction should be minimized.

Some captive populations have been subjected to accidental 'introgression', the process whereby genes of one species are introduced into the population of another. As a result, several wild species have been contaminated to varying degrees with genes from domestic stock. Certain captive lines of Przewalski's horse now carry domestic horse (*Equus caballus*) genes, and the European bison is universally contaminated with cattle (*Bos taurus*) genes. A related problem concerns hybridization between natural subspecies, as a result of early breeding programmes not recognizing the importance of maintaining genetic pedigrees. Thus, current zoo populations of tigers (*Panthera tigris*), lions (*Panthera leo*) and Indian elephants (*Elaphas maximus*) are mixtures of various geographical subspecies. Ordinarily, it would be important to keep such subspecies separate, but for these species at least this possibility has now been lost.

Behavioural factors may determine the success of both captive breeding and reintroduction programmes. It will be important to maintain the cohesion of social groups, particularly on release into the wild. This may mean, for example, penning animals at the reintroduction site and not releasing them until social interactions have stabilized. Readjustment to the natural environment will be easier for species with innate behaviour patterns. Conversely, species in which important behaviours are culturally transmitted may have more difficulty in adjusting back to natural conditions because these learned behaviour patterns are most likely to have been lost during prolonged captivity. In some species, even quite elaborate behaviour patterns such as predator avoidance may be artificially taught to captive-bred animals.

The reintroduction of animals back into the wild is not a trivial exercise. Stanley-Price (1989) describes the issues that need to be considered. Undoubtedly the most important is to understand the reasons causing the original collapse of the natural population. Project planners should be confident that the adverse factors responsible have been removed and that sufficient suitable habitat remains. Release strategies may be complicated, involving consideration of the numbers to be liberated, the optimal timing, any pre-release accustomization and whether the release stock should be spread across more than one site. Some programmes use 'probe' releases, where a small number of individuals are liberated and their fates monitored closely; this is a prudent strategy if the causes of the original extinction are not known or if it is suspected that adverse factors (e.g. predators or disease) may still be operating. Finally, all reintroductions, whether successful or otherwise, should be carefully monitored, so that lessons can be learnt from what are essentially field experiments; this important principle has been repeatedly overlooked in the past (see example in Chapter 15).

Translocations

Similar principles to those outlined above apply to translocations, where stock is transferred from one site to another but without an intermediate captive phase. A distinction can be drawn between 'true introductions' into sites where the species has no known historical precedence, 'reintroductions' where an attempt is made to re-establish a local population that has gone extinct, and 'augmentations' where individuals are released into an existing population to boost numbers, increase genetic heterogeneity or adjust demography (e.g. age structure or sex ratio).

Translocations are likely to be of maximum benefit to species with limited powers of dispersal that are specific to naturally or artificially fragmented habitats. Circumstances likely to promote the success of such ventures include a high potential rate of population increase and a high number and genetic diversity of founders. Experience with birds and mammals shows that scarce or sensitive species are hardest to translocate and that protection and management of the recipient habitat enhances the likelihood of success. Interestingly, however, translocations into the core of a species' historical range appear to be strikingly more successful than those into sites near or beyond the margin. This conforms with the theoretical gradient of declining mean net rate of population increase from the centre to the edge of a species' range. In practical terms, it means that it is very important to locate receiving sites appropriately within the existing or historical geographical range. Furthermore, the density and dynamics of the donor population appear to affect the outcome; translocations from high density populations that are increasing have empirically better chances of success compared to those from lower density populations where absolute numbers are stable or declining (Griffith *et al.*, 1989).

Invertebrates offer several advantages over vertebrates as potential subjects for both captive breeding programmes and translocation experiments. Reintroductions are likely to have a better chance of success in invertebrates, because behaviour patterns are less flexible and therefore less likely to be lost during generations of captivity. Their small size allows the maintenance of comparatively large captive populations and their high reproductive potential facilitates rapid population increase in suitable post-release habitats. Measured against this, rare species tend to have very exacting habitat requirements which may take considerable research to unravel, be difficult to mimic in captivity and difficult to reinstate in the wild.

Sequence of events

Full recovery of a threatened species back to a self-sustaining population level is potentially a lengthy and intricate process. The following list outlines some key steps.

1 First, we need to identify the problem. Accurate estimates will be needed of the current population level and the changes over time that have aroused concern. Here, the importance of long-term monitoring cannot be overstated, to help distinguish between minor short-term population fluctuations and long-term trends.
2 Wherever possible, the factors causing the adverse population trend should be identified, bearing in mind that they may be multiple, subtle and interlinked. Proceeding beyond this step without adequate knowledge of such primary causes would be unwise, except in extreme circumstances where extinction is imminent.
3 The degree of threat posed by each factor needs to be assessed, bearing in mind that past threats may no longer apply and future ones may emerge following particular actions. The list of threats should be graded for severity and possible actions should be identified to nullify or minimize the effect of each.
4 A full review of all appropriate actions should be assembled, with clear statements of the potential benefits and dangers of each. This will involve evaluation of a variety of alternative actions, including different types of *in situ* management, *ex situ* conservation, reintroduction programmes and other strategies.
5 Once a decision has been taken on which actions are required to ameliorate or reverse current downward population trends, responsibilities for action have to be assigned, either to individuals or to organizations. These may change as the programme progresses.
6 A formal 'Recovery Programme' should be agreed that maps out the path to full recovery of the population. This will become the source of reference for all those involved and must state clearly each component in the programme, the costings, sources of financial (and other) support, the full intended sequence of events, time scales for each action, targets (usually expressed in terms of population levels), landmark criteria by which success or failure in reaching targets is to be measured, the interdependencies between actions, responsibilities for each component and the procedures by which results are to be reported back.
7 Undoubtedly one of the most important components in any recovery programme is careful monitoring of the population itself. At the most basic level, this will focus on numbers, but may need also to include the collection of demographic, and perhaps even genetic, information about the population. Emphasis has already been placed on the importance of long-term data, the value of which increases disproportionately with the timespan covered.
8 Finally, the programme should be reviewed periodically, to allow adjustments to be made as circumstances change. In this context, there may be several iterations between steps (3) and (7), before the population or species can be considered 'safe'.

A way forward?

Like it or not, we cannot possibly hope to conserve all the species in imminent danger of extinction. Choices must be made between many more deserving cases than we will have the resources to rescue. Hitherto, considerable attention has been given to high-profile species with significant public appeal, undoubtedly at the expense of other less obvious and attractive ones. Mammals, birds and selected plants traditionally have been the primary focus of conservationists' attention; amphibians, reptiles and invertebrates have been correspondingly undervalued, despite the latter group's domination of the world's biodiversity. Should we be prioritizing particular taxonomic or ecological groups of organisms for particular attention? Are there 'umbrella species' that require broad habitat management strategies under which a range of other, perhaps less conspicuous, species can shelter? Can we identify 'keystone' species that are vital links in major food webs, the loss of which would cause a 'chain reaction' of extinctions amongst other species in the community?

One promising approach to this dilemma may be to identify geographical concentrations of biodiversity, which can become the focus of conservation attention, thus helping to conserve maximal diversity with a fixed amount of resources. Such a strategy depends critically on two assumptions: (1) that concentrations of diversity coincide geographically for different taxonomic groups, and (2) that the important rare species are also concentrated in these biodiversity 'hotspots'. Prendergast *et al.* (1993) have tested these assumptions on selected groups of flora and fauna in Britain and found disappointingly poor associations on both counts. However, Britain is a small and highly modified island and therefore may not be typical of other larger land masses. The hypothesis urgently needs to be tested at other scales. This strategy may be our best hope for retaining the maximum global biodiversity.

References

Berger, J. (1990) Persistence of different-sized populations: an empirical assessment of rapid extinctions in bighorn sheep. *Conservation Biology*, **4**, 91–8.

Boyce, M.S. (1992) Population viability analysis. *Annual Review of Ecology and Systematics*, **23**, 481–506.

Cadbury, C.J., G. Elliott and P. Harbard (1988) Birds of prey conservation in the UK. *RSPB Conservation Review*, **2**, 9–16.

Caughley, G. (1994) Directions in conservation biology. *Journal of Animal Ecology*, **63**, 215–44.

Clarke, B.C., J.J. Murray and M.S. Johnson (1984) The extinction of endemic species by a program of biological control. *Pacific Science*, **38**, 97–104.

Dempster, J.P. (1991) Fragmentation, isolation and mobility of insect populations. In N.M. Collins and J.A. Thomas (eds) *The conservation of insects and their habitats*. London: Academic Press, pp. 143–53.

Ehrlich, P.R. and E.O. Wilson (1991) Biodiversity studies: science and policy. *Science*, **253**, 758–62.

Falk, D.A. and K.E. Holsinger (eds) (1991) *Genetics and conservation of rare plants*. New York: Oxford University Press.

Fiedler, P.L. and S.K. Jain (eds) (1992) *The theory and practice of nature conservation, preservation and management*. New York: Chapman & Hall.

Gilpin, M.E. and M.E. Soulé (1986) Minimum viable populations: the process of population extinction. In M.E. Soulé and B.A. Wilcox (eds) *Conservation biology, an evolutionary-ecological perspective*. Sunderland, MA: Sinauer Associates, pp. 13–34.

Greig-Smith, P. (1983) *Quantitative plant ecology*, 3rd edn. Oxford: Blackwell.

Griffith, B., J.M. Scott, J.W. Carpenter and C. Reed (1989) Translocation as a species conservation tool: status and strategy. *Science*, **245**, 477–80.

Halliday, T. (1978) *Vanishing birds*. New York: Holt, Rinehart & Winston.

Hanski, I. and I.P. Woiwod (1993) Spatial synchrony in the dynamics of moth and aphid populations. *Journal of Animal Ecology*, **62**, 656–68.

Harrison, S. (1991) Local extinction in a metapopulation context: an empirical evaluation. *Biological Journal of the Linnean Society*, **42**, 73–88.

Harrison, S. (1994) Metapopulations and conservation. In P.J. Edwards, R.M. May and N.R. Webb (eds) *Large-scale ecology and conservation biology*. Symposium of the British Ecological Society, **35**, 111–28.

Harrison, S., D.D. Murphy and P.R. Ehrlich (1988) Distribution of the bay checkerspot butterfly, *Euphydryas editha bayensis*: evidence for a metapopulation model. *American Naturalist*, **132**, 360–82.

Hobbs, R.J. (1992) The role of corridors in conservation: solution or bandwagon? *Trends in Ecology and Evolution*, **7**, 389–92.

Hutchings, M.J. (1987) The population biology of the early spider orchid, *Ophrys sphegodes* Mill. I. A demographic study from 1975–1984. *Journal of Ecology*, **75**, 711–27.

Hutchings, M.J. (1991) Monitoring plant populations: census as an aid to conservation. In F.B. Goldsmith (ed.) *Monitoring for conservation and ecology*. London: Chapman & Hall, pp. 61–76.

Jenkins, M. (1992) Species extinction. In B. Groombridge (ed.) *Global biodiversity*. London: Chapman & Hall, pp. 192–205.

Krebs, C.J. (1989) *Ecological methodology*. New York: Harper & Row.

Kunin, W.E. and K.J. Gaston (1993) The biology of rarity: patterns, causes and consequences. *Trends in Ecology and Evolution*, **8**, 298–301.

Lawton, J.H., S., Nee, A.J. Letcher and P.H. Harvey (1994) Animal distributions: patterns and processes. In P.J. Edwards, R.M. May and N.R. Webb (eds) *Large-scale ecology and conservation biology*. Symposium of the British Ecological Society, **35**, 41–58.

Newton, I. (1979) *Population ecology of raptors*. Berkhamsted: Poyser.

Oostermeijer, J.G.B., R. van't Veer and J.C.M. den Nijs (1994) Population structure of the rare, long-lived perennial *Gentiana pneumonanthe* in relation to vegetation and management in the Netherlands. *Journal of Applied Ecology*, **31**, 428–38.

Pollard, E. (1991) Synchrony of population fluctuations: the dominant influence of widespread factors on local butterfly populations. *Oikos*, **60**, 7–10.

Prendergast, J.R., R.M. Quinn, J.H. Lawton, B.C. Eversham and D.W.Gibbons (1993) Rare species, the coincidence of diversity hotspots and conservation strategies. *Nature*, **365**, 335–7.

Pulliam, H.R. (1988) Sources, sinks, and population regulation. *American Naturalist*, **132**, 652–61.

Quinn, R.M., J.H. Lawton, B.C. Eversham and S.N. Wood (1994) The biogeography of scarce vascular plants in Britain with respect to habitat preference, dispersal ability and reproductive biology. *Biological Conservation*, **70**, 149–57.

Rabinowitz, D. (1981) Seven forms of rarity. In H. Synge (ed.) *The biological aspects of rare plant conservation*. Chichester: Wiley, pp. 205–17.

Ratcliffe, D.A. (1980) *The peregrine falcon*. Calton: Poyser.

Schemske, D.W., B.C. Husband, M.H. Ruckelshaus, C. Goodwillie, I.M. Parker and J.G. Bishop (1994). Evaluating approaches to the conservation of rare and endangered plants. *Ecology*, **75**, 584–606.

Simberloff, D. (1988) The contribution of population and community biology to conservation science. *Annual Review of Ecology and Systematics*, **19**, 473–511.

Soulé, M.E. (1987) *Viable populations for conservation*. Cambridge: Cambridge University Press.

Soulé, M.E., M. Gilpin, W. Conway and T. Foose (1986) The millenium ark: how long a voyage, how many staterooms, how many passengers? *Zoo Biology*, **5**, 101–13.

Stanley-Price, M.R. (1989) *Animal re-introductions: the Arabian oryx in Oman*. Cambridge: Cambridge University Press.

Templeton, A.R. (1986) Coadaptation and outbreeding depression. In M.E. Soulé (ed.) *Conservation biology: the science of scarcity and diversity*. Sunderland, MA: Sinauer Associates, pp. 105–16.

Thomas, J.A. (1980) Why did the large blue become extinct in Britain? *Oryx*, **15**, 243–7.

Thomas, J.A. (1984) The conservation of butterflies in temperate countries: past efforts and lessons for the future. In R.I. Vane-Wright and P.R Ackery (eds) *The Biology of Butterflies*. London: Academic Press, pp. 333–53.

Thomas, J.A. (1991) Rare species conservation: case studies of European butterflies. In I.F. Spellerberg, F.B. Goldsmith and M.G. Morris (eds) *The scientific management of temperate communities for conservation*. Oxford: Blackwell Scientific, pp. 149–97.

Waite, S. and M.J. Hutchings (1991) The effects of different management regimes on the population dynamics of *Ophrys sphegodes* Mill.: analysis and description using matrix models. In T.C.E. Wells and J.H. Willems (eds) *Population ecology of terrestrial orchids*. The Hague: SPB Academic Publishing, pp. 161–75.

Warren, M.S. and R.S. Key (1991) Woodlands: past, present and potential for insects. In N.M. Collins and J.A. Thomas (eds) *The conservation of insects and their habitats*. London: Academic Press, pp. 155–211.

Warren, M.S., C.D. Thomas and J.A. Thomas (1984) The status of the heath fritillary butterfly, *Mellicta athalia* Rott., in Britain. *Biological Conservation*, **29**, 287–305.

Webb, N.R. and L.E. Haskins (1980) An ecological survey of heathlands in the Poole Basin, Dorset, England in 1978. *Biological Conservation*, **17**, 281–96.

Western, D. and M.C. Pearl (eds) (1989) *Conservation for the twenty-first century*. Oxford: Oxford University Press.

Wilcox, B.A. (1986) Extinction models and conservation. *Trends in Ecology and Evolution*, **1**, 46–8.

Wilcox, B.A. and D.D. Murphy (1985) Conservation strategy: the effects of fragmentation on extinction. *American Naturalist*, **125**, 879–87.

CHAPTER 12 Conservation of habitats

KEITH J. KIRBY

Introduction

'No man is an island', wrote John Donne, the English author (1573–1631), reflecting on the interdependence between individuals and the society in which they live. While some species may be preserved in zoos, botanic gardens or seed-banks (see Chapter 15), conservation must include maintaining the relationships between species, and between species and their physical environment.

In many countries nature conservation increasingly means considering how species have adapted to and may continue to fit into the cultural landscapes that we have created; the managed and much reduced forest area, the agricultural grassland and rivers controlled to reduce the risks of flooding. The structure and composition of cultural landscapes are largely determined by the patterns of human activity, principally farming (Box 12.1) (Birks *et al.*, 1988; Whitney, 1994) with often only fragmented and relict semi-natural community types (see Chapter 13).

Rarely can the whole landscape be conserved; it is either too expensive or would interfere too much with other land uses (forestry, farming, recreation, etc.). Therefore areas are selected in the hope that they contain the most important species and ecological processes (see Chapter 3).

Common species and those well adapted to human modified landscapes may need no special conservation measures at present but agricultural changes and introduced pests or diseases can rapidly alter the situation. Improved cereal cultivation techniques and herbicides mean that some arable weeds in Britain are now so rare that they are the subject of special conservation programmes. The chestnut blight disease in North America eliminated a major forest tree as a dominant canopy species over much of its range (Whitney, 1994). Ways must be found to maintain species and processes, the links between them and their relationships to surrounding or indeed far-distant areas by influencing land-use policies and practices across the whole landscape, not just within protected areas.

This chapter looks at what is usually meant by 'habitat' and some related terms. We consider the circumstances where habitat conservation is most likely to be appropriate, the use of habitat surveys and inventories, the need or otherwise for management to maintain the condition of habitat patches, and possible mechanisms for putting conservation ideas into practice. Case studies are described mainly from Britain where there is a high population density and 'cultural landscape'.

Habitats and niches – the concepts introduced

All species of plants or animals survive in a limited range of environmental conditions (see Chapter 8). Most are further constrained in where they actually occur through competition, predation and other interactions between species. The *habitat* for a particular population of an organism is used in this chapter to define or describe the conditions under which it usually occurs, taking account of both physical and biological constraints on its distribution. The related concept of *niche* (Elton, 1966) involves also the role that the species is playing in the system (predator, herbivore, decomposer, etc.). If habitat is like a person's address, then niche describes their job or profession. Other terms such as biotope and formation may be used in ways that overlap with the habitat (see Glossary).

Box 12.1 Cultural landscape elements in Britain

Element in the modern landscape	Vegetation in former natural landscape	Factors that created or now maintain element
Heathland	Acid woodland	Controlled fire, grazing and cutting
Meadows	Neutral–calcareous woodland	Cutting and grazing
Coppice woodland	Natural high forest	Deforestation of surroundings and subsequent regular cutting of remaining fragments

Some parts of the cultural landscape are very ancient. Much heathland in Britain dates back at least 3000 years and woods may have been managed by coppicing for as long a period (Rackham, 1986).

Like an address, the habitat of a species population may be defined with varying degrees of precision, although ultimately it must be expressed in terms of temporal and spatial scales appropriate to the organism concerned. A beetle species may, for example, be described at one level of habitat definition as being found in woodland, at a finer level as a species of old oak forest, at a third level its habitat is rot-holes within oak trees. Many invertebrates and amphibians (such as frogs and toads) have differing requirements in their juvenile and adult lives.

Humans assess a landscape from our particular scale and perspective. Even so there may be different interpretations of what is seen: a child of four sees a sand dune in a very different way to a fifty-year-old botany professor. Some species differ in how they use what appear to us to be fairly homogeneous 'habitats' while others occur in what to us are very distinct patches. For example, in Britain the stinging nettle (*Urtica dioica*) occurs in woodland communities, in fens and in highly disturbed urban sites because the key factor it needs is soil rich in phosphate which is associated with a diverse range of other conditions.

Habitats must also be defined in a particular geographical and historical context. Several butterfly species on the edge of their climatic range are restricted to grassland with very short turf in Britain. Their larvae need particularly warm conditions for their development and so the food plants must not be shaded by tall grasses; in warmer climates the cooler conditions among taller grasses are sought (Box 12.2) (Thomas, 1991).

A species may also be absent from a particular geographic area through accidents of history, giving opportunities for other, less competitive, species to take its place. For example the lack of predatory mammals has contributed to the success and

Box 12.2 Habitat requirements for butterflies on the northern limit of their range in Britain

Species	Habitat requirements	
	In Britain	In continental Europe
Lysandra bellargus and *Hesperia comma*	Rare, restricted to short turf and often to south-facing slopes	Widespread and common, and not so restricted in site type, where the climate is 2–3°C hotter in central Europe
Violet-feeding fritillaries	Restricted to open sunny areas in woods, such as immediately after a coppice cut	Common in semi-shaded woods in the Dordogne in southern France
Maculinea arion	Requires very short turf (<2 cm) (because of its ant associate, see Box 12.3) on south-facing slopes	In south-east Sweden, where spring and summer temperatures are warmer, it uses 2–5 cm swards, and in southern Europe 20–30 cm swards

Source: Thomas (1991).

abundance of flightless and ground-nesting birds on many remote islands such as New Zealand

Use of the habitat concept in practical conservation

Even if it is not possible to define the habitat for a species precisely, for many it can be described in broad terms as woodland, grassland or bog, etc. This provides a starting point for practical conservation – at least this range of habitats must be maintained within the landscape. This is also often the only level at which there is much information on the extent and distribution of habitats, and in particular good historical data on changes in their extent. Such data are often used to fuel conservation campaigns by focusing on losses of or damage to habitats. For example, campaigns to save American old-growth forest stress that it occupies less than 1% of the original forest cover in many states (Whitney, 1994).

Habitat-based conservation is easy for non-specialists to understand and support; the main weakness lies in deciding at what level habitats are defined. Even if some patches of habitat are conserved, some important species may not be present in those particular areas because they require some feature (such as old trees in a forest) that is not present or a mosaic of different vegetation types. For example, many insects in their larval form feed on dead wood but their adult stages feed on nectar. Although thought of as woodland species, they are more likely to thrive where there is a combination of old trees and relatively open sunny conditions with many flowers.

Advantages and disadvantages of habitat-based conservation

Many natural landscapes have an easily recognizable pattern to the vegetation that they contain. In savanna there may be gallery forests associated with the main waterways; in boreal forests tree-covered land is interrupted by open bogs. Any conservation programme for the savanna or boreal forest must include the distinct habitats provided by the gallery forest or bogs respectively because they contain many species not found in the surrounding landscape.

However, habitat-based conservation is applied more often in cultural landscapes where natural or semi-natural vegetation patches and their associated fauna survive within a landscape that is of relatively low value for nature conservation. Each patch tends to be dominated by one or other major habitat type – woodland, grassland, wetland, heathland – and mosaics or transition zones between types are scarce. The type reflects the history of land use in the area as well as underlying environmental conditions. For example, wetland areas or rocky pastures may survive because they were too difficult to farm; in parts of Britain, Denmark and Germany heathlands occur – the vestiges of a farming system that is now abandoned. The frequency and abundance of these relict habitats and their associated species are not necessarily representative of the original landscape.

The compartmentalism trap

Defining distinct habitats and the boundaries of the habitat patches that they form is a powerful tool in conservation biology. However, all these habitat patches and their limits are partly artificial creations of ecologists, planners and conservationists. Many species use more than one vegetation type and individuals may use more than one patch in any landscape. There is a tendency to place more value on the conservation of species that are restricted to the core of a patch because these may be rarer, less mobile and have more restricted requirements than those found at the edge. However, edge species, for example tall herb communities of scrub and forest margins (Ellenberg, 1988), and those of mosaics also need to be conserved. Concentrating on habitat patches rather than on the whole landscape also limits the scope for conservation at a wider scale. Defending individual habitat patches, particularly those that are nature reserves, is essential but is only a holding operation; it may slow the rate of species loss within the patches but not prevent it. We need to move beyond the siege-walls of site defence and look at the relationships between patches and the surrounding countryside.

Scales and levels of habitat differentiation

Animals respond to different cues in the landscape (Figure 12.1). A herd of fallow deer (*Dama dama*) may use small blocks of woodland and some of the intervening land with little regard to the botanical or structural variation present in the woods. A small

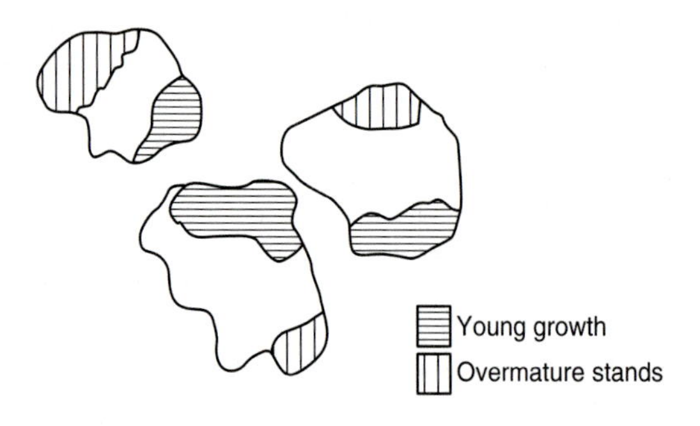

Fig. 12.1 Spatial scales and differentiation of habitats in woods. Three woods used by deer, containing patches of young growth suitable for breeding nightingales, and overmature stands with a rich deadwood fauna.

songbird such as the nightingale (*Luscinia megarhynchos*) is most abundant in areas with dense undergrowth that provide favourable conditions for nesting. Beetles requiring rotting wood are restricted to the stands with large old trees.

The relationship between habitat patches and their use by individuals and populations is different. All the deer form one herd but use all the woods; there may be regular interchange of individuals with neighbouring herds using other woods. The nightingales are part of a single population, but the individuals in it occur in only one or other of the woods. The deadwood beetles have colonies confined to individual trees and there is little or no interchange between populations from separate woods.

The importance of microhabitat differentiation

Habitat specialization occurs on even finer scales with, for example, different invertebrates associated with different decay zones within rot-holes in a tree trunk. Plant distribution in grassland can vary over the space of a few metres, between the tops of old plough ridges and the bottom of the adjacent wetter furrow (Figure 12.2; Harper and Sagar, 1953) or between the top of an anthill and the surrounding grassland (King, 1977). Annual plants and short-lived

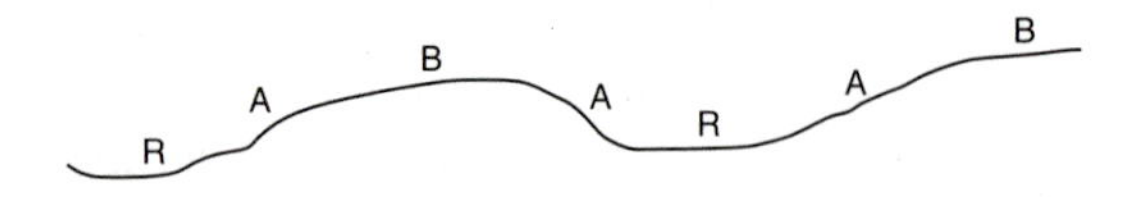

Fig. 12.2 Microhabitat variation in grassland. Consistent differences were found in the distribution of three species of buttercup across a series of ridges and furrows: *Ranunculus bulbosus* (B), *R. acris* (A), *R. repens* (R).

Box 12.3 The extinction of the large blue butterfly in Britain: an interaction of plant structure, summer temperatures and an ant

The large blue (*Maculinea arion*) was confined to short turf sites in southern Britain. Its eggs are laid on *Thymus praecox*, but after a short period of feeding the larvae drop off and have to be 'adopted' by an ant, *Myrmica sabuleti*, in whose nest the larvae then spend the next nine months as predators on the ants.

Myrmica sabuleti only forms dense colonies in Britain where the turf is less than 2 cm high; if the grass grows to 5 cm the spring soil surface temperatures decrease by 5°C and the ant dies out. Hence the butterfly was also confined to short turf sites.

By the mid-1970s all but one of the sites for the large blue had been abandoned as grazing land by farmers and the rabbit population was insufficient to keep the grass grazed so short. Soon afterwards all the colonies of the butterfly had died out.

Reinstatement of grazing on some sites has led to the recovery of the ant populations and there has been a successful reintroduction to these sites of the large blue from Swedish stock.

Source: Thomas (1991).

perennials were found to be more abundant on the anthills, where the soil surface is disturbed by the activity of the ants.

Small differences in the height of the grass alter the humidity and temperature regimes at ground level and have contributed to species extinctions (Box 12.3). Elsewhere, reduction of grazing pressures may lead to the development of a tall tussocky structure rather than a close-grazed one. Small mammal populations may then expand because there is more cover and food for them and in turn may attract raptors that feed on them.

Examples of other temporary microhabitat resources include dung and carrion in grasslands and deadwood in forests. The dynamics of occurrence of these microhabitats may be driven by natural processes or, particularly in agricultural landscapes, by management.

Even where a habitat is apparently present all the time, its suitability for growth and reproduction of a species may still vary. The place where a species thrives in a dry year may not be the same as in a wet one. For example in one study in New York State, the rare fringed gentian (*Gentianopsis crinita*) grew best and produced the largest plants in one part of a field, but died out from there in drought years. Recolonization occurred from what were suboptimal areas in growth terms in normal years (Robertson, 1992). Failure to include sufficient habitat variation within conserved patches to allow for such differences in survival may lead to species extinction in unfavourable years.

Keeping microhabitat variation in perspective

No two woods are the same; no two trees are precisely identical; and even on a single tree there is variation. Different lichens occur on the outermost twigs compared to those on the main trunk; some birds feed among the fine twigs, while others hunt along the main branches and trunks (Ellenberg, 1988; Edington and Edington, 1972). Around the base of the tree run-off from the trunk may lead to soil acidification in some directions with corresponding changes in the vegetation.

Not all microhabitats can be created or maintained in each habitat patch because the conditions that create some may destroy others. The short turf needed by the ant host for the large blue butterfly (Box 12.3) is lost if grazing is relaxed to favour the large tussocks used by small mammals. A key feature of habitat conservation is to recognize the important features and species in a site and to build on those strengths. Priorities must be set for conservation both locally and nationally and refined in the light of survey results.

Habitat survey methods

The design of all habitat surveys should be made only within clear objectives and a clear statement of the aims. Surveys and evaluation of habitats for nature conservation purposes take place at at least three levels. There is a need to identify what habitats are present and whether all are of equal value (are woods more important than grassland or arable fields?). There is evaluation to determine which examples of a particular habitat are most worthy or in need of conservation in a particular region. Finally, there may be the need for research and monitoring to understand how habitat patches function, whether they are changing and if that is desirable from a conservation perspective.

The survey methods used must separate the different habitat types and establish their extent and distribution and the relationships between them. Further information on the composition and structure of the habitat, the variety of species present and the occurrence of rare or endangered species may be needed to assess the comparative value for nature conservation of different patches.

From knowledge of the needs of particular species and the history of a site, decisions must be made as to how the land or water should be managed in future. Maintenance of hay-meadow communities in northern Europe depends on particular management regimes being continued: signs of active management might thus be considered beneficial. In near-natural forest types on the other hand evidence of human impact would be a negative feature in site evaluation.

Survey methods are almost always a compromise between the ideals of good ecological science and the practicalities of getting enough information for sufficient sites within the available budget and time-scale. Frequently the experience of the surveyor must act as a substitute for more objective, but time-consuming measures.

Broadscale identification and mapping of habitats

Defining the extent, location and composition of habitats is a first step towards an evaluation of their significance for nature conservation. If the results are to be compared with pre-existing data that have already been classified into types, the existing classification may need to be part of the new survey. Otherwise a habitat classification may be one of the results of the survey. Many habitats can be distinguished by non-specialists (woodland, grassland, sand-dunes) and be shown on large-scale maps. Distinctions between different grassland types on the other hand may be less obvious. The variations in species composition are as great, but to many people all grasses look the same.

Surveys can be carried out by people walking the ground and mapping what they see, as for example in the American land survey (Whitney, 1994). Such direct habitat mapping is still widely used, particularly in small or local surveys. Many details in the structure and composition of the vegetation and signs of animal life can be recorded. Direct mapping relies, however, on good access to the areas concerned and that access may not be possible for physical or legal reasons. Mapping is also time-consuming and may require large teams of surveyors. Maintaining recording standards between and within teams then becomes difficult.

The habitats present can also be assessed from aerial photographs and satellite images (see for example Plate 6). The advantages of remote sensing over field mapping are that large areas can be considered relatively quickly, interpretation and analysis can be more easily standardized, with satellite imagery much of the data processing may be automated, and the information can be frequently updated by new flights or satellite passes.

Habitat mapping from aerial photographs depends on the scale and clarity of the photograph and on the skill of the person interpreting the picture. Ground truthing of the interpretation (checking that what seems different on the photograph matches a change on the ground) is required. More detail can be obtained if colour prints are available, if there are summer and winter views of the same area, or if two photographs of the same scene, taken from slightly different positions, are viewed through a stereoscope. This device merges the two images to give a three-dimensional effect from which the heights of trees or variations in topography may be deduced. Old aerial photographs can be used to assess habitat change, for example the rapid changes in land cover that have occurred in many countries over the last 50 years (Box 12.4).

Aerial photographs may not be available for some areas, or they may be out-of-date, or parts of the picture may be hidden by cloud. Some of these problems can be overcome with satellite imagery. Various satellites regularly record land cover data using a range of wavelengths within the electromagnetic spectrum, which can be translated into visual images. The size and type of habitat patch that can be distinguished depends on the spatial resolution of the satellite system and also on the degree to which different habitats can be unambiguously separated according to their reflectance of different types of radiation. There may, for example, be difficulty in separating marsh vegetation from dry tall rough grassland or in separating deciduous woodland from a mixture with coniferous trees. Discrimination between habitats can be improved by comparing areas of known land cover with their satellite images and then searching for other areas with a comparable signal. Nevertheless there remain difficulties in reconciling estimates of the extent of and change in land cover types from different surveys because of differences in cover type definition and recognition.

Box 12.4 Use of remote sensing for habitat mapping to assess changes in the land cover of Scotland

The land cover of Scotland was stratified into broad classes using Landsat Multispectral Scanner prior to sampling. Sample squares (each initially 5 km × 5 km, later reduced to 2.5 km × 2.5 km) were selected within the strata to provide a sample coverage of 7.5% overall (464 squares in total). Aerial photographs from the 1940s were compared with those from 1977–78 for each square.

The photographs were at 1:10 000 scale and features down to 0.1 ha were split between eight major areal categories each of which was subdivided to give a total of 36 types and a further five linear types.

Estimates of change for the major areal groups and linear features over the c.30-year period are as follows:

Type	1940s	1977–78	Change
Major areal groups (km^2)			
Grassland	24 348	23 858	−491
Mire	20 191	18 500	−1 691
Heather moorland	15 377	12 636	−2 741
Arable	8 246	8 098	−148
Woodland and scrub	4 544	8 974	4 429
Open water	1 572	1 612	40
Built-up land	1 604	2 066	462
Other	2 042	2 183	141
Linear features (km)			
Hedgerows	42 819	26 874	−15 946
Tree-lines	19 334	17 600	−1 734
Running natural waters	123 904	122 728	−1 176
Canalized waters	56 784	71 718	14 935
Unsurfaced tracks	31 344	39 240	7 896

The results demonstrate the major loss of mires, heather moorland and grassland (particularly unimproved grassland), and an increase in (mainly coniferous) woodland. The hedgerow loss was mirrored elsewhere in Britain. The increases in built-up land (including roads and tracks) and in canalized waters reflect the trend towards a more managed and less 'wild' countryside.

Source: Tudor *et al.* (1994).

Box 12.5 The decline in the deadwood fauna in British woodland over the last 5000 years

Peats and inorganic silts preserve not only a record of the past flora as both fossils and macrofossils, but also insect remains. These may be identifiable to species level and by comparing their current distribution and habitat requirements conclusions may be drawn about the past structure and composition of the landscape.

At least 30 beetle species that are now extinct in Britain have been identified from Holocene deposits. A disproportionately high number of these are known to be associated with dead wood or other habitats found in old-growth type forests. Wetland species have also declined. There is evidence for major reductions in the range of other species, some of which survive in only a handful of sites, sometimes only in a handful of trees.

This reinforces work on the contemporary fauna in forests in Scandinavia on the vulnerability of 'old growth' specialists (Samuelsson *et al.*, 1994).

Broadscale surveys of past land cover

In northern Europe and in North America, much of the original natural vegetation has been cleared or modified by agriculture. However, one of the objectives for habitat conservation is the restoration of species and communities which were there in the past. Evidence for past natural vegetation (and some of the associated fauna) comes from archaeological and historical remains and peat deposits, and by analogy with such remnant near-natural areas as do survive, for example old-growth forests (Birks *et al.*, 1988; Whitney, 1994). For parts of North America there are also documentary accounts of the vegetation before European settlement. These data are combined with the current distribution patterns of habitat types and species in relation to geology and climate to produce potential vegetation maps.

Knowledge of past or potential vegetation cover and its structure may help to identify which habitats and species have suffered most loss and reduction in their range or have increased during historical times (Box 12.5).

Broadscale field surveys and habitat inventories in Britain

Remote sensing cannot eliminate the need for some direct surveying either for ground truthing or to provide more detail on habitat patches. In Britain, during the 1980s, the Nature Conservancy Council (the then government conservation agency) combined aerial photographic interpretation with quick, superficial ground surveys. These 'Phase 1' surveys mapped habitats at 1 : 10 000 scale using finer divisions than are possible from remote sensing, and also provided a very crude evaluation of the likely importance of different patches in nature conservation terms based on the surveyor's impressions (NCC, 1990). At the time, virtually all semi-natural areas were threatened by intensive agriculture, commercial forestry and urban developments.

Surveyors used photographs to assess the broad patterns of variation in the landscape and to identify areas likely to be of some interest which were in turn checked on the ground from paths and roads and by scanning areas with binoculars from vantage points. Ninety 'habitat types' were recognized in the surveys, the most abundant species in each habitat patch were usually listed, and for many patches of semi-natural vegetation a short description was produced.

The maps and supporting legends created by Phase 1 surveys were a cumbersome way of presenting and storing data. Subsequent development and use of computer databases and spreadsheets make it easier to summarize large volumes of data in readily retrievable forms. Geographic Information Systems (GIS) allow different elements within the landscape to be overlain and relationships can be sought between the occurrence of species or habitats and soils, topography or land use.

Habitat inventories in Britain have also helped to focus attention on the need to change land-use policies at local and national level (see Chapter 6). In the early 1980s ancient broadleaved woodland was being lost or severely degraded under the then current forestry policies. It was recognized that these woods should receive special treatment, but the policy change could only be brought about once there was a mechanism for identifying such sites and listing them across the whole country – the Ancient Woodland Inventory (Box 12.6). The inventory showed that the threats to these important woods were as widespread as had been previously estimated from sample surveys and that far less than had been thought was protected in nature reserves or other special conservation sites. The inventories can also be used to target new woodland establishment to complement existing ancient woodland patterns.

Box 12.6 Ancient woodland inventories in Britain

Background

Peterken (1974) and Rackham (1976) in regional studies highlighted the importance for nature conservation of sites which had existed as woodland, at least back to the medieval period (AD 1600), even if it could not be proved that these woods were primary in the sense of never having been cleared. These studies could not be duplicated everywhere because the procedure was time-consuming and the sources were missing in some areas. Therefore in 1981 the Nature Conservancy Council (NCC) started a project to produce an inventory of ancient woods for the whole country, using a simpler approach. It was decided to include sites for which the data were uncertain or ambiguous in the first instance; this it was hoped (and it has proved to be the case) would stimulate more detailed work on such sites. The inventories remain therefore subject to revision as more information becomes available.

Building up the inventories

Possible ancient woodland sites were identified from their appearance on maps produced in the 1970s, 1930s and the first half of the nineteenth century, plus other readily available historical data. Their current state, whether semi-natural or replanted with introduced species, was determined from aerial photographs and field surveys (where available). Data were assembled on the location and extent of sites over 2 ha for each county and summary statistics produced on site size, recent losses in cover or changes in woodland composition, and variations in ancient woodland occurrence across the country (Spencer and Kirby, 1992).

Subsequent uses

The inventories are used by the Forestry Authority and local government to assess the impact of forestry proposals, plans for new housing, etc. They have stimulated interest in ancient woodland conservation among local communities and provided a resource for those researching land use and landscape change.

European habitat classification

Inventories covering several countries require even clearer agreement on habitat definition than do those just within a country. In Europe, the CORINE biotopes listing (Box 12.7) is one system of classifying areas for conservation purposes which includes types based on geographic and structural characteristics as well as purely phytosociological ones (European

Box 12.7 Extract form the CORINE* classification for European woodland

Five or more levels are involved in parts of the classification:

- Level 1
  Coastal and halophyte communities
  Non-marine waters
  Scrub and grassland
  Woodland
  Bogs and marshes
  Rocky habitats
  Agricultural land and highly artificial communities
- Level 2 (for woodland only)
  Broadleaved deciduous
  Native coniferous
  Mixed woodland
  Alluvial and very wet forests and brush
  Broadleaved evergreen
- Level 3 (for broadleaved deciduous only)
  Beech forests
  Oak–hornbeam forests
  Fraxinus excelsior woods

* CORINE = CoORdination of INformation on the Environment

Communities Commission, 1991). There is also a database with details of the distribution of habitats across Europe. The information is mainly from protected sites (which biases the data) and the coverage from different countries varies considerably. However, as the classification system and the site inventory are improved they will provide a European context for country conservation programmes.

Detailed plant surveys

Detailed surveys are needed where there is a choice as to which sites are protected or how the site should be treated to ensure that its features of importance are maintained. Initially these are likely to be vegetation surveys, because they are usually quicker and easier to do than animal surveys.

Plant data may be collected in various ways depending on the extent of the area to be surveyed and the nature of the vegetation. One requirement may be a plant species list which should be supported by herbarium specimens if the flora is not well known, or if the surveyor is inexperienced. Species lists may be used as one measure of biodiversity but it is important to recognize that the number of species found is affected by the intensity of the survey, by seasonality, by the experience of the surveyor and even sometimes by the weather (Kirby *et al.*, 1986).

Plant lists are more informative if the abundance of the species concerned and their distribution are indicated (a subjective assessment is better than none at all). Plant species data can also be collected in ways that provide more objective measures of plant abundance (Kent and Coker, 1992). Sample plots are commonly used with the size varying according to the type of vegetation (from 1 m × 1 m plots for studies in grassland to plots of a hectare or more in tropical rain forest).

Sample plot data provide the frequencies and distributions of individual species and can be analysed to bring out the similarities between plots, for example using computer programs such as TWINSPAN (Hill, 1979) or CANOCO (ter Braak, 1987). Although the groupings of plots produced by such analyses may provide a way of classifying the vegetation, it is worth trying to relate results to those from existing classifications and hence sites previously described. Description of the vegetation may also include maps.

The structure of the vegetation and other features that contribute to microhabitat diversity should also be noted, either through simple listing of attributes, through the construction of indices such as the foliage-height index as used in many bird studies (James and Wamer, 1982), or through detailed scale drawings of features, such as canopy profiles and the locations of individual trees.

Collecting data on all plant species may be very time-consuming. An alternative is just to record those plants thought to be indicators of particular conditions. Certain vascular plants, for example, are strongly associated with a long continuity of woodland cover (Peterken, 1974). Other plants may be used as indicators of the site conditions: the presence (or an increase in the abundance) of species intolerant of grazing or of a high water table may be used to assess the impact of management of the site or its surroundings (Ellenberg, 1988).

The levels and stages described above are summarized in Table 12.1.

Table 12.1 Different levels of and stages in plant surveys

1 Divide landscape into habitat patches.
2 List plant species and make subjective assessment of abundance.
3 Record species presence/cover in quadrats or through plotless sampling.
4 Combine 2 and 3 with descriptions of vegetation structure and mapping.
5 Analyse and interpret 2–4 in terms of vegetation types or indicator species.

Surveys of animals

Surveys of animals involve a wide range of methods, depending on the group concerned and the habitat type. Invertebrate surveys in woodland, for example, might include pitfall traps for ground-dwelling species, interception traps for those flying in the space below the canopy, direct observation of large, easily identified butterflies and releasing controlled amounts of an insecticide into the canopy to sample the species in the canopy (for details see texts such as Southwood, 1978). Intensive trapping may be needed to establish which small mammals are present, but for birds the same sort of information may be obtained from recording their calls (Ralph and Scott, 1981). In short visits to an area recording indirect signs, e.g. tracks, dung, nests, or structure of the vegetation, may be as informative as rare direct observations. Some indirect methods can be made quantitative, for example by counting the number of leaf mines or spiders' webs or dung pellets of deer found in sample plots.

Particular animals may be selected as indicators of the occurrence or quality of particular habitat features (that is on the assumption that if conditions are suitable for the indicator other species will similarly benefit). For example, brown trout (*Salmo trutta*) are used as an indication of the quality of forest streams because they are sensitive to acidification. More generally a range of invertebrates may be used as indicators for different types of river pollution (Spellerberg, 1991).

Sites and their surroundings

The land surrounding a site or a patch of habitat influences many of the species that occur within it. Forest-nesting birds may use adjacent open land for feeding; amphibians may breed in a pond but spend part of their adult life in a nearby field. There may be herbicide drift from adjacent agricultural land, alterations to the water table by drainage, or increased predation of birds at the edge of the habitat patch. Where important species or communities might be affected the nature of the surrounding land should be recorded, both in absolute terms and as a proportion of the perimeter. Species in the centre of the patch, however, particularly if they are sedentary, may be little affected by the nature of the surroundings.

Survey, evaluation and management

Rarely is it possible to protect all examples of a particular habitat. Conservation bodies, owners and managers need guidance as to which are the more important sites to conserve or the more important areas within a site. The issues are explored in Spellerberg (1992) and in Chapter 14.

The initial survey and subsequent monitoring studies are also important in deciding what the appropriate management regime should be and whether the current management is likely to maintain the important species and communities on that site.

Management or minimum intervention

In most countries, nature conservation is the primary objective on only a small part of the land. Where nature conservation needs to be integrated with other land uses, some management of habitats may be unavoidable. For example, grassland may have to be grazed by cattle because the farmer who owns it runs a dairy farm, reeds in a wetland may be cut for thatch, or some of the trees in a wood may be felled for sale.

Where conservation is the main objective of land use, two differing philosophies have developed. On the one hand there are landscapes where much of the landscape is still predominantly natural or tending towards a more natural state (frequently forest). In large areas natural disturbance regimes such as fire, windstorm, disease, flood, etc., may maintain the full diversity of conditions in one block at all times. At any one point some species may become extinct, but there are always other patches within recolonization distance. The emphasis is therefore towards restricting human intervention, or at least limiting it to a low level. Some North American national parks, for example the Great Smoky Mountains, are in forests that were heavily logged in the last century but may now revert to wilderness over large areas. There may be attempts to reduce existing human activities, for example by resettling people who live within the reserve boundaries.

The other extreme is illustrated by much of England, where the habitats important for conservation are mostly in small isolated patches. They cannot be left alone because they are too affected by the land beyond their boundaries; they are not large enough to show the full range of disturbance patterns in each site and are too isolated to be able to rely on recolonization from adjacent patches; and many are examples of arrested or diverted succession maintained by former agricultural practices. These issues are explored further in turn.

Boundary conditions

The management of adjacent land may have deleterious effects on the reserve so the internal reserve management must be altered to counteract them. If a highly invasive species starts to spread in from adjacent land the reserve manager may be forced to control that species on the reserve. Conversely species from the reserve may spread on to adjacent land (or predate on farmers' crops and stock) leading to a need to limit that species on the reserve as part of a 'good neighbour' policy. Natural processes, for example wild fires, may be acceptable in principle but have to be stopped from spreading to managed forest or houses.

Patch dynamic factors

For many species the particular microhabitat that they require comes and goes at various places according to growth and disturbance phenomena. A beetle that depends on fallen logs may be able to pass through several generations within one log before it decays completely, but eventually the population must move on to another tree. Species of bare sand on heathland may need to move to new, freshly created patches as heather growth gradually recovers on the original one. The minimum area for these species to have the potential to survive indefinitely is therefore that which always contains their microhabitat in a suitable condition and where the microhabitat patches are close enough for the species to move easily between them (Pickett and White, 1985).

As an example, windstorms, or tree deaths from disease, might affect 3% of a wood per year sufficiently to create a gap. If gap-dependent species need 0.5 ha of gap to survive on the site then the wood must cover at least $0.5 \times 100/3 = 16.7$ ha (the minimum patch dynamic area). In reality, a much larger area would be required since it is unlikely that natural processes would produce exactly 0.5 ha each year. In some years it would be much more and in others no gaps at all might be formed.

Where forests are now only a small part of the countryside, the minimum patch dynamic area for species of open glades in the original natural forests may far exceed the size of the surviving woods, and little movement can occur between woods because they are too far apart. Gap species might not be able to survive now in unmanaged stands. Survival is, however, possible under forestry systems that increase the rate of gap formation or make it more regular in occurrence such that the need for large areas to allow for stochastic variation is reduced. The common practice throughout much of Europe and also in Japan of coppicing woods (regular cutting of the trees at 10–30 year intervals with most of the regrowth coming from the stumps) did both; not surprisingly open stage species flourished. In Britain half of the 16 butterflies that are largely confined to woodland are edge or young growth specialists, while 60% of the 125 rare woodland moths are associated with transitional woodland habitats (Fuller and Warren, 1991). The decline of coppice management over the last century has led to widespread losses of these species, although there are instances where the decline has been reversed after the coppice regime has been restored (Thomas, 1991).

Thus in landscapes where habitat patches are small and isolated management may help maintain patch dynamic cycles.

Arresting or diverting succession

Many habitats (meadows, heaths and some forms of wetland and wood pastures) in the cultural landscapes of Europe, highly valued for their wild plants and animals, were created or are maintained by regular grazing or cutting. Some of the species assemblages may not have occurred in the virgin landscape; others may have been present but were rare and transitory. If the management that created or maintained these cultural habitats is stopped, successional processes take over and the site and species assemblages will change.

Whole landscapes may alter quickly. In Britain, much grassland has changed to scrub or woodland in the last 40 years following a steep decline in rabbit populations. The abandonment of traditional farming in the meadow–forest zones of southern Europe is increasing the woodland at the expense of grassland. In North America, much farmland in the eastern states of the USA has reverted to forest since the end of the nineteenth century as farmers moved to more productive ground further west (Whitney, 1994).

Habitat creation for conservation

Habitat creation by humans on a large scale started when agricultural and forestry practices altered the composition and structure of the original landscapes in a permanent way. Recent developments in habitat creation for conservation purposes differ in that now there is a deliberate intention to try to mimic natural or semi-natural communities and structures. This includes creating wetland for birds, restoring herb-rich grassland, and introducing ground flora species to newly planted woodland (Buckley, 1989; Howell and Jordan, 1991). Some schemes involve translocation of species from existing (usually threatened) sites to the new one.

Concern has been expressed at these practices, often centring on the following issues:

1 The poor success rate of many schemes represents an inefficient and ineffective use of scarce conservation resources.
2 Even if successful the new habitat usually lacks the more specialist species and features associated with long-established examples.
3 Successful habitat creation may be used as an argument for destroying existing high-value areas.
4 The plants used in creation schemes may be of introduced or inappropriate genotypes which could spread to nearby semi-natural examples of the habitat.
5 The translocation of species may make it difficult to determine past distribution patterns and thus to assess human impact on farms, in forests and in towns.
6 There is inadequate monitoring of what has been done and its success or failure.

These are important ecological concerns and these and others are discussed further in Chapter 16.

Conservation mechanisms – ideals into practice

Those promoting conservation must always appreciate the implications for others of what they seek to achieve, as well as making others aware of the conservation case in a way that is as positive and encouraging as possible. Building support amongst land managers, for example through educational programmes, is a keystone of good conservation. Public support may, however, need to be backed by legislation (prohibiting or promoting particular practices); by grants and incentives particularly where positive action from a private landowner is needed; by designation of special areas (which may be backed by extra legislation or incentives); and by sympathetic ownership (private or public bodies and private individuals who wish to promote conservation). Good examples of all of these can be found, as, however, can places where they have all failed.

A recurring theme is the conflict of interest between the immediate interests of the individual which may be best served by damaging or destroying a patch of habitat (to make it more productive) and those of society which values the communities and species present. This is repeated at larger scales, as the clash between a region that wishes to see its large areas of semi-natural habitat made more productive to generate income and jobs for its inhabitants, and the state as a whole that values the relatively undeveloped region for its high wildlife content. There may be international conflicts with, for example, the British government's decisions to site roads through sites of high conservation value being queried by officials of the European Union (see example of Twyford Down in Chapter 1).

The different mechanisms may overlap on any one piece of ground. A site identified as of high nature conservation value may also be within a protected landscape or be owned by a conservation organization. This is beneficial when two weak mechanisms reinforce the overall level of protection received by a site, but may also waste resources (Colman *et al.*, 1993).

Conclusions

Habitat-based conservation is a way of structuring conservation efforts such that the range of conditions across a landscape may be represented in a series of relatively small patches. It is commonly developed in areas where much of the landscape has been modified by intensive agriculture or commercial forestry.

Inevitably wide-ranging species and those that depend on mosaics and transitions between habitats may not be favoured by this approach. The small-scale variations that occur within what at a human scale are homogeneous habitats must also be recognized.

Survey methods must include both broad-ranging assessments of the extent of different habitat types across a landscape and more detailed descriptions of individual habitat patches. They must provide data on how habitats are changing in extent and composition and provide the basis for determining appropriate management.

Active habitat management is essential in many sites if their interest and character are to be maintained. However, opportunities for minimum intervention should also be sought.

There is a wide range of mechanisms for getting conservation implemented, but the main hurdle to be overcome is the conflict between individual losses resulting from the implementation of conservation measures and the wider (often unquantified) benefits.

References

Birks, H.H., H.J.B. Birks, P.E. Kaland and D. Moe (1988) *The cultural landscape: past, present and future.* Cambridge: Cambridge University Press.

Buckley, G.P. (1989) *Biological habitat restoration.* London: Belhaven Press.

Colman, D., J. Fround and L. O'Carroll (1993) The tiering of conservation policies. *Land Use Policy*, **10**, 281–92.

Edington, J.M. and M.A. Edington (1972) Spatial patterns and habitat partition in the breeding birds of an upland wood. *Journal of Animal Ecology*, **41**, 331–56.

Ellenberg, H. (1988) *Vegetation ecology of central Europe*, 4th edn. Cambridge: Cambridge University Press.

Elton, C.S. (1966). *The pattern of animal communities.* London: Chapman & Hall.

European Communities Commission (1991) *CORINE biotopes – the design, compilation and use of an inventory of sites of major importance for nature conservation in the European Community.* Luxembourg: European Communities Commission.

Fuller, R.J. and M.S. Warren (1991) Conservation management in ancient and modern woodlands: responses of fauna to edges and rotations. In I.F. Spellerberg, F.B. Goldsmith and M.G. Morris (eds) *The scientific management of temperate communities for conservation.* Oxford: Blackwell Scientific, pp. 445–72.

Harper, J.L. and G.R. Sagar (1953) Some aspects of the ecology of buttercups in permanent grassland. *Proceedings of the British Weed Control Conference*, **1**, 256–65.

Hill, M.O. (1979) *TWINSPAN – a FORTRAN program for arranging multivariate data in an ordered two way classification of the individuals and the attributes.* Ithaca, NY: Cornell University, Department of Ecology and Systematics.

Howell, E.A. and W.R. Jordan III (1991) Tall-grass prairie restoration in the North American Midwest. In I.F. Spellerberg, F.B. Goldsmith and M.G. Morris (eds) *The scientific management of temperate communities for conservation.* Oxford: Blackwell Scientific, pp. 395–414.

James, F.C. and N.O. Wamer (1982) Relationships between temperate forest bird communities and vegetation structure. *Ecology*, **63**, 159–71.

Kent, M. and P. Coker (1992) *Vegetation description and analysis – a practical approach*. London: Belhaven Press.

King, T.J. (1977) The plant ecology of anthills in calcareous grasslands. I. Patterns of species in relation to anthill in Southern England. *Journal of Ecology*, **65**, 235–56.

Kirby, K.J., T. Bines, A. Burn, J. Mackintosh, P. Pitkin and I. Smith (1986) Seasonal and observer differences in vascular plant records from British woodlands. *Journal of Ecology*, **74**, 123–31.

NCC (1990) *Handbook for Phase 1 habitat survey*. Peterborough: Nature Conservancy Council.

Peterken, G.F. (1974) A method for assessing woodland flora for conservation using indicator species. *Biological Conservation*, **6**, 239–45.

Pickett, S.T.A. and P.S. White (1985) *The ecology of natural disturbance and patch dynamics*. London: Academic Press.

Rackham, O. (1976) *Trees and woodland in the British landscape*. London: Dent.

Rackham, O. (1986) *The history of the countryside*. London: Dent.

Ralph, C.J. and J.M. Scott (1981) *Estimating numbers of terrestrial birds*. Studies in avian biology 6, Lawrence, KS: Allen Press.

Robertson, H.J. (1992) *A life history approach to the study of plant species rarity*: Gentianopsis crinita *in New York State*. Unpublished Ph.D. thesis, Cornell University, Ithaca, NY.

Samuelsson J., L. Gustafsson and T. Ingelog (1994) *Dying and dead trees: a review of their importance for biodiversity*. Uppsala: Swedish Threatened Species Unit.

Southwood, T.R.E. (1978) *Ecological methods*. London: Chapman & Hall.

Spellerberg, I.F. (1991) *Monitoring ecological change*. Cambridge: Cambridge University Press.

Spellerberg, I.F. (1992) *Evaluation and assessment for conservation*. London: Chapman & Hall.

Spencer, J.W. and K.J. Kirby (1992) An inventory of ancient woodland for England and Wales. *Biological Conservation*, **62**, 77–93.

ter Braak, C.J.F. (1987) The analysis of vegetation–environment relationships by canonical correspondence analysis. *Vegetatio*, **69**, 69–77.

Thomas, J.A. (1991) Rare species conservation: case studies of European butterflies. In I.F. Spellerberg, F.B. Goldsmith and M.G. Morris (eds) *The scientific management of temperate communities for conservation*. Oxford: Blackwell Scientific, pp. 149–98.

Tudor, G.J., E.C. Mackey and F.M. Underwood (1994) *The national countryside monitoring scheme: the changing face of Scotland 1940s to 1970s*. Edinburgh: Scottish Natural Heritage.

Whitney, G.G. (1994) *From coastal wilderness to fruited plain*. Cambridge: Cambridge University Press.

CHAPTER 13 Communities and ecosystems

S. McINTYRE, G.W. BARRETT and H.A. FORD

Introduction

In this chapter we define and describe communities and ecosystems and some of the human activities that threaten them. We aim to present a unified picture of types of modifications that are occurring worldwide. Finally, we shall discuss how conservation strategies need to take into account both the nature of the destruction, and the evolutionary history of the affected ecosystems. The 1980 World Conservation Strategy identified three major objectives for environmental management: (1) maintaining essential ecological processes and life-support systems; (2) the preservation of genetic diversity; and (3) the sustainable utilization of species and ecosystems. The achievement of these objectives requires our wise stewardship of ecosystems.

Biological communities represent a higher level of organization than populations or species and have properties that are not possessed at lower levels. Examples include number of species, amount of biomass or total rate of photosynthesis. These are termed collective properties and can be simply derived by aggregating the units that make up the community. Certain collective properties such as diversity indices may be of particular value to conservation as they provide various means of describing numbers and relative abundance of species in ecosystems (Spellerberg, 1991).

Beyond the simple aggregation of components, ecosystems have other features that result from interactions between communities and their environment (O'Neill *et al.*, 1986). Ecosystem processes are examples of emergent properties that are not predicted from the simple scaling up of community components. We often only identify the importance of ecosystem processes when human activities perturb them, for example by interfering with hydrological cycles through tree clearing or irrigation, or by disrupting predator/prey relationships through the use of pesticides. These processes are important, because all species (including ourselves) depend on them for survival.

The concept of community and ecosystem

Collective and emergent properties

We have recognized that an assemblage of species may exhibit collective properties (e.g. community productivity, species diversity, species composition). These properties could be observed in any collection of interacting species; for example Krebs (1978) described a community as 'any assemblage of populations of living organisms in a prescribed area or habitat'. This is a very loose definition that makes no assumptions about the types of interactions between populations or the structure of the community. Similarly, the description and classification of communities often does not recognize them as being any more than collections of co-occurring species (Whittaker, 1978).

Nonetheless, community ecologists have recognized that many assemblages of organisms have evolved into configurations having particular emergent properties, e.g. stability – the tendency for the community to return to a particular state, and resistance – the degree to which a community avoids moving from that state (Begon *et al.*, 1990). Communities in so-called 'fragile ecosystems' are likely to have low resistance and low stability because they tend to be easily and permanently affected by perturbations. The structure of 'evolved' communities is the product of coevolution and can be predicted by the physical and biological environment.

Evolved communities differ from random collections of organisms in that they contain species that have coexisted under particular environmental conditions for long periods. This coexistence has been sufficiently prolonged for individual species to have adapted to environmental pressures created by the other organisms, including predation and herbivory. If they had not adapted, they would have been excluded from the community.

Conservation focuses on 'evolved' communities

When we consider problems of conservation of communities and ecosystems, we need to distinguish between more or less random collections of organisms and tightly evolved communities. Firstly, the stability of evolved communities would be expected to be greater than that of randomly occurring species assemblages (at least under their 'normal' environmental regime) making them easier to recognize as distinct entities. Secondly, the organisms within an evolved community often depend upon each other and therefore conservation of individual species may depend on maintaining a collection of other species. Finally, highly evolved communities may have (but not necessarily have) developed higher species richness which has tended to give them a higher perceived conservation value (see Chapter 3).

However, it would be incorrect to assume that all communities need to be stable to achieve conservation significance. Stability is not a necessary feature of ecosystems, and the stability of communities can change with the spatial scale being considered and the way in which stability is assessed. Assumptions about stable equilibrium states have led to mistakes in nature conservation and will be discussed later in the chapter.

Community or ecosystem?

The concept of the ecosystem encompasses communities (the organisms), their physical environmental properties (resilience, etc.) and processes (successional change, etc.). However, for the purposes of this discussion at least, the distinction between communities and ecosystems is not important. Communities and their physical environment are inextricably linked, so much so that communities may be defined by their environment, e.g. sandy desert, chalk, rocky shore or alpine communities. Examples of other ways of identifying or classifying communities include the use of one or more of the following:

- dominant species (e.g. *Astrebla* grassland, redwood forest)
- physical structure (e.g. fauna of tree hollows and of forest or woodland)
- location (e.g. arctic communities, Amazonian rainforest community)
- trophic structure (e.g. food chain with bald eagle at its apex)
- other organisms (e.g. bird communities of deciduous forests, microbial communities of cattle rumina)

Hierarchies of classification occur and these may have arbitrarily defined boundaries. No particular approach to classification can claim to be correct (Whittaker, 1978) and a community can be defined at any physical scale or level within a hierarchy (Begon *et al.*, 1990).

Objectives and approaches that have been applied to community and ecosystem conservation

For some years, a major aim of community conservation has been to maintain species richness by maintaining habitat diversity within an ecosystem. This was based on ecological studies that showed higher species richness where the structural complexity of the ecosystem was highest. For example, in some North American forests there are more bird species along the forest edges than in the interior of the forest and it was thought that species richness would be enhanced if forests were broken up so as to increase the amount of edge. Others argued that the rarer species that used the forest interior habitat would be disadvantaged in these circumstances (Temple and Cary, 1988).

Conservationists and ecologists have on occasions focused on threatened species, not only to avert imminent extinctions, but as an approach to the conservation of entire communities. The rationale is that by catering for the needs of certain threatened species (usually by setting aside large reserves), the more tolerant organisms will be automatically protected (Soulé, 1987). In some regions, it is questionable whether this assumption is correct, as there are cases when management for some species on a regional basis may not be appropriate for the majority of species (see later sections).

Without doubt, the dominant approach to ecosystem conservation in practice has been the setting aside of protected areas such as nature reserves.

Although individual reserves may be substantial in area, communities still tend to be fragmented so that on a regional scale only small fragments are protected. While reserves will continue to be important, and adequately protect many species, they are far from the total solution. This is becoming more widely recognized – even quite large reserves do not support viable populations of bears and wolves in North America (Belovsky, 1987). Moreover, opportunities for the expansion of reserve networks are decreasing rather than increasing, as land is allocated to other uses. Attention is now being shifted towards intervening areas.

Conservation in areas between reserves must be integrated with other land uses, and the goals may therefore need to be more humble. However, they are no less important in the longer term (see Chapter 17). Compared to the setting aside of reserves, managers of multi-use areas will require a greater breadth of knowledge of community dynamics, a philosophy that can accommodate a range of perspectives and a pragmatic approach to community conservation (see Chapter 14 for more details).

Processes leading to the degradation and loss of communities and ecosystems

Two types of evolved communities

The concept of conservation rests on the notion that the activities of people often reduce the variety of species and lead to undesirable changes or losses of communities. While the story is not as simple as this, it is helpful to recognize the existence of *natural* communities as those which have evolved in the relative absence of human influence. However, it is also necessary to recognize species assemblages that have been exposed to humans over an evolutionary time scale and that may therefore have adapted to, or even become dependent on, specific human activities. These we shall term *semi-natural* communities. Both these terms refer to evolved communities, i.e. they contain species that have coevolved with each other. Artificial communities are those that result from novel combinations of species, e.g. a garden containing plants from different regions, an aquarium in a living room or a wheat field in a region that has been cultivated for a few centuries or less. We will not consider artificial communities further.

The concept of exogenous and endogenous disturbance in conservation biology

The concept of disturbance is crucial to the discussion of community conservation, as disturbances can determine the persistence and composition of communities. Rykiel (1985) considered a disturbance as 'a cause that results in a perturbation' – a sufficiently imprecise explanation to cover the wide range of definitions used by various authors. Rykiel stresses the importance of the relativity of disturbance, in which the disturbed community moves away from a recognized reference state. The resulting change is a perturbation. Thus disturbance is in the eye of the beholder, or the community.

Distinguishing between natural and human-induced disturbances as well as between novel and long-standing disturbances is a major issue for conservation biology. Consequently, we will make the distinction between an *endogenous* disturbance as that to which the community has been repeatedly exposed through evolutionary time, and *exogenous* disturbance which tends to be novel (Fox and Fox, 1986). Endogenous disturbances tend to be 'natural' while exogenous disturbances are generally human-induced. An important exception, and one that forms a major message of this chapter, is that certain human activities may constitute an essential endogenous disturbance for the maintenance of semi-natural communities.

Examples of both disturbance types are given in Table 13.1, and it will be apparent that one type of activity could constitute an exogenous or an endogenous disturbance, depending on the type of community. In terms of actual management, the boundaries between natural and human-induced disturbances may blur, as it is possible to substitute human disturbances for natural disturbances in some cases. However, the distinction between endogenous and exogenous disturbances remains critically important to the conservation of ecosystems. Conservation usually aims to maintain endogenous disturbances and minimize those that are exogenous.

Human activities as exogenous disturbances

The most extreme impact of human populations on biotic communities results in their complete destruction and replacement with settlements, agriculture and other forms of intensive resource exploitation. Outright destruction of communities has been identified as the major cause of species loss on a worldwide basis. In Australia as is the case elsewhere, the

Table 13.1 Endogenous and exogenous disturbances that are associated with specific examples of natural and semi-natural ecosystems

Ecosystem type	Endogenous disturbances	Exogenous disturbances
Natural		
Intermittent wetlands	Wetting/drying cycles Herbivory Grazing	Permanent drainage Permanent flooding Nutrient enrichment Grazing (higher levels) Cultivation
Coral reef	Cyclones Predation Tides	Nutrient enrichment Trampling, boat anchors Dredging Fishing, shell collecting Oil spills
Coastal heathlands (Australia)	Fire (natural and anthropogenic) Drought Herbivory	Protection from fire Nutrient addition Clearing Weed invasion
Semi-natural		
Coppiced woodlands (Great Britain)	Regular harvesting of wood Grazing	Protection from tree cutting Clearing
Traditionally managed farmland (Europe)	Rotational cropping Grazing Organic sources of fertilization	Annual cultivation Continuous cropping No tillage farming Heavy use of inorganic fertilizers Herbicide use Improved cleaning of crop seeds Abandonment of some crops

clearing of vegetation for agriculture is the most important past and present threat to native birds (Recher, 1990).

Where communities are not destroyed outright, the effects of exogenous disturbances lead to losses of species and, frequently, their replacement by exotic species (McIntyre and Lavorel, 1994). These changes may also be accompanied by alterations to ecosystem processes. Take the example of a coral reef which remains mostly intact but attracts many sightseers. The visitors themselves may directly cause damage through trampling and boat anchorage, as well as selectively removing edible and attractive species from the reef. The provision of tourist facilities nearby will attract many more people and may result in pollution and nutrient enrichment of surrounding waters. Finally, the resulting alterations to the environment may alter competitive relationships between coral reef species and lead to irruptions of native or introduced species of marine life. Any alterations that affect the survival of the corals themselves (e.g. the growth of algae) disrupt the entire reef building process and could ultimately destroy the reef.

Although generally considered to be negative in effect, exogenous disturbances may inadvertently favour organisms through the creation of habitats, removal of predators and competitors or through an increase in resources. Many of the benefiting species may come to be regarded as pests, but sometimes rare species are assisted. The construction of sewerage and saltworks has created major habitats for shorebirds (Lane, 1987), an outcome that has important lessons for conservation management. Although such developments may simultaneously destroy the original habitats, there is scope in new developments for more creativity in the provision of substitute habitats.

Human activities as endogenous disturbances

In contrast to exogenous disturbances, those that are endogenous in nature may be essential for maintaining community structure and function. Of particular importance to conservation is the widely acknowledged role of disturbance in maintaining species richness (Petraitis *et al.*, 1989). In the case of semi-natural communities, some of the endogenous disturbances are human activities. This situation seems to have developed only when there has been extended and relatively stable coexistence between people and ecosystems. Thus aboriginal burning has been recognized as a factor maintaining species richness and structural characteristics of some Australian ecosystems (Pyne, 1991). Similarly, a long association with agricultural cultivation in parts of Europe has led to a dependence of crop weed species on certain cultural practices (Hilbig, 1982). Many European birds (previously of woodland habitats) have also come to rely on farmland, with its mosaic of habitats and variety of food resources (O'Connor and Shrubb, 1986).

There is nothing inherently different about the types of disturbance associated with endogenous versus exogenous regimes. The difference is entirely to do with the evolutionary context of the community in question. Consequently, the soil cultivation which may be necessary to maintain various wildflower species in Europe has caused significant loss of species in Australian vegetation that had not been exposed to large-scale cultivation previously. An exogenous disturbance could become endogenous if it prevails over an evolutionary time scale.

A significant event that has transformed human activity from being predominantly endogenous to exogenous is the industrial revolution. Radical changes in technology and associated increases in human populations have meant that the intensity and extent of disturbances are far greater. Although adaptation has become evident in some organisms, for example industrial melanism in the English peppered moth (Kettlewell, 1956), changes are often too rapid for most species in the communities to adapt. Species that cannot rapidly adapt to the disturbance or that are not pre-adapted to them are eliminated. Even apparently human-adapted communities such as those of arable land have been adversely affected by non-traditional management resulting from rapid technological change such as herbicide use and continuous cropping. These changes have led some (e.g. Orr, 1991) to consider the industrial revolution as the event that has finally forced the issue of conservation onto our society – we can no longer ignore the massive effects of our technologies.

Spatial patterns of exogenous disturbance

Of four categories of disturbance identified by Rykiel (1985), destruction is the most severe. Where destruction is of a similar scale to that of the communities, complete removal or fragmentation can occur. When destruction occurs at very small (fine-grained) spatial scales, or when other forms of disturbance occur at any scale, the result is ecosystem modification rather than total loss. These types of disturbance (which can be used to classify both endogenous and exogenous disturbances) are described in Table 13.2, together with examples from human activities.

Because the scale at which destruction can occur is a continuous variable, there is also a continuum between fragmentation and modification. Recognizing such a continuum, authors such as Lord and Norton (1990) have described both coarse- and fine-grained destruction as forms of fragmentation. However, we have chosen to describe communities that are affected by fine-grained destruction or any of the three other forms of disturbance described in Table 13.2 as *modified* (cf. *unmodified*). Conversely,

Table 13.2 Examples of human activities that result in disturbance of biotic communities. The four types of disturbance listed are those recognized by Rykiel (1985)

Disturbance type	Human activities
Destruction (reduction of total biomass)	Clearing, fire, use of machinery; non-selective pesticides and toxic pollutants
Discomposition (selective changes to populations)	Hunting, fishing, logging and other forms of harvesting; selective destruction resulting from pollutants or pesticide use; introduction of exotic animals and plants
Interference (inhibition of matter, energy or information exchange processes)	CO_2 pollution (global warming), dam construction (inhibiting fish migration), irrigation (alteration of hydrological cycles); eutrophication (increased resources available for growth, altering competitive interactions)
Suppression (prevention of natural disturbance)	Fire protection, flood mitigation, control of key species

modified or unmodified communities that have been destroyed over substantial parts of the landscape are *fragmented* (cf. *intact*).

The concept of habitat fragmentation is one that has dominated thinking in conservation biology. The fragments themselves become modified by smaller-scale or lower-intensity disturbances and by cross-boundary influences of the surrounding area. Exogenous disturbances clearly can be imposed at different spatial scales and intensities, often simultaneously in ecosystems. At one extreme (coarse-grained), the degree of destruction determines the pattern of the landscape. Lower intensities or finer patterns of disturbance determine the degree of modification of intact or remnant ecosystems.

We should clarify our use of the terms semi-natural and modified as they relate to biotic communities. Semi-natural communities and ecosystems have experienced human management on an evolutionary time scale and certain human activities are part of the endogenous disturbance regime (compared to natural communities in which humans have had little or no role in their history). Modified communities are those that have been affected to a greater or lesser extent by exogenous disturbances. The type of disturbance that is exogenous to the community will vary depending on the evolutionary history of that community. Therefore intact semi-natural and natural communities can be in a modified or unmodified state depending on whether they have experienced exogenous disturbances.

Variegation and fragmentation as transitions from intact to relict communities

A diagrammatic depiction of possible transitions between intact, unmodified communities and modified relicts is presented in Figure 13.1. Two intermediate states are recognized in which the spatial pattern of destruction differs. In *variegated* ecosystems, the communities still form the landscape matrix although there may be areas where they have been destroyed (McIntyre and Barrett, 1992). *Fragmented* ecosystems have experienced a considerable degree of destruction so that the communities occupy a small proportion of the landscape. The boundaries between the fragments and the surrounding 'hostile' areas tend to be abrupt. In variegated landscapes, there are more gentle gradients of modification and intermediate stages of modification are better represented. Both variegation and fragmentation could be considered as nodes on a continuum of destruction, although with certain land uses it is possible to transform an intact community directly to a fragmented one (Figure 13.1).

More intensive land uses such as urbanization and cropping, where the original habitat is replaced, tend to result in fragmented landscapes. Variegated ecosystems may arise when there is exploitation of animals and plants within the community itself, such as the grazing of rangelands or timber extraction from forests.

Community classification for conservation

Eight community/ecosystem types

In conservation management, it is important to recognize the different circumstances of communities and ecosystems, and to match appropriate strategies to them. We recognize eight major types of community that provide the range of situations in which conservation strategies need to be applied. This typology takes into account disturbance history and current spatial patterns of community destruction as described in previous sections (see Table 13.3 for summary of terms used). Thus semi-natural and natural communities combined with the four spatial configurations (intact, variegated, fragmented, relict) result in eight combinations. Although the categories are arbitrarily delineated along two gradients, they provide a convenient framework for the discussion of conservation strategies. In Figure 13.2, we have presented the eight categories in two dimensions, together with examples of communities and ecosystems.

Management priorities – a brief summary

Each community or ecosystem will have its own specific issues that need to be addressed and which take into account history, current condition, land use and socio-economic circumstances. Nonetheless, it is useful to set objectives that can guide management and the eight community types will vary in this respect.

The spatial pattern of destruction of the community, and its extent, is the primary factor determining the appropriate strategy. In the case of variegated landscapes, there is a need to maintain the community matrix and integrate potentially damaging activities in such a way as to avoid further deterioration. This integration could be achieved by limiting, modifying

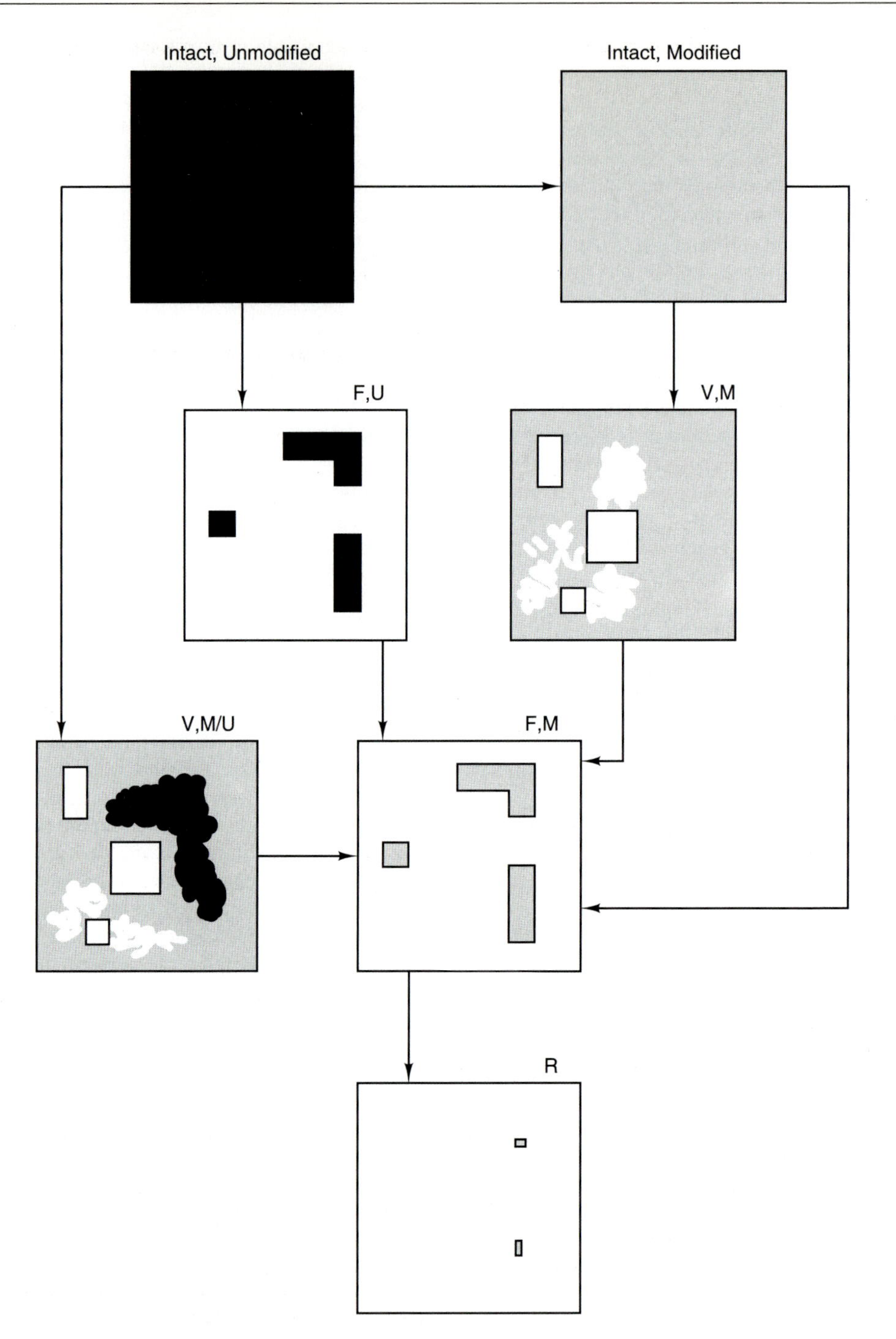

Fig. 13.1 Diagrammatic plans of landscapes showing possible transitional states and pathways that result from the effects of exogenous disturbances. Intact communities are converted to variegated (V) and/or fragmented (F) ecosystems and may eventually be destroyed to the extent that only small relicts (R) remain. Regardless of spatial configuration, the communities may be subjected to varying degrees of modification (U = unmodified, M = modified).

Table 13.3 Summary of terms used to describe ecosystems and communities and the changes induced by exogenous disturbances. The types of disturbance described in Table 13.2 apply to both exogenous and endogenous disturbance

Ecosystem/community type	
Natural	Ecosystem/communities which have evolved in the virtual absence of human influence
Semi-natural	Ecosystem/communities which have been exposed to human activity over an evolutionary time scale and that may therefore have adapted to, or even become dependent on, specific activities
Disturbances	
Endogenous	Disturbances which the ecosystem/community *has* been repeatedly exposed to through evolutionary time
Exogenous	Disturbances which the ecosystem/community *has not* been exposed to through evolutionary time
Changes induced by exogenous disturbances:	
(1) *Landscape scale, spatial patterns*	
Intact[a]	Ecosystem/community is largely unaffected by landscape-scale destruction
Variegated[a]	Ecosystem/community still forms the landscape matrix but is variously affected by fine- and coarse-grained exogenous disturbances
Fragmented[a]	Ecosystem/community has been destroyed over much of the landscape but persists in isolated fragments
Relict	Ecosystem/community largely destroyed with only tiny areas containing elements of the modified communities remaining
(2) *Changes within ecosystems/communities (fine-grained)*	
Unmodified	Natural or semi-natural ecosystem/community that *has not* been affected by exogenous disturbance
Modified	Natural or semi-natural ecosystem/community that *has* been affected by exogenous disturbance

[a] Ecosystems/communities in these spatial configurations may be modified or unmodified.

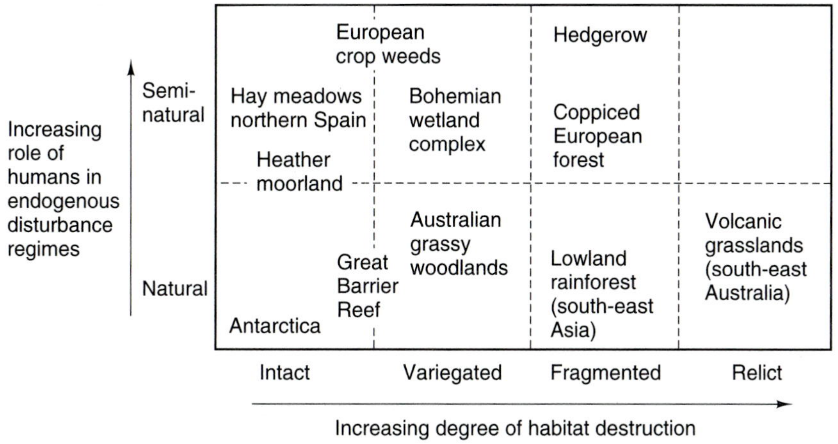

Increasing role of humans in endogenous disturbance regimes

	Intact	Variegated	Fragmented	Relict
Semi-natural	European crop weeds; Hay meadows northern Spain; Heather moorland	Bohemian wetland complex	Hedgerow; Coppiced European forest	
Natural	Antarctica; Great Barrier Reef	Australian grassy woodlands	Lowland rainforest (south-east Asia)	Volcanic grasslands (south-east Australia)

Increasing degree of habitat destruction

Fig. 13.2 Examples of communities and ecosystems and their classification according to disturbance history and spatial patterns of destruction.

Table 13.4 Management strategies for eight community types

Community type	Priority	Secondary strategies
Natural, Intact	Protection – minimize direct effects of exogenous disturbance	Control off-site disturbances (e.g. pollution); control invading species
Natural, Variegated	Integration – maintain landscape matrix by controlling effects of exogenous disturbances	Manage for maximum diversity by stratifying land uses; protect significant areas for conservation
Natural, Fragmented	Rehabilitation – minimize and treat effects of exogenous disturbance (e.g. control exotic species)	Buffer existing remnants from disturbances; extend area of remnants; increase connectivity
Natural, Relict	Reconstruction – recreate community around remnant species	Protect from exogenous disturbance; species reintroduction to increase diversity
Semi-natural, Intact	Maintain endogenous disturbance regime	Protect from exogenous disturbance; diversify endogenous disturbance regime to maintain diversity
Semi-natural, Variegated	Integration – maintain landscape matrix by controlling effects of exogenous disturbances	Manage for maximum diversity by stratifying land uses; maintain endogenous disturbance regimes; protect significant areas for conservation
Semi-natural, Fragmented	Protect from effects of exogenous disturbance; maintain endogenous disturbance regimes	Adjust disturbances to maximize diversity in fragments; extend area of remnants; increase connectivity
Semi-natural, Relict	Reconstruction – recreate community around remnant species	Reinstate endogenous disturbance; protect from exogenous disturbance; species reintroduction to increase diversity

and locating damaging activities in such a way as to control their impact. Fragmented landscapes also require reintegration of the community remnants, but as the community matrix has been lost, the priority must be given to rehabilitation. This is required to maintain remnant condition, and in some cases to connect the fragments. This may require major efforts in weed control and habitat enhancement. In the case of relict vegetation, we are referring to situations where no unmodified community remains; only small areas containing some of the constituent species remain. On the fertile volcanic plains west of Melbourne, Australia, the native grasslands were rapidly settled for livestock production after European invasion. The effects of exogenous disturbances were extremely destructive and now only tiny fragments of a once extensive community remain. In cases such as these, the focus needs to be on habitat reconstruction and care of particular species, and may include habitat creation at sites where none of the community components are present. Habitat creation and enhancement are discussed in Chapter 16 and will not be considered further.

The degree of naturalness of a community is the second major factor that needs to be taken into account and determines the need for protection versus active management. Table 13.4 presents a summary of the management priorities identified for the eight community types, which are discussed in more detail below.

Examples of ecosystems and strategies for their management

Natural, intact

Identifying naturalness

There are relatively few examples of totally natural ecosystems, and even fewer that have been left relatively intact. Antarctica may be the most conspicuous example. However, the term 'natural' has been more loosely used to include the activities of indigenous people, or through failing to recognize their influence. Consequently countries such as Australia have been considered to contain ecosystems that were natural at the time of European settlement, ignoring the profound effects of at least 40 000 years of Aboriginal occupation on the country's biota (Taylor, 1990). In choosing to retain our proposed anthropocentric concept of naturalness, we are suggesting a re-examination of which ecosystems are really natural. Rather than change the definition, we need to look critically at the circumstances under which communities may have evolved.

Where human influences have been restricted to hunting and gathering, there is a greater tendency to consider ecosystems as natural. For example, forests in East Africa have been exploited by humans for more than 2000 years, yet the rainforests are considered by Bjorndalen (1992) to need protection rather than active human management. Similarly, although hunting and trapping is traditional in central African rainforests, it appears to influence mammal populations. Also, increasing hunting pressure can cause impoverishment of the fauna (Carpaneto and Germi, 1992). However, we do not know how much the rainforest community depends upon low levels of hunting to maintain biodiversity, and therefore total protection from human influence may be undesirable.

In a similar way, deciduous forests in North America are considered natural communities. However, Whitcomb (1987) attributed the occurrence of those forest birds which were adapted to broken forest or edge habitats to disturbances during pre-Columbian times. Whether such disturbances were anthropogenic is unclear and illustrates the problems of determining the endogenous disturbance regime of a community. Australia has many ecosystems where Aboriginal management does not seem to be a major endogenous disturbance, but in most cases we do not really know, e.g. in coral reefs, wetlands, rainforests, mangroves, tall open forests and saltmarsh.

Management priorities for intact natural ecosystems

The key issue is whether anthropogenic management needs to be continued to maintain the integrity of the community. If it does not, the main priority in the management is to avoid human activities that threaten to modify or destroy the community. The usual response in this situation is the creation of conservation reserves. Usually, conservation reserves are located in areas that are not particularly attractive for economic development, with the decision being politically, more often than ecologically, based.

Establishment of reserves can go a long way towards protection but there are other issues. The integrity of conservation reserves may be threatened by a wide range of factors related to population, visitor pressure and widespread effects of pollution and resource gathering. A by-product of population pressure is the building of roads that can fragment otherwise intact systems and serve as a conduit for more people and disturbances. Andrews (1990) identified the following consequences of road construction through natural areas: habitat loss and alteration, alteration of hydrology and siltation of aquatic habitats, edge and barrier effects on fauna, road kills, weed invasions and greater human access. Hence the focus of management of intact natural areas is often the control of human behaviour, whether visiting, poaching or encroaching with economic developments.

A secondary activity is the control of pest plants and animals. Invasions are most often associated with disturbances (Hobbs and Huenneke, 1992). In countries with clearly defined native and exotic species such as New Zealand and Australia, exotic invasions are a common indicator of modification of natural or near natural systems. Exotic invasions are capable of obliterating entire communities; for example the introduction of tree snakes has almost eliminated land birds from the island of Guam. However, the control of exotics may not be a black and white issue. Whereas the most damaging species need to be removed, some species have apparently benign effects on the ecosystem and others have become valued by the public.

No matter how extensive they are, natural ecosystems are influenced by global-scale disturbances. For example, the thinning of the ozone layer is likely to affect Antarctic phytoplankton even though the region is isolated from the source of pollution (El-Sayed *et al.*, 1990). Similarly, ozone depletion and human-induced climate change could deleteriously affect the survival of coral reef ecosystems, although the effects of other disturbances such as sedimentation, pollution and eutrophication may be more immediate and increase susceptibility to climatic stresses (Smith and Buddemeier, 1992). Off-site disturbances can be manifested in other ways. Approximately 2 million shorebirds migrate between Australia and northern Asia annually (Lane, 1987) and although well protected in both regions, they are increasingly exploited and disturbed in their stopover sites in south-east Asia. Such examples highlight the need for environmental management at all scales.

Natural, variegated

Forestry areas as variegated ecosystems

When natural communities become variegated, they are influenced by one or more land uses that affect the ecosystem in a variable and patchy manner. These land uses may exploit elements of the community, which has therefore been able to survive over much of the landscape, albeit with modification. In the case of low-intensity forestry which relies on natural regeneration, selective or non-selective removal of trees over limited areas may produce a patchwork of regrowth of different ages across the landscape. Plants and animals that can survive in regrowth are seldom disadvantaged in these environments, while species that depend on mature forest may become threatened by loss of key resources such as tree hollows, e.g. the northern spotted owl (*Strix occidentalis caurina*) in coniferous forests of North America and Leadbeater's possum (*Gymnobelideus leadbeateri*) of south-eastern Australia, which requires mature eucalypt forests.

The habitat characteristics of particular areas will change as trees mature and others are removed. The landscape can be envisaged as a shifting mosaic of habitats of varying suitability for the biota. The aims of management in variegated ecosystems, generally, are to:

1 maintain the integrity of the communities so that they continue to form the habitat matrix;
2 retain the range of modification states that are adequate to support most (if not all) of the biotic community.

This assumes that there is a tendency for land use to intensify and increasingly dominate the community and reduce biodiversity. For species with particular ecological requirements such as the spotted owl and Leadbeater's possum there may be real problems in satisfying requirements, despite efforts to retain mature forests. But for the majority of the biota, these strategies should control resource exploitation sufficiently well to enable their survival.

Rangelands as variegated ecosystems – managing temperate grassy ecosystems in eastern Australia

The ecosystem for which the term 'variegated' was developed is that supporting grassy woodlands in temperate eastern Australia (McIntyre and Barrett, 1992). These areas are subject to livestock grazing of native grasslands, with relatively little sown pasture or cropping land. In this sense they are similar to grazed rangelands throughout the world. The grassy woodlands of the New England Tablelands of eastern Australia will be used to illustrate the phenomenon of variegation in some detail and discuss management issues of variegated landscapes.

At the time of European invasion in the 1830s, the region supported hunter–gatherer Aboriginal populations. Their use of fire appeared to have maintained open grassy communities with widely spaced eucalypts as the dominant vegetation and some denser eucalypt forests. This endogenous disturbance has now been overwhelmed by the exogenous effects of pastoral settlement. Although the grassy woodland matrix still exists, it is overlain by grazing, clearing, soil disturbance, nutrient and water enrichment and exotic introductions occurring at different intensities and spatial scales. Many mammal species have declined or become locally extinct as a result of these changes. Today only about one-third of the tree cover remains, forming a diffuse mosaic of scattered trees, woodland and patches of forest.

The decline of trees is not just the direct result of clearing and grazing, but a complex phenomenon of insect defoliation leading to death, related to the use of phosphate fertilizers and intensified grazing since the 1950s (Lowman and Heatwole, 1992). All components of the community appear to be involved: the larvae of defoliating beetles benefit from the extra forage growth of fertilized and sown pastures upon which they feed. This puts increasing defoliation pressure on the remaining trees by adult beetles. Fertilization and sowing of exotic pastures permits heavier stocking rates, which prevent tree regeneration. Birds can remove up to half of the tree leaf-eating insects in healthy woodland (Ford, 1989), but they too are adversely affected by the loss of trees and native understorey. Mammals, reptiles, insects and spiders also play a part in controlling dieback, but they too are affected by habitat destruction and modification. There appears to be a threshold of community modification beyond which tree populations may enter a positive feedback loop of decline.

Although not closely coupled with tree survival, the native herbaceous species are also affected by exogenous disturbances (McIntyre and Lavorel, 1994). Soil disturbance, fertilization and intensive grazing lead to significant declines in these species, but it is a question of degree, and there is a small core of native species that can tolerate extreme modifications. The least modified, species-rich habitats are very restricted in area. Although pastures are modified, and therefore have a lower species richness of native and rare plants, they represent a huge biological resource, occupying over 90% of the region.

The woodland bird community responds to ecosystem changes in a similar manner to that of plants. Exogenous disturbances lead to the progressive disappearance of woodland birds. Variation in the bird community appears to be linked to structural features in the habitat, e.g. the population density of shrubs and trees. There are also specific links between bird and plant species, e.g. seed eaters depend on particular grasses and nectarivores on certain trees and shrubs. When a range of eucalypt types and ages is present (including mature trees with hollows; fallen and standing dead timber) and there are native shrubs in the understorey, the full complement of woodland birds might be expected to occur. Where the woodland is reduced to a stand of mature trees with little or no understorey or no fallen timber, the bird community is limited, and in cases where only scattered trees remain, the woodland community is replaced with grassland birds (Figure 13.3).

Most woodland birds in New England (Australia) do not require extensive areas of intact woodland. There, approximately 17 species are dependent upon areas of woodland that are larger than 400 ha in area, but these birds are marginal in distribution on the Tablelands, tend to be forest rather than woodland species and are common elsewhere in their range (Barrett *et al.*, 1994). Conventional conservation practice would focus on these 17 forest species. In a region where most of the land is privately owned and managed as rangeland, demanding large areas is unrealistic. Fortunately large reserves occur to the east of the region.

In a variegated landscape, as much priority should be given to managing the habitat matrix with its more tolerant plant and animal species as is given to the areas of high quality, or extensive, habitat required by rare species. Using this strategy, progression towards a completely fragmented landscape can be arrested, and the long-term survival of a reasonably representative biota is more likely. The disadvantage of taking a broad approach to conservation at the level of the community is that rarer species may be neglected. Perhaps a stronger emphasis on matrix management is justified on the basis of the greater conservation gains that could be achieved for the efforts made, and prevention of additional species becoming threatened.

Ecosystem management in rangelands is primarily the responsibility of the land holders. General principles and strategies have been developed around a theme of incorporating a range of intensities of management – or disturbance – into the farm operation (Barrett *et al.*, 1994; McIntyre, 1994) and can be incorporated into general farm planning. In the New England Tablelands, as in other parts of rural Australia, there is growing awareness about land degradation, leading to the formation of community-based land care groups. With adequate government support and regional coordination such groups can be an effective way to disseminate information, build community support and reduce ecosystem destruction and modification. Currently, these groups are primarily concerned with planting rather than protection of remnant vegetation. However, there is a developing awareness of the value of native fauna and flora. There is also a need for a regional approach that goes above this level of management.

Secondary conservation strategies for variegated landscapes

We have emphasized management of the landscape matrix in the belief that integrating existing land use with survival of native vegetation and wildlife is realistic and viable. More conventional approaches that only focus on rare species and the management of isolated reserves further encourage fragmentation by devaluing 'lower quality' habitat such as logged forest and native pastures.

Because the influences of exogenous disturbances tend to dominate variegated landscapes, endogenous disturbances have a different context and may no longer be relevant or even produce undesirable effects. For example, the effects of fire are different in grazed compared to ungrazed vegetation. A pasture that has been grazed and unburnt for some decades may have already lost its fire-dependent species. The sudden imposition of a fire regime (even though it may have been an endogenous disturbance) may simply cause further loss of grazing-tolerant, fire-sensitive species without any further gains in diversity. Only when significant areas of relatively unmodified ecosystems are managed for conservation will endogenous disturbance regimes assume major importance.

Natural, fragmented

In a variegated ecosystem, the landscape is relatively permeable to the movement of many organisms. In a

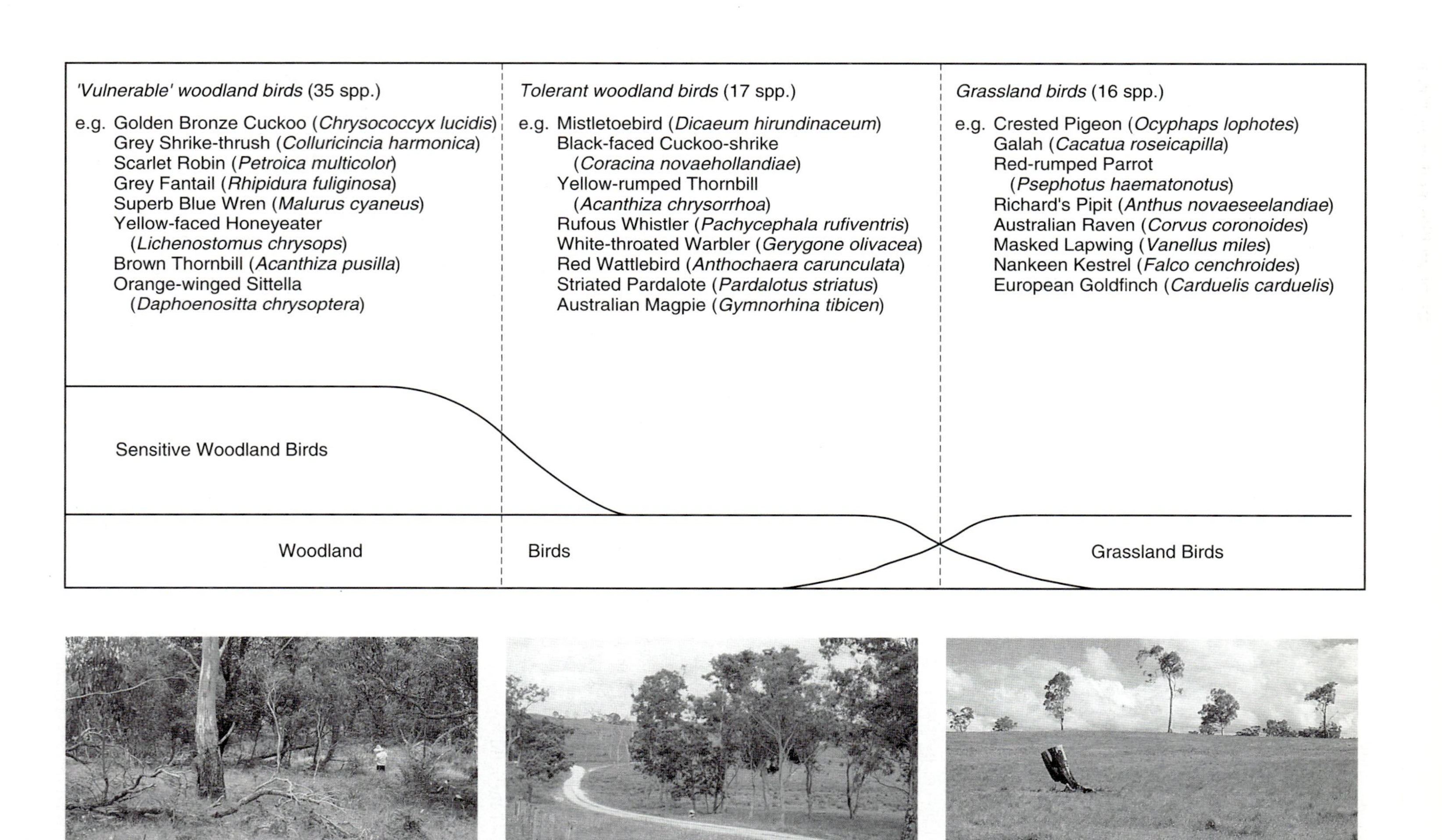

Fig. *13.3* Examples of bird species associated with grassy eucalypt woodlands subjected to increasing level of exogenous disturbance and consequent habitat modification.

fragmented ecosystem, a far greater proportion of species may have their movement restricted by surrounding hostile environments. The genetic and demographic effects of these restrictions have dominated early thinking in conservation biology. There are many effects of fragmentation on natural communities (Chapter 2) and there are numerous examples of communities throughout the world that have become fragmented, although the same community type may be intact, fragmented or variegated, depending on the locality. The processes leading to fragmentation commonly stem from industrial, urban or intensive agricultural development. In Western Australia, for example, woodland and heathland communities are fragmented by wheat cropping areas. In North America, deciduous forests are fragmented by urban areas, while in Brazil, tropical rainforest is being fragmented by farming, settlement and industry. Lowland rainforest in peninsular Malaysia has been highly fragmented by plantations of rubber and oil palm. Although many birds occupy the plantations, few are rainforest species, and this habitat appears to be inhospitable to many rainforest birds.

Management strategies for these landscapes have endeavoured to buffer and extend existing remnants, particularly through the use of connecting corridors. However, some authors have questioned the extent to which fauna actually uses corridors, and have also identified potential biological disadvantages, including their role in the spread of exotic species and diseases.

Nonetheless, it can be accepted as a general principle that reducing the effects of exogenous disturbances on the remaining fragments is important for the integrity of the remaining natural community. This can be achieved by intensive protection and management to control exotic plants and animals. In some cases it may be desirable and feasible to reintroduce locally extinct species if immigration is not possible. An effective strategy in the longer term might be to buffer the fragments with areas of relatively benign land use such as parkland, plantations or reserves of various kinds. More usefully, extending the area of fragments using techniques to encourage regeneration and perhaps planting will assist the protection and possibly enhance the viability of remnants. Extending fragments in such a way as to increase connectivity between them is still likely to be a useful strategy, although perhaps it should be secondary to buffering existing remnants.

Semi-natural, intact

A number of semi-natural communities have evolved with European agriculture, but there are relatively few examples where traditional methods are still practised as a dominant land use. The hay meadows of northern Spain are an exception. The species-rich flora is adapted to endogenous disturbances in the form of seasonal grazing, annual cutting and sometimes manuring and/or irrigation. Here the biota is threatened by socio-economic circumstances that threaten to depopulate the region and reduce the importance of cattle production (García, 1992). For conservation, therefore, the management priority must be to maintain these farming practices. Within the traditional management framework are a range of possible approaches that will have major effects on composition of the communities. It is important that these relationships be understood in order to provide the appropriate traditional management combinations that will support the widest range of species in the community.

There are also many examples of traditional agroforestry systems in tropical regions that involve the use of natural and semi-natural ecosystems (Orians and Millar, 1992). From what we know of western traditional systems, it is likely that they may also play a role in supporting regional biodiversity, but this aspect has received little attention.

Semi-natural, variegated

The Třeboň Biosphere Reserve of the Czech Republic in Central Europe is a semi-natural landscape that has been exploited by humans since the thirteenth century. Jeník and Květ (1984) give an account of the region and describe the effects of exogenous disturbances that have recently threatened the ecosystems. The region has a complex mosaic of woodland, peatland, wetland, small lakes and arable land – a mixture of natural and semi-natural communities. About 700 years ago, the natural landscape was altered by the development of an extensive canal and fishpond system. A pattern of land use linking agriculture, pisciculture and forestry evolved and largely persisted until the middle of this century. This has enabled a stable, diverse complex of habitats to develop which support a significant fauna and flora.

Since 1948, the application of intensive technologies has increased agricultural production and the yields of fish and wood. Unfortunately, the changes began to affect species richness adversely and to endanger the stability of the landscape as a whole. For example, excess fertilizer and waste from pig and duck farming has polluted underground and surface waters. Drainage of wetlands and meadows has affected the biota within them and the intensification of pisciculture has lowered species richness in fishponds, through the effects of eutrophication and disturbance.

As in the example of natural variegated ecosystems discussed above, the management of the Třeboň Biosphere Reserve needs to focus on limiting the extent and intensity of intensive production (exogenous disturbances). Jeník and Květ recognized degrees of compatibility of lower-intensity land use with protection of nature and this reflects the long-term coexistence of traditional farming and the biota. Whether endogenous disturbances associated with traditional management need to be consciously reinstated may not become apparent until the major effects of exogenous disturbance are controlled.

There are similarities between the Třeboň Biosphere Reserve and the British rural landscape which is a patchwork of forest, grasslands, crops and hedges supporting a rich fauna and flora. This landscape has also been affected by intensive production methods. Indeed, many of the threatened and declining species in Europe are associated with semi-natural habitats. Great bustard (*Otis tarda*), corncrake (*Crex crex*) and more recently corn buntings (*Miliaria calandra*) are all species of birds that have suffered from the intensification of farming.

Semi-natural, fragmented

European forests were farmed for centuries by the system known as 'coppice with standards' (Orians and Millar, 1992). Elements of this system are a few full-sized trees (standards) for timber production with other trees being routinely cut to ground level (coppiced) or several metres above the ground (pollarded) with livestock grazing beneath. This management supports the richest assemblage of fauna of all western European terrestrial ecosystems. These forests, together with other managed ecosystems such as hedgerows and crop weed communities, have been seriously affected by destruction and other exogenous disturbances (including neglect). In the case of cropping, the traditional rotational systems with low inputs have been replaced by continuous cropping with high fertilizer and pesticide use. As a consequence, arable habitats contain amongst the highest numbers of endangered species in Europe (Hilbig, 1982).

These ecosystems are perhaps the most demanding in their conservation requirements as they require some degree of protection as well as specific management to maintain the integrity of the communities. Except in the case of some traditional cropping rotations, it is probably not useful for management to be strictly prescriptive. Even if it is well understood which endogenous disturbances should be maintained (or re-established) in a particular remnant community, it may not be optimal to apply it blindly, particularly if the disturbance has been withheld for some time. For example, to reintroduce coppicing into a forest that has developed a high forest structure may favour some bird species, e.g. nightingale (*Luscinia megarhynchos*) and garden warbler (*Sylvia borin*), at the expense of other significant species, such as woodpeckers and the wood warbler (*Phylloscopus sibilatrix*), that have established in old-growth habitat (Fuller, 1982). Light-demanding, understorey plants that rely on coppicing may have already disappeared from the locality and not be able to recolonize the site. Clearly, management needs to be responsive to the particular circumstances, with the intention of maximizing diversity or retaining sensitive species. Initially this should focus on maintaining the elements of the community that are present. Subsequently more fine-tuned management may allow for reintroduction of species and other aspects of rehabilitation.

Conclusions

We have stressed how important it is to identify the historical processes that influence the structure of communities today. This does not mean that management has to be restricted to maintaining these processes. Rather, we need to consider what sort of communities we want and, by recognizing both past and present influences, decide how we are going to impose management that will conserve or produce the desired communities.

The more fragmented and modified an ecosystem is, the more need there is for creativity in setting conservation goals and developing management strategies. But, regardless of the state of the community, there are a number of guiding principles that need to be followed. This is why the management strategies identified for the community types vary in emphasis rather than in actual content (Table 13.4). Depending on the extent and pattern of destruction, the degree of modification and the importance of human-induced endogenous disturbance, each site will have its specific priorities, but the common aims are: (1) to reduce the impact of exogenous disturbances; (2) to retain or improve the spatial integrity of the community; (3) to retain or increase species and habitat diversity; and (4) to maintain or reinstate endogenous disturbance regimes.

Managers that translate these principles into practice need to take a more pragmatic approach to conservation than is sometimes advocated. For example, if we give priority to the rarest species in a region, large tracts of undisturbed habitat may have to be reserved. In rural areas, such a conservation goal may be currently unrealistic, nor may it be necessary to achieve conservation of the great majority of species, as in the case of the New England region of Australia.

In the case of variegated and fragmented communities, the full complement of species may no longer be present, and some of the distinctive environmental conditions of the original ecosystem may no longer prevail. In these cases, it seems inevitable that we will have to accept that slightly different communities will evolve out of these changing conditions. Managers will still need to retain important features. Where possible, vital ecosystem processes must be protected. For example, in much of southern Australia, the excessive clearing of eucalypt forests and woodlands has changed the hydrology and led to severe soil salinization. The return of tree cover in this case is a realistic conservation goal for the present, while the return of all species in the original communities may not be. The maintenance of most of the species assemblage is also a realistic goal. Catering for each species individually is unlikely to be possible, but it is feasible to consider providing a range of land management types and disturbance intensities. It is important to determine the best patterns of management to provide suitable habitat for most species.

Finally, it is worth reflecting on the differences between what we know and what we actually do in conservation practice. Although the general ecological principles for maintaining biological diversity have been developed over the last 20 years, loss of species and communities continues unabated. It is now widely recognized that without community involvement and cooperation, conservation management plans will be ineffective. Progress has been slow and because so many of these projects are long-term, their progress is often difficult to measure. In some respects, achieving conservation gains through community participation will be the greatest challenge for conservation – one for which there are no simple solutions, except perhaps persistence, goodwill and encouragement.

References

Andrews, A. (1990) Fragmentation of habitats by roads and utility corridors: a review. *Australian Zoologist*, **26**, 130–141.

Barrett, G.W., H.A. Ford and H.F. Recher (1994) Conservation of woodland birds in a fragmented rural landscape. *Pacific Conservation Biology*, **1**, 245–56.

Begon, M., J.L. Harper and C.R. Townsend (1990) *Ecology: Individuals, populations and communities*, 2nd edn. Cambridge, MA: Blackwell.

Belovsky, G.E. (1987) Extinction models and mammalian persistence. In M.E. Soulé (ed.) *Viable populations for conservation.* Cambridge: Cambridge University Press, pp. 35–57.

Bjorndalen, J.E. (1992) Tanzania's vanishing rain forests – assessment of nature conservation values, biodiversity and importance for water catchment. *Agriculture, Ecosystems and Environment*, **40**, 313–34.

Carpaneto, G.M. and F.P. Germi (1992) Diversity of mammals and traditional hunting in central African rain forests. *Agriculture, Ecosystems and Environment*, **40**, 335–54.

El-Sayed, S.Z., F.C. Stephens, R.R. Bidigare and M.E. Ondrusek (1990) Effect of ultraviolet radiation on Antarctic marine phytoplankton. In K.R. Kerry and G. Hempel (eds) *Antarctic ecosystems. Ecological change and conservation.* Berlin: Springer-Verlag, pp. 379–85.

Ford, H.A. (1989) *Ecology of birds – an Australian perspective.* Sydney: Surrey Beatty.

Fox, M.D. and B.J. Fox (1986) The susceptibility of natural communities to invasion. In R.H. Groves and J.J. Burdon (eds) *Ecology of biological invasions: an Australian perspective.* Canberra, Australia: Australian Academy of Science, pp. 57–66.

Fuller, R. J. (1982) *Bird habitats in Britain.* Calton: Poyser.

García, A. (1992) Conserving the species-rich meadows of Europe. *Agriculture, Ecosystems and Environment*, **40**, 219–32.

Hilbig, W. (1982) Preservation of agrestal weeds. In W. Holzner and N. Numata (eds) *Biology and ecology of weeds.* The Hague: Junk, pp. 57–9.

Hobbs, R.J. and L.F. Huenneke (1992) Disturbance, diversity and invasion: implications for conservation. *Conservation Biology*, **6**, 325–37.

Jeník, J. and J. Květ (1984) Long-term research in the Třeboň Biosphere Reserve, Czechoslovakia. In F. Di Castri, F.W.G. Baker and M. Hadley (eds) *Ecology in practice. Part I: Ecosystem management.* Dublin: Tycooly International Publishing, pp. 437–59.

Kettlewell, H.B.D. (1956) Further selection experiments on industrial melanism in the Lepidoptera. *Heredity*, **10**, 287–301.

Krebs, C.J. (1978) *Ecology: the experimental analysis of distribution and abundance*. New York: Harper and Row.
Lane, B.A. (1987) *Shorebirds in Australia*. Melbourne: Nelson.
Lord, J.M. and D.A. Norton (1990) Scale and the spatial concept of fragmentation. *Conservation Biology*, **4**, 197–202.
Lowman, M.D. and H. Heatwole (1992) Spatial and temporal variation in defoliation of Australian eucalypts. *Ecology*, **73**, 129–42.
McIntyre, S. (1994) Integrating agricultural land-use and management for conservation of a native grassland flora in a variegated landscape. *Pacific Conservation Biology*, **1**, 236–44.
McIntyre, S. and G.W. Barrett (1992) Habitat variegation, an alternative to fragmentation. *Conservation Biology*, **6**, 146–7.
McIntyre, S. and S. Lavorel (1994) Predicting richness of native, rare and exotic plants in response to habitat and disturbance variables across a variegated landscape. *Conservation Biology*, **8**, 521–31.
O'Connor, R.J. and M. Shrubb (1986) *Farming and birds*. Cambridge: Cambridge University Press.
O'Neill, R. V., D.L. DeAngelis, J.B. Waide and T.F.H. Allen (1986) *A hierarchical concept of ecosystems*. Princeton: Princeton University Press.
Orians, G.H. and C.I. Millar (1992) Forest lands. *Agriculture, Ecosystems and Environment*, **42**, 125–40.
Orr, D. (1991) Biological diversity, agriculture and the liberal arts. *Conservation Biology*, **5**, 268–70.
Petraitis, P.S., R.E. Latham and R.A. Niesenbaum (1989) The maintenance of species diversity by disturbance. *The Quarterly Review of Biology*, **64**, 393–418.
Pyne, S.J. (1991) *Burning bush: a fire history of Australia*. New York: Henry Holt.
Recher, H.F. (1990) Wildlife conservation in Australia: state of the nation. *Australian Zoologist*, **26**, 5–11.
Rykiel, E.J. (1985) Towards a definition of ecological disturbance, *Australian Journal of Ecology*, **10**, 361–6.
Smith, S.V. and R.W. Buddemeier (1992) Global change and coral reef ecosystems, *Annual Review of Ecology and Systematics*, **23**, 89–118.
Soulé, M.E. (1987) Introduction. In M.E. Soulé (ed.) *Viable populations for conservation*. Cambridge: Cambridge University Press, pp. 1–10.
Spellerberg, I.F. (1991) *Monitoring ecological change*. Cambridge: Cambridge University Press.
Taylor, S.G. (1990) Naturalness: the concept and its application to Australian ecosystems. *Proceedings of the Ecological Society of Australia*, **16**, 411–18.
Temple, S.A. and J.R. Cary (1988) Modelling dynamics of habitat-interior bird populations in fragmented landscapes. *Conservation Biology*, **2**, 340–7.
Whitcomb, R.F. (1987) North American forests and grasslands: biotic conservation. In D.A. Saunders, G.W. Arnold, A.A. Burbidge and A.J.M. Hopkins (eds) *Nature conservation: the role of remnants of native vegetation*. Chipping Norton, NSW: Surrey Beatty, pp. 163–76.
Whittaker, R.H. (1978) *Classification of plant communities*. The Hague: Junk.

CHAPTER 14

Protected areas: where should they be and why should they be there?

R. L. PRESSEY

Introduction

Protected areas are not a new idea. Even the establishment of the world's first national park in 1872, Yellowstone in the north-western United States, is relatively recent. The practice of setting aside land from unrestricted exploitative use is actually many centuries older and has had many motivations apart from nature conservation. These include religion, protection of commercial infrastructure like waterways and harbours, and safeguarding of resources for use by the privileged (Leader-Williams *et al.*, 1990; Grove, 1992). Two things are new today. First, the increase in number and extent of protected areas throughout the world has been very fast in recent decades. In the 1970s and 1980s, more area was given protected status than in all the previous decades combined (Reid and Miller, 1989). Second, human populations are growing at an unprecedented rate, along with human requirements for food, materials, energy, and places to put waste products. In other words, the pressure on biodiversity is also unprecedented.

Is the mounting pressure on biodiversity balanced by all the new protected areas? Unfortunately not. Protected areas are still relatively restricted in area – about 6% of the world's land surface. More importantly, like the major threats to biodiversity, they are very unevenly distributed. There has been a tendency for protected areas to be established in different places than those affected by major threats to biodiversity (Pressey, 1994), although there are also some overlaps. To judge the progress of a system of protected areas by the hectares it occupies is therefore to miss the point. Just as important as hectares is the effectiveness with which protected areas do their job. This depends on their goals, locations, management, the types of threats they are meant to offset, the nature of their boundaries, and their size and shape.

This chapter is about the effectiveness of protected areas. Effectiveness is by no means guaranteed when a protected area has been established. Ehrlich (1988) concluded that reserves and other conservation measures 'are largely the tactics of a small and beleaguered army fighting rearguard actions on battlegrounds of the enemy's choosing' and that 'the tactical successes of the conservation movement can be appropriately evaluated only against a backdrop of total and continuing strategic disaster'. Is this too pessimistic a view? The answer depends not on hectares but on effectiveness. This chapter begins by discussing some basic but important issues. What are protected areas and what are they for? It then covers two broad questions about effectiveness: where should protected areas be located, and what factors determine the future of the pieces of biodiversity that have been included in protected areas? For each of these questions, there are ideal answers that assume that the conservation of biodiversity is of the greatest importance to everyone. Then there are the real answers that emphasize that biodiversity is a low priority for many people, including some who make decisions on new protected areas. The future of the planet's biodiversity depends mainly on the balance struck by one species, *Homo sapiens*, between the ideal and the real.

What are protected areas and what are they for?

The basic purpose of protected areas is to separate the components of biodiversity from the things that threaten their persistence. Threats to biodiversity

Table 14.1 IUCN protected area categories

Category	Description
Ia	Strict nature reserve: managed mainly for science
Ib	Strict nature reserve: managed mainly for wilderness protection
II	National park: managed mainly for ecosystem protection and recreation
III	Natural monument: managed mainly for conservation of specific natural features
IV	Habitat/species management area: managed mainly for conservation through management intervention
V	Protected landscape/seascape: managed mainly for landscape or seascape conservation and recreation
VI	Managed resource protected area: protected area managed mainly for the sustainable use of natural ecosystems

Source: IUCN (1994).

Table 14.2 Management objectives in relation to IUCN protected area categories: 1 = primary objective; 2 = secondary objective; 3 = potentially applicable objective

Management objective	Ia	Ib	II	III	IV	V	VI
Scientific research	1	3	2	2	2	2	3
Wilderness protection	2	1	2	3	3		2
Preservation of ecosystems, habitats, species and genetic diversity	1	2	1	1	1	2	1
Maintenance of environmental services	2	1	1		1	2	1
Protection of specific natural/cultural features			2	1	3	1	3
Tourism and recreation		2	1	1	3	1	3
Education			2	2	2	2	3
Sustainable use of resources from natural ecosystems		3	3		2	2	1
Maintenance of cultural/traditional attributes						1	2

Source: Modified from IUCN (1994).

can be grouped under the term 'threatening processes'. There are many kinds of threatening processes and they differ in their importance in space and time. Examples are clearing for agriculture, chemical pollution from mines, and the introduction of herbivores such as goats or rabbits. Such threats are probably familiar, but not all threats come so directly from the hands of humans. Natural processes can threaten biodiversity if their pattern or timing is altered. In Australia, much of the vegetation around Sydney is adapted to fires but it can be burnt too often for some species to persist. If fire becomes too frequent, say because of vandalism or prescribed burning to reduce the risk of wildfires around towns, some plants that take several years to set seed after fire can become locally extinct (Auld, 1987). In the same environment, fire could be devastating for the fauna in a small patch of habitat if the patch were completely burnt and isolated from sources of recolonization.

'Protected area' is more than a fashionable term for nature reserve. It is deliberately broad because not all protected areas are dedicated only for nature conservation. The establishment and management of protected areas covers a spectrum of arrangements from strict reservation solely for nature conservation to various combinations of conservation and extractive use such as logging or grazing by domestic stock (Soulé, 1991). This range of protection measures is critically important for nature conservation. Without it, there are only two states of nature: reserved or not. This assumes that the elements of biodiversity in reserves are safe and those outside are doomed, both of which are unrealistic. Furthermore, unless the reserved areas in such a scenario occupied nearly all of the globe, there is no chance that they would be adequate to protect biodiversity.

The spectrum of protection measures is formally recognized in the IUCN classification of protected areas (Table 14.1). One of the aims of this new classification is to reduce confusion over the many different terms used internationally to describe types of protected areas. Australia, for example, uses 45 different names and the United States National Parks Service is responsible for 18 different types of areas. Other important aims of the new categories are the improvement of global and regional accounting for protected areas and provision of a framework for data collection and communication. Protected areas are meant to be assigned to categories based on their primary management objective listed in Table 14.1. All have additional objectives apart from these primary ones (Table 14.2).

A feature of the 1994 classification is a new category – VI: managed resource protected area. This is defined as an 'area containing predominantly unmodified natural systems, managed to ensure long term protection and maintenance of biological diversity, while providing at the same time a sustainable flow of natural products and services to meet community needs' (IUCN, 1994). There are several requirements that must be satisfied before such an area is listed, including the proportion of its total area that remains in a natural state. Determining whether extractive use in such an area is sustainable or not is likely to be difficult in many situations. However, the importance of the new category is that it recognizes the contribution that areas like native production forest or semi-arid rangelands with introduced livestock can make to nature conservation.

IUCN (1994) stressed that the numbering of categories does not reflect importance for conserving biodiversity – a combination of all categories is necessary – but the numbering system does indicate a gradation of human intervention. According to IUCN (1994), categories I–III are mainly concerned with the protection of natural areas where direct human intervention and modification of the environment are limited. Categories IV–VI, and particularly the last two, generally involve greater intervention and modification. This chapter takes a slightly different view. 'Strict' protection for nature here includes categories I–IV. Category IV is included under strict protection because the primary goal of management is still nature conservation. Management intervention is not to allow commercial or traditional extractive uses but to maintain certain natural conditions or processes.

In many environments, management intervention might be essential to protect the natural values for which protected areas were originally dedicated. This is because nature is not static (see Chapter 12). Erecting fences and signs without actively managing the area can amount to neglect of natural values if, for example, certain disturbance regimes due to fire or flooding are essential for the persistence of some species or communities. Two important category IV protected areas are Haleji Lake Wildlife Sanctuary in Pakistan and Selous Game Reserve in Tanzania (IUCN, 1994). Channels in Haleji Lake are regularly dredged and reedbeds cut annually to maintain value for waterfowl. Miombo woodland is maintained in Selous Game Reserve with a scientifically based programme of fire management. Another example of the importance of management intervention comes from Barro Colorado Island in Panama. Patches of cleared forest re-established on the island after reservation and one of the results was a decline in butterfly species. The species that were lost relied on food plants that occurred in clearings but were absent or uncommon in mature forest. In this case, there was suitable habitat outside the reserve, but the results emphasize the importance of active intervention to produce particular habitats in many reserves, particularly those that do not have the benefit of surrounding natural or semi-natural habitats.

Even with protected areas defined as broadly as in Table 14.1, they will not do the job alone. The conservation of biodiversity will rely to a large extent on the 'unreserved matrix' (Franklin, 1993), all the pieces of natural or semi-natural habitat that occur outside formally declared protected areas. These range from large tracts of intact vegetation in private or public ownership to small patches of habitat such as hedgerows, strips of native vegetation along roadsides and railways, and urban parks or gardens. The idea of regional-scale planning or ecosystem management (Noss and Cooperrider, 1994) concerns an overall conservation strategy that integrates the management of all natural and semi-natural areas, whether or not they are formally listed and categorized as protected areas.

The remainder of the chapter deals with the effectiveness of protected areas. Two things are critically important: the locations of protected areas in relation to the components of biodiversity that need protection, and the 'viability' of those components. Viability is a complex issue. The term refers to the ability of the components of biodiversity, both patterns and processes, to persist after areas have been declared protected. It relies on many factors, some within and some outside reserves. Table 14.3 gives an outline of these factors and the following sections enlarge on the outline.

Effectiveness of protected areas in relation to the goal of representativeness

The notion of representativeness is of fundamental importance in conservation planning. It refers to protected areas as ways of sampling the natural environment. Just as study plots, perhaps 20 m wide and 50 m long, in patches of forest give a sample of the plants and animals living in the forest, protected areas sample the range of natural diversity in a region. A system of protected areas is representative of a region if it samples the range of natural variation in that region.

The difficulty of achieving the goal of representativeness depends on the level at which biodiversity is defined for the sampling exercise. If a country is divided into 20 ecosystems, then 20 protected areas could do the job, less if they were located to straddle the boundaries between ecosystems. If the country

Table 14.3 Factors that determine the viability of biodiversity in protected areas

Representation targets for reserves

1 Number of occurrences of each species or plant community. Multiple occurrences can serve as insurance, say against the loss of one or more occurrences due to fire or disease. It can also relate to the geographical spacing of occurrences, perhaps in an attempt to sample the likely but undescribed variation in species composition within the geographic range of a plant community. The same approach could be used to sample the genetic variation within a species.
2 The relative extent of each plant community or ecosystem. This can relate to the need to cover as much internal variation as possible, in which case 20% is probably better than 5%, particularly if the 20% is made up of smaller pieces that are spread geographically. It can also depend on the status or vulnerability of a community. If only a small part of a once extensive community remains after clearing, the entire remaining area might need to be included in reserves. If an ecosystem is highly vulnerable to severe alteration for intensive cropping, a larger proportion will need to be reserved than that of an ecosystem that is not threatened, at least in the short term.

Security

Dedication of a reserve does not guarantee that it will serve its purposes. There might be no real commitment on the part of the government to prevent exploitation of the reserve. Its natural diversity could therefore be degraded, despite official listing of the area as protected. This emphasizes the importance of the distinction between documenting the size, number and management categories of protected areas and assessing the effectiveness of those areas for their stated purposes (IUCN, 1994).

Size, shape and connectedness

These are important criteria for the viability of biodiversity in protected areas but only a few general rules are reliable with current knowledge. Large size will increase the number of species that persist in a protected area, allow larger animals to persist there, and buffer internal habitat better from external influences. More compact shapes will also buffer interiors better. Many other issues, such as single large versus several small, and the pros and cons of corridors to connect protected areas are unresolved. The best approach will vary with circumstances.

Types of threatening processes (see Table 14.5 for detail)

Some threatening processes are widespread extractive uses of the land such as clearing, grazing and logging. These are relatively easy to offset by strict reservation. Other processes, such as encroachment by exotic plants and changed fire regimes, are difficult to exclude from protected areas without intensive management. For other processes, formal protection can have little effect. Many impacts on protected areas, such as air pollution and destruction of complementary habitats, cannot be offset by management of the protected areas themselves.

Management and monitoring of reserves

Legal boundaries and signs that say 'Nature Reserve' do not secure the biodiversity of protected areas. Active management is often necessary to ensure that protected areas achieve their purposes, for example by enforcing regulations on hunting. Management is also necessary to maintain the natural patterns and processes for which the protected area was dedicated. Monitoring complements management by gauging the effectiveness of particular approaches.

Types of boundaries and buffer zones

The boundaries of protected areas 'leak' to varying extents by allowing in threatening processes. This depends on the types of threatening processes (Table 14.5) but also on the nature of the boundaries. For example, some impacts on protected areas from outside activities can be excluded or reduced if the boundaries are aligned with watersheds. Another approach to preventing or minimizing leaks of threatening processes is to establish buffer zones around the protected areas that are managed to allow some extractive use while protecting the core zone from major outside impacts.

were subdivided into major plant communities, perhaps 200 in total, more protected areas would be needed. If all known species were to be represented, the task would be much harder, not only because there are many more species in a country than plant communities, but also because, in many countries, the location records of many described species are old and imprecise. Representing all species, including the undescribed ones, presents obvious difficulties. The same problem of lack of knowledge applies to the goal of representing all biological variation at the level of separate populations and genes. There is another problem with the goal of representativeness, too. Elements of biodiversity are being progressively lost from the world before they can be effectively protected and, very often, before they can even be described.

Despite these difficulties, the goal of representativeness is vital for the conservation of biodiversity. The greatest hope for biodiversity is the protection, *in situ*, of as many of its elements as possible as soon as possible. The fact that the goal of representativeness will not be fully achieved should not prevent the most strenuous efforts to maximize success. The goal is now widely endorsed as a fundamental basis for conservation planning and can be summarized in two ways:

- protected areas in the broad sense (IUCN categories I–VI) should be as representative as possible of all levels of biodiversity;

- strict reserves (IUCN categories I–IV) should be as representative as possible of those elements of biodiversity that are least likely to persist under any significant levels of extractive use.

These two points raise another question: when is an ecosystem or species represented? Is one occurrence of each species in protected areas enough? Is one occurrence of an ecosystem in protected areas enough? The answer to both questions is 'probably not' but firm alternative guidelines are difficult to derive. Replication of species, ecosystems or other features and representation of substantial parts of their ranges is desirable. There are three main reasons: as insurance against disasters, to cover undescribed geographical variation, and to recognize that some elements of biodiversity need more protection than others (Table 14.3). Only two guidelines are presently reliable in all cases: more replicates are better than less, and larger proportions of ranges are better than smaller ones. Compromises are necessary with competing demands for money and space. Conservation planners have to determine the representation goals for particular regions based on what they know about the region's biodiversity and what they think might be achievable. Once this has been done, they are in a position to judge the adequacy of a region's protected areas and the necessary number and extent of new ones.

Effectiveness of protected areas in relation to their location

It might seem blindingly obvious to say that protected areas should be located so that they best cover the components of biodiversity that need protection. But this often does not happen. There are two serious outcomes: limited resources for nature conservation are often misplaced, and components of biodiversity are lost unnecessarily. This section reviews the ways in which protected areas can be put in the wrong places, discusses the reasons for this problem, and proposes some key principles for choosing the locations of new protected areas to maximize their contribution to nature conservation. The discussion draws from research on strict protected areas (IUCN categories I–IV – called 'reserves' here for convenience) but the findings are applicable to protected areas in general.

Past locations of reserves – tactical successes, strategic problems

Outlined here are two case studies that illustrate past reasons for establishing reserves in parts of New South Wales, Australia. The dedication of reserves in both regions has been *ad hoc* in the sense that the selection of sites usually lacked a regional perspective on threatening processes and conservation priorities. The case studies illustrate the disadvantages of *ad hoc*, rather than strategic, decisions on the location of reserves.

Case study 1: north-eastern New South Wales

The first national park in this region of about 80 000 km^2 was dedicated in 1931. The conservation reserve system now includes areas totalling more than 520 000 ha or nearly 7% of the landscape. These are spectacular and biologically rich but how representative are they? This question is easiest to answer in relation to physical environments defined by slope and fertility. These give a region-wide picture of biodiversity at a broad level that is otherwise difficult to obtain. An analysis of reserve coverage in relation to environments shows that reserves are concentrated in the steep and infertile parts of the region (Figure 14.1). These environments have least potential for clearing for cropping, grazing, and urban development. The main reason for this pattern of reservation is that reserves have been established as cheaply as possible. This has meant that they have been largely converted from land owned by the government ('Crown' land) which has had the least potential for intensive extractive use. The land with most promise for clearing has been released for private ownership and is therefore expensive to buy for nature conservation. Apart from reservation, the other main use of the Crown land is commercial forestry, so it was inevitable that logging and nature conservation would be seen as competing uses of the more rugged and infertile parts of the region.

Reserves in this region are the leftovers from clearing and grazing. The implications for nature conservation are serious. The environments that are largely in private ownership are very poorly reserved and have had their natural vegetation largely removed and fragmented. Their associated biodiversity has been vanishing while the reserve system has expanded in other areas. Most of the remnant vegetation still has no effective protection. The continuing emphasis of the conservation lobby groups is on the wild and scenic parts of the region remaining on Crown land,

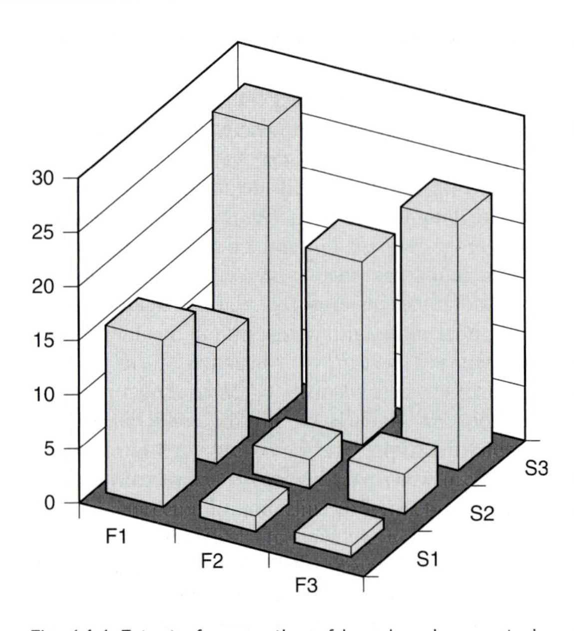

Fig. 14.1 Extent of reservation of broad environments in north-eastern New South Wales. The vertical axis shows the extent of reservation as percentage areas of nine environmental units; the nine units are defined by combinations of slope and fertility. Slope categories are S1: <3° (flat), S2: 3–10° (moderate), and S3: >10° (steep). Fertility categories are F1: low fertility (leucogranite, coarse-grained sedimentary, Quaternary sand), F2: intermediate fertility (granite, acid volcanics, Quaternary alluvium, fine-grained sedimentary, serpentine, limestone), and F3: high fertility (basic igneous). From Pressey (1995).

and largely in State Forests. So logging continues to be emphasized as a threatening process even though clearing and grazing are the most widespread threatening processes in the region and despite the fact that clearing is the most dramatic of impacts on biodiversity. A new focus for conservation efforts is needed that gives much more emphasis to habitat protection and reconstruction in environments that are most vulnerable to serious alteration from clearing and grazing.

Case study 2: the Western Division of New South Wales

The Western Division is an administrative region in New South Wales covering about 320 000 square kilometres of semi-arid and arid rangelands. The most extensive land use in the region is sheep grazing but increasing areas are being cleared for cropping along major rivers and around the eastern and southern margins with relatively high rainfall. Because of constraints on funding and prevailing opposition to nature conservation as a land use in past decades, the establishment of reserves has been opportunistic. The first reserve in the region was Round Hill Nature Reserve, investigated because the lease had been surrendered. The dedication of subsequent reserves was dominated by the availability of expiring or surrendered leases and interests in particular sites or environments. The total gazetted area now covers some 8660 square kilometres or 2.6% of the region. The pattern of reservation has been less strongly biased towards the least productive land than in the north-east of the State. This is partly because virtually all of the Western Division is useful to some extent for grazing and partly because of a less biased distribution of available land (Plate 5 depicts Sturt Nature Park, NSW).

One way of testing the representativeness of the existing reserve system is to find out how many of the region's ecosystems are within reserves. About one-third of all the mapped environments are represented in reserves to some extent. This identifies the gaps in the existing reserve system. Another analysis is more revealing of the contribution of the existing reserves to a future reserve system that would represent all environments. Using a computer program, it is possible to estimate the minimum area of pastoral holdings, ignoring the existing reserves, that is necessary to represent each environment to some extent (Figure 14.2). This is equivalent to the pre-1960 situation. It is then possible to start the same analysis with the existing reserves in 1993 to find out what total area, including the reserves plus additional pastoral holdings, is then necessary to represent every environment. The answer, which might seem counter-intuitive, is somewhat larger. This result is an important one for future reserve selection. It shows that the existing reserves have represented ecosystems inefficiently relative to the computer analysis. The computer analysis is very efficient because, once a representation target is set, it identifies a set of potential reserves that are highly complementary in the ecosystems they contain. The existing reserves have not been selected to best complement previous ones so a greater total area is needed to represent all land systems in 1993 than was needed before 1960.

In a world with limited resources for nature conservation, this increase in required area is a problem. If the required area continues to increase, there might not be enough resources to represent all the environments in the region. The likelihood of this happening is indicated by the continued increase in

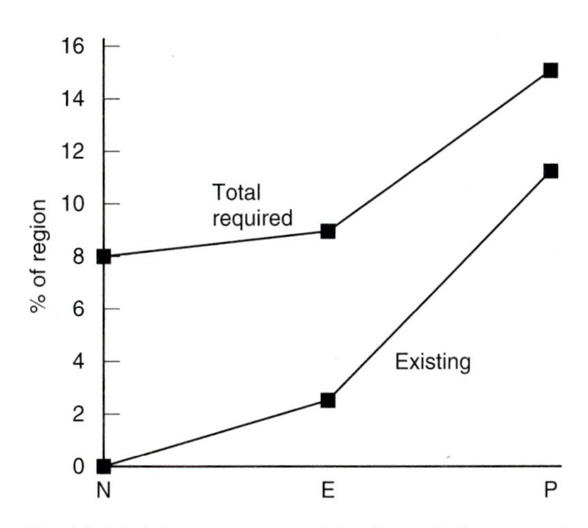

Fig. 14.2 Existing reserves and total required reserves to represent 5% of the total area of every natural environment in the Western Division of New South Wales; areas expressed as percentages of the whole Western Division. Three reservation scenarios are shown: the pre-1960 situation with no existing reserves (N); starting with all existing reserves in 1993 (E); and starting with all existing reserves and all *ad hoc* proposals for additional reserves (P). Modified from Pressey and Tully (1994).

required area when the analysis is started with all existing reserves plus all proposed reserves currently being considered. The proposed reserves have also been selected in an *ad hoc* way. The problem is, in fact, more serious than some environments being missed by the reserve system. If this process continues, there will have been no explicit decisions made on which ecosystems are to be left unreserved. Some of them could be the ones most in need of protection from clearing and damage from stock grazing, in which case the purpose of reservation, to separate elements of biodiversity from threatening processes, will not have been achieved.

How typical are these case studies of the rest of the world?

These case studies illustrate that the number and total area of reserves in a region can be a poor indicator of the effectiveness with which the region's biodiversity has been protected. The location of the reserves in relation to the features that most need strict protection is a much more telling measure of effectiveness. So too is the extent to which the reserves of a region complement, rather than duplicate, one another once a representation goal has been set. The case studies demonstrate that tactical efforts to dedicate individual reserves are of limited effectiveness unless they are part of a larger strategic picture. The larger picture should consist of a regional view of threatening processes, the elements of biodiversity most at risk from these, and the ways in which new reserves can best complement existing ones. To what extent are the case studies generally applicable?

In his book *National Parks: the American experience*, Alfred Runte wrote a chapter called 'Worthless lands' (Runte, 1979). In this chapter he built on his earlier writings by citing examples of famous national parks in the United States that had been dedicated, he said, not primarily to conserve nature, but because they were demonstrated to be relatively useless for any major commercial land uses. This became known as the 'worthless lands thesis'. Runte was criticized by other authors for this view and a vigorous debate followed. In 1994, however, the core of his arguments stands firm – authors from all over the world have come to the same conclusion about the location of strict reserves (Pressey, 1994). The residual nature of the reserve system in north-eastern New South Wales is a reasonable picture of the worldwide pattern of strict reservation. The 'worthless lands thesis' is a tenable generalization – reserves are dedicated not so much for what they are but for what they are not. There are, of course, exceptions but these are sufficiently uncommon to prove the rule rather than undermine it.

Similar analyses from elsewhere, such as South Africa (Rebelo and Siegfried, 1992) and Norway (Sætersdal *et al.*, 1993), have indicated the same problem. It seems reasonable to conclude that any purpose of reservation other than efficient representation of biodiversity will usually make the achievement of a representative reserve system more expensive and, as a result, less likely. This is not an argument against these other purposes but for recognizing that they come with a potential cost for biodiversity. Effective conservation planning should weigh that cost against the perceived benefits.

Reasons for *ad hoc* reservation

Regional or national reserve systems that are fully representative of ecosystems or vegetation types have been explicit goals of conservation planning for much of this century – since the 1930s in the United States (Noss and Cooperrider, 1994) and at least since the 1940s in New South Wales (Strom, 1979). Many *ad hoc* decisions on the locations of reserves have been made in spite of these goals, not because of their absence. Understanding the reasons for *ad hoc* reservation is a first step towards making reservation a more strategic and effective process.

There are three main reasons for *ad hoc* reservations:

1 Political pragmatism. An early aim of the Fauna Protection Panel, the agency that preceded the National Parks and Wildlife Service in New South Wales, was to protect samples of all the State's natural systems, as they were then known. The political situation was such that 'the pattern of faunal reserve dedication which subsequently emerged was unfortunately a scramble for whatever was offering' (Strom, 1979). The constraints faced by Strom – conflicting land uses, resistance to locking up valuable natural resources, and an electorate that was not discerning about protected areas – have applied for the whole history of the reserve system. If conservation progress can be sold as extra hectares, where is the political advantage in dedicating expensive, contentious reserves on freehold land when cheaper ones will do? This has been a serious constraint on the best efforts of conservationists for decades.

   The same types of problems apply much more generally, of course. Political pragmatism can amount to blatant misuse of large amounts of public money to gain political advantage with little or no benefit for biodiversity. Such misuse is by no means limited to environmental portfolios but tends to be scrutinized less rigorously than in others. Voters tend to be less discerning about the establishment of protected areas than about health or employment issues, for example. If a health minister diverted money from upgrading cardiac wards to repainting several hospitals to match his or her new car, it is likely that the reaction in the media and in Parliament would be less than favourable. But environment ministers can and do allocate funds to protected areas with similar care and effectiveness without censure.

2 Failure to take regional perspectives on the significance of areas or their need for protection. Everyone has favourite natural areas and it is not surprising that some people lobby for the conservation of those areas. But what is the net benefit of using limited resources to conserve a well-known area if other more valuable areas with more urgent need for protection are lost as a consequence? This possibility is not so far-fetched. Many conservation decisions are made for parochial reasons and not in the best interests of biodiversity. This has been one of the major motivations for more systematic approaches to setting conservation priorities (Chapter 3). Ten or twenty years ago, regional perspectives were harder to obtain. Today, with geographic information systems and refined techniques for conservation assessment, they are readily available. Despite this, many reserve proposals remain *ad hoc*.

3 Conflicting agendas of lobby groups and government agencies. There are many reasons for establishing new reserves and many ideas about conservation priority. Unfortunately, these are not always complementary. Wilderness, old growth, endangered species, under-represented plant communities, and recreation all compete for limited conservation resources. Wins for one can mean losses for others. Resolution of the issues can be complex and, at the political level, can boil down to the slickest promotion and the optimal electoral impact. Reservation will not become more strategic until the potential conflicts between demands for reserves can be identified and resolved by the proponents themselves. Resolution is necessary at a regional level and should allocate coordinated mixes of protection measures for various purposes and locate sites that can satisfy more than one requirement.

Three key principles for effectively locating reserves

If the effectiveness of biodiversity protection is to be maximized, some basic guidelines are necessary for the placement of reserves. Three principles for effectively locating reserves are outlined below. These are necessary, but not sufficient, for the protection of biodiversity. Their potential benefits will only be realized with the support of changes in conservation policy (Chapter 4).

Vulnerability

The principle of vulnerability is listed first because it will often determine where, when and how the others are applied. It has two applications:

- to focus resources, conservation planning, and restoration efforts on strategically important parts of the landscape where threatening processes are depleting biodiversity most rapidly;
- to focus strict reservation, as a subset of the *in situ* measures, on those environments, communities or taxa that are least likely to persist under any form of extractive use.

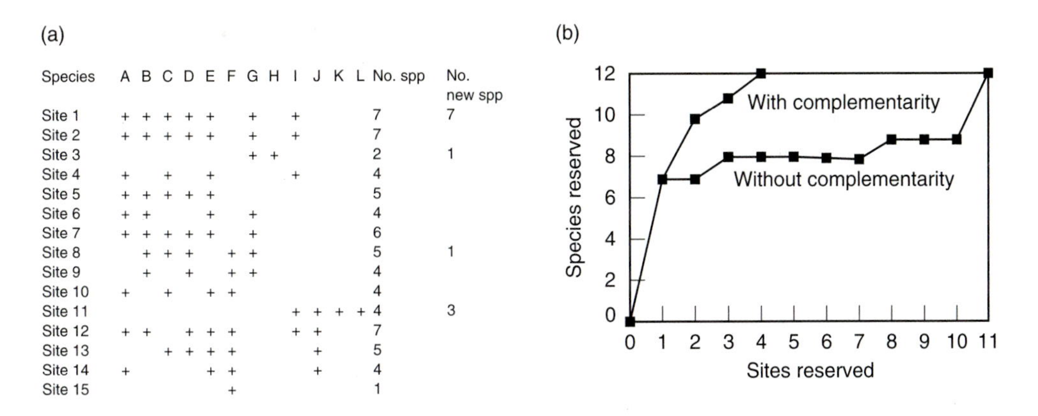

(a)

Species	A	B	C	D	E	F	G	H	I	J	K	L	No. spp	No. new spp
Site 1	+	+	+	+	+		+		+				7	7
Site 2	+	+	+	+	+		+		+				7	
Site 3							+	+					2	1
Site 4	+		+		+				+				4	
Site 5	+	+	+	+	+								5	
Site 6	+	+			+		+						4	
Site 7	+	+	+	+	+		+						6	
Site 8		+	+	+		+	+						5	1
Site 9		+		+		+	+						4	
Site 10	+		+		+	+							4	
Site 11									+	+	+	+	4	3
Site 12	+	+		+	+	+			+	+			7	
Site 13			+	+	+	+				+			5	
Site 14	+				+	+				+			4	
Site 15						+							1	

Fig. 14.3 Reservation of sites in order of species richness with and without complementarity: (a) reservation without complementarity ranks sites in order of species richness, regardless of overlaps in species; reservation with complementarity begins with a representation target (every species at least once) and then selects sites that progressively add the most unrepresented species; (b) cumulative number of species represented with number of sites reserved from (a), with and without complementarity.

The vulnerability of individual taxa, with Red Data Books and lists of threatened species, is a familiar idea. The same idea can be applied to parts of the landscape. An example at the level of whole countries or large regions is Myers' hotspots analyses (Myers, 1988), highlighting parts of the globe deserving the most urgent attention. Regional- or local-scale assessments are also possible (Pressey, 1995). Application of this principle does not force conservation to react constantly to threats – it allows conservation to anticipate threats and to put protection measures in place before the habitats of concern are lost. Without this anticipation, resources could easily be spent on conserving apparently valuable areas that do not need protection, at least in the short term. Applying vulnerability is a way of keeping options open for conservation. Options will remain for areas that are not imminently threatened but will continue to close for those areas that are. If vulnerability is not emphasized in conservation planning, the 'worthless lands' approach to reservation (Runte, 1979) is more likely to prevail and patterns of reserve coverage will continue to resemble that in Figure 14.1.

Complementarity

Complementarity simply means the extent to which reserves complement one another in the features they contain (Pressey *et al.*, 1993). Once the representation target has been set for the species, communities or ecosystems in a region, it is easy to find out if a proposed reserve contains many features that are absent or under-represented in the existing reserve system. If it does, it is highly complementary. If it duplicates features that are already satisfactorily represented in the reserve system, it has low complementarity and is not an effective addition to the system. The consequences for conservation are straightforward. If priorities for reservation are allocated, for example, according to the number of species in sites, without taking complementarity into account, the number of sites that have to be reserved to represent all species is larger than if complementarity is considered (Figure 14.3). *Ad hoc* decisions on new reserves ignore complementarity and increase the total number and area of reserves needed in a region to achieve a representation target (Figure 14.2).

Complementarity therefore leads to the efficient allocation of resources to protect the biodiversity of any region. The cost of achieving a representation target, in terms of money, staff, and opportunity costs for other land uses, is minimized and the likelihood of achieving the target maximized. This efficiency, while important, needs to be applied judiciously, like any principle. There will often be a need to depart from strict efficiency of representation to achieve other important conservation goals. These could include amalgamating individual reserves into large ones to improve viability (Table 14.3 and below) or avoiding some of the most efficient selections to secure other areas that are less disturbed. The important thing is that the potential costs of such departures from efficiency are recognized and balanced against the intended benefits.

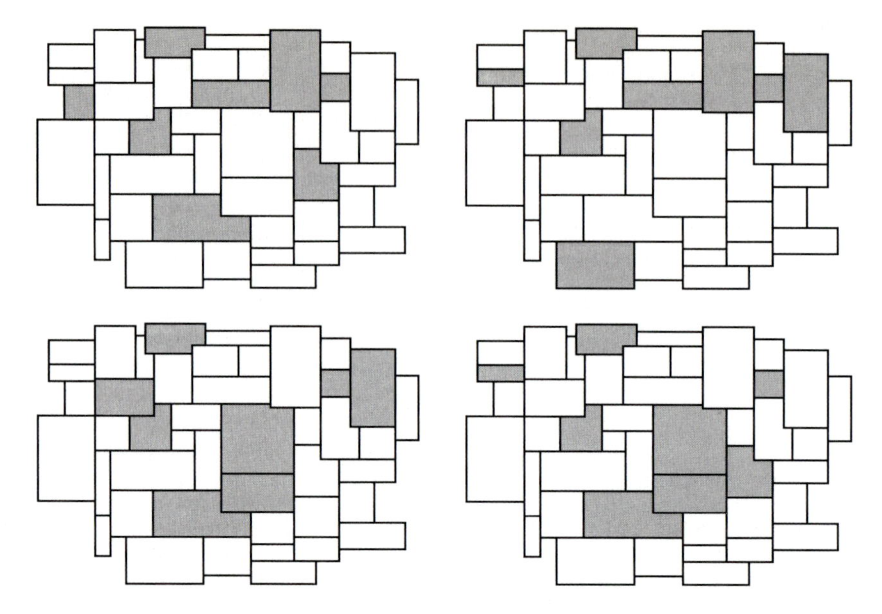

Fig. 14.4 Four alternative reserve networks composed of pastoral holdings in a hypothetical region, all of which could adequately represent all the environments in the region.

Flexibility

In most regions, it is possible to achieve a representation goal by constructing many slightly different reserve networks. The four reserve networks in Figure 14.4 could all achieve the same representation goal, even though they differ somewhat in the specific sites they contain and the precise areas of natural environments they cover. This flexibility in the location of reserves is very important in conservation planning. It opens up options for planners to achieve goals other than the representation of biodiversity, such as maximizing the adjacency or proximity of individual reserves. It also means that there is usually some scope for negotiation between different interest groups on the final configuration and content of a reserve network. Recent approaches for conservation planning have been specifically designed so that this flexibility can be explored (Bedward *et al.*, 1992).

The extent of flexibility can be illustrated for a group of actual pastoral holdings in the Western Division of New South Wales. These constitute a small environmental region in their own right and total several thousand square kilometres in area. The numbers of possible combinations of individual holdings can be very large, as can the numbers of representative combinations (Table 14.4a). These representative combinations are alternative reserve networks. All achieve the basic goal of representing every natural environment at least once.

The data in Table 14.4a can be used not only to illustrate flexibility but to say something useful about individual pastoral holdings. Take, for example, the 81 combinations of four holdings that are representative. Do individual holdings occur in these 81 representative networks with the same frequency or are some more common than others? As it happens, individual holdings occur in representative combinations with very different frequencies – some in all the alternative reserve networks and some in very few (Table 14.4b). This is important information for conservation planning. It means that the holding occurring in all representative combinations is an essential component of any representative network and that, if it is destroyed, the representation goal can no longer be achieved. This holding is irreplaceable. Table 14.4b also indicates that there is a gradation of irreplaceability of pastoral holdings from very high to very low, referred to elsewhere as 'shades' of irreplaceability (Pressey *et al.*, 1994a).

When a range of alternative reserve networks is mapped (Figure 14.4), the overlaps between networks in the sites reserved are related to their irreplaceabilities (Figure 14.5). Mapping the shades of irreplaceability for a region in this way provides a guide to the options for reserving individual sites and the consequences of losing those sites. It indicates

Table 14.4 Flexibility and irreplaceability in a small environmental region in the Western Division of New South Wales, environmental province 3b of Morgan and Terrey (1992)

(a) Numbers of representative combinations of pastoral holdings (representing each natural environment at least once)

Number of sites in combination	Number of possible combinations	Number of representative combinations
1	29	
2	406	
3	3 654	2
4	23 751	81
5	118 755	1 120
6	475 020	9 020
7	1 560 780	50 540
8	4 292 145	214 166
9	10 015 005	718 732
10	20 030 010	1 965 148

Source: Pressey *et al.* (1994b).

(b) Percentage of the 81 representative combinations of four holdings in Table 14.4(a) in which each holding occurs

% occurrence in representative combinations	Number of holdings
100.0	1
63.0	1
50.6	2
39.5	1
4.9	15
2.5	9

Source: Pressey *et al.* (1994b).

where conservation attention should be focused and where compromises can be made.

Effectiveness of protected areas in relation to security, design, threats, management and boundaries

Assume that good decisions have been made on the location of protected areas in a region. They have been chosen to implement sensible representation goals that recognize the need to replicate the samples of some communities and species. Steps have also been taken to represent the range of likely variation within communities and species. The planners have ensured that the protected areas are as complementary as possible and that priority and strictest protection have been given to the features most susceptible to extractive uses. Possible alternative networks of protected areas were considered before the final decisions were made. What else has to be done?

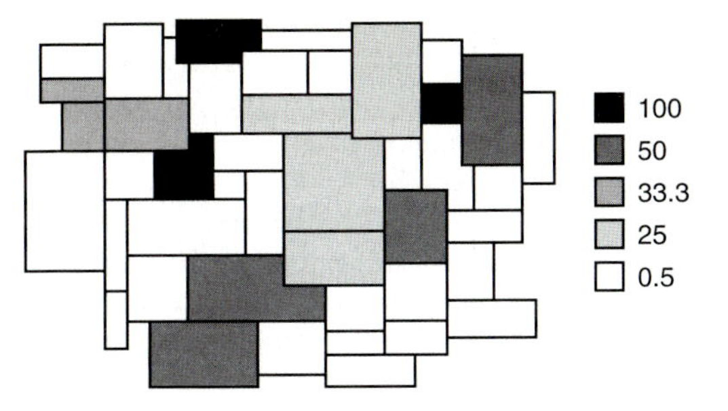

Fig. 14.5 Levels of irreplaceability of pastoral holdings in the same hypothetical region as in Figure 14.4. Note that the degree of overlap between alternative networks in Figure 14.4 depends on the irreplaceability of the individual holdings; totally irreplaceable holdings occur in all four networks in Figure 14.4.

The protection of biodiversity does not end with the dedication of a protected area – it begins there. Basically, protected areas are nothing more than pieces of paper. They are lines on maps and statements in government gazettes supported by pieces of legislation, which themselves are no more than pieces of paper. The persistence of biodiversity within the areas dedicated to protect natural diversity needs much more than words and paragraphs. Apart from representation targets, there are other important factors that determine the effectiveness and viability of protected areas once they have been established. These are listed in Table 14.3 and discussed below.

The effectiveness of protected areas in separating elements of biodiversity from threatening processes depends primarily on the legal and financial support they receive to ensure protection of the features they contain. Unfortunately, in both the terrestrial and marine environments, there are examples of 'paper parks' – areas that are protected on paper but which have no effective management, legal title or clear boundaries. The destructive pressures on these areas can become so great that the protected areas themselves are regarded as threatened from the very processes they are intended to offset, as in parts of Central America (Barzetti, 1993) and south-east Asia (Parr *et al.*, 1993). There are also many documented cases of reserve status being legally removed by governments to make natural resources available for exploitation (Aiken, 1994). Soulé (1991) has advocated an actuarial approach to predicting the life expectancies and ongoing effectiveness of protected areas for their specified roles according to biogeographic, social and political factors. The results would guide such considerations as the need for replication, the emphasis given to alternative *in situ* protection measures, and the importance of backup *ex situ* approaches.

Location is one critical aspect of the pattern of protected areas in the landscape. Others include the size, shape and connectedness of the protected areas. All these issues have been dealt with at length by other authors and several major issues have emerged. Other things being equal, a large area of habitat will support more species of a particular taxonomic group than a small one, and some species with large area requirements will only persist in large protected areas. A patch of natural habitat experiences 'edge effects' from adjoining land uses or habitats. Large protected areas therefore tend to have less disturbed interiors. Examples of edge effects are increased insolation and wind and drift of sprayed herbicides at a cleared boundary (Saunders *et al.*, 1991) and altered species composition due to increased colonization by native species from adjacent secondary forest (Janzen, 1983). There is no overall guide as to whether a large area of habitat will support more species than several smaller ones of the same size, so this choice has to be applied on a case-by-case basis, and more smaller areas can help to replicate samples of species or environments. As for shape, a reasonable generalization is that protected areas with irregular and elongated shapes (high perimeter:area ratios) will suffer proportionally more from edge effects. The connectedness of protected areas, or the extent to which they are linked by natural or semi-natural habitat, has both benefits and drawbacks and also needs to be considered carefully for specific planning problems.

Given strong legal support for protected areas and some resources for management, their effectiveness, or the viability of the natural features they contain, depends largely on the threatening processes that are operating in particular places (Tables 14.3, 14.5) and this is closely tied to the management of the areas and the nature of their boundaries. Of all the processes that threaten biodiversity, the impacts of clearing, forestry, and grazing by domestic stock are most effectively offset by protected areas, and particularly by strict reserves. Marine equivalents to these processes include trawling and long-line fishing. These activities have well-defined spatial limits and are easy to demarcate on the ground or at sea. In most cases, the legal support for protected areas is sufficient to exclude these processes. In other cases, patrolling and monitoring by management staff are necessary to prevent incursions.

The effectiveness of protected areas in offsetting other processes, like the impacts of exotic plants, feral animals, altered fire patterns, hunting and illegal collecting, depends very much on the resources available for weed and feral control, patrolling and prosecuting, and exclusion of fire from outside areas. Active management is also necessary to regulate the pattern and timing of natural processes such as fire, sedimentation of wetlands, or grazing by native herbivores. Generally, as protected areas become smaller, the extent of management intervention to maintain their desired characteristics must increase (White and Bratton, 1980). This is because natural processes can more often be left to run their own course in large, unaltered areas. Management and monitoring are closely linked in a feedback loop. A management action is in many ways an experiment to see what works best to achieve a desired goal. Monitoring is necessary to test how well the goal has been achieved. The results of the monitoring will then determine whether and how management should be changed.

Protected areas are generally ineffective in controlling a range of other processes (Table 14.5). If the flow of a major river is largely diverted for irrigation use, establishing a reserve on part of the floodplain will have no effect on the unnatural drying of wetlands. This problem could be ameliorated to some extent by local management, for example by installing control structures on wetland basins to manipulate water levels. Similarly, rising saline groundwater that kills native vegetation might originate hundreds of kilometres away because of excessive clearing of trees in the recharge zone of the groundwater basin. Airborne chemical pollution also has no respect for reserve boundaries. Visibility in the Grand Canyon in Colorado can be impaired by air pollution from external sources to the point where views disappear (Freemuth, 1991). The location and nature of the boundaries of protected areas are important for many reasons (Theberge, 1989) but very clearly for issues relating to hydrology. Hydrological impacts such as sedimentation from upstream erosion and waterborne chemical pollutants can be eliminated if boundaries are aligned with those of watersheds, although sedimentation problems can

Table 14.5 Examples of threatening processes and the effectiveness of protected areas in offsetting them

Threatening process	Effectiveness of protected areas[a]
Clearing for cropping, grazing, housing or mining	Generally high
Forestry activities	Generally high for strict reserves
Grazing by domestic stock	Generally high for strict reserves
Introduced plants	High to zero, depending on management
Feral herbivores and carnivores	High to zero, depending on management
Changed fire regime	High to zero, depending on management
Hunting/illegal collecting	High to zero, depending on management
Off-road vehicles	High to zero, depending on management
Erosion/sedimentation	High to zero, depending on management and location of boundaries
Chemical pollution – waterborne	High to zero, depending on location of boundaries
Changed water regime	Generally low to zero
Salinization due to rising groundwater originating outside the reserve	Zero
Chemical pollution – airborne	Zero
Impacts of damaging activities on surrounding areas and the atmosphere	Zero
Human population growth outside protected areas	Zero

[a] The tabulated comments are necessarily general because effectiveness depends to varying extents on local conditions and management initiatives.

originate from accelerated internal erosion, perhaps from access tracks and other developments.

Protected areas also generally have no influence on the impacts of human activities on the surrounding land or sea, unless adjacent areas are zoned for sympathetic use, as in biosphere reserves (Dyer and Holland, 1991). This has obvious implications for those elements of biodiversity that are not adequately represented in reserves. Perhaps less obviously, it also has important implications for the reserves themselves. Both Yellowstone and Everglades National Parks in the United States are very extensive but their biodiversity is declining because of clearing and water control works outside their boundaries (Noss and Cooperrider, 1994). Reserves can change from being part of a sea of natural habitat to islands in a sea of cleared land. As this change progresses their biodiversity is reduced by size and edge effects (Saunders *et al.*, 1991). Another important effect of isolation is 'truncation of the habitat matrix' (Terborgh, 1992) or a reduction in access to suitable habitats. Access by some species to habitats outside the reserves is prevented. The species for which these external habitats provide essential seasonal resources will therefore eventually disappear from the reserve. Other examples of impacts on reserves from events in surrounding areas are global and regional climate change exacerbated by clearing and burning of vegetation (Ehrlich and Ehrlich, 1992).

Outside activities are therefore important influences on protected areas for several reasons: the effective size of the protected areas is reduced by destruction of external habitat, the physical and biological interactions between the protected areas and their surrounds is altered, and the boundaries of protected areas 'leak' with respect to some impacts of outside activities such as pollutants, weeds or reduced water flows. These are major reasons for ecosystem- or regional-scale management (Noss and Cooperrider, 1994) that attempts to coordinate human activities right across a natural region to maximize the overall benefits to biodiversity. Part of ecosystem management would be the establishment of buffer zones around protected areas. These are areas outside the boundaries of the core protected area that are managed sympathetically to minimize the impacts of outside activities. For example, they could be managed to exclude domestic stock and so prevent the spread of disease into a national park, they could be burnt in a pattern that would buffer the park from fires originating in settled areas, or they could be used for selective logging while buffering the park from intensive agriculture. While doing all these things, buffer zones increase both the effective size of the protected area and the likelihood that all the life requirements of protected organisms will be provided in this larger area.

Reserves also have no effect on external human population growth but the implications for the 'protected' biodiversity might also be very serious. Among the more intractable problems are incursions by people needing food, fuelwood or living space, for example in India (Ward, 1992) and Malaysia (Aiken, 1994). It was because of this pressure and the intensifying external threats outlined above that Ehrlich (1988) predicted the strategic defeat of the conservation movement, despite tactical successes such as the establishment of protected areas. This view cannot be dismissed and, given the increasing impacts of outside activities on protected areas in economically developed nations, cannot be regarded as accurate only for developing countries. A slightly more optimistic assessment would be to see protected areas as sandcastles on a beach with a rising tide. Some will be safe because they are high up, some will be washed away entirely, and some will need very careful and intensive management if they are to persist. Scientists like Michael Soulé have been talking for some years about a demographic winter, the nadir of biodiversity at the zenith of the world's human population. This hammers home an unpleasant fact: biodiversity will only decline from here on. We cannot stop this decline, only slow it, and the best way of slowing it is to reduce the number of people on the planet.

Conclusions

So are individual conservation actions only short-term successes that address the symptoms and not the cause of long-term disaster, as Ehrlich (1988) proposed? Ehrlich's comment is important because it questions the worth of the most important and fondly-held approach to conserving biodiversity – establishing protected areas. The continuing decline of biodiversity in the face of increasing human populations is unarguable. All the threatening processes that protected areas are meant to offset have one root cause – people – and protected areas are themselves threatened by increasing numbers of people. But the scenario is not one of utter destruction. Many elements of biodiversity will survive, as long as human population growth peaks and, hopefully, begins to decline.

Questioning the worth of protected areas also does more than highlight the ultimate cause of the decline in biodiversity. It raises difficult and unresolved issues about the effectiveness of protected areas that are hidden by assessments based only on hectares. The issues raised in this chapter present many intellectual and practical challenges to conservation planners. The extent to which the challenges are met will help to determine how much biodiversity the human race takes with it into the future. However, the most serious impediment to effective conservation planning is not lack of knowledge. It is the lack of effective use of the knowledge now available. Just as the ultimate threat to biodiversity is human numbers, the ultimate block to effective conservation planning is the low priority given to biodiversity conservation on political and economic agendas. New knowledge in itself will not change this situation unless it is used to close the gap between science and environmental policy.

References

Aiken, S.R. (1994) Peninsular Malaysia's protected areas' coverage, 1903–92: creation, rescission, excision, and intrusion. *Environmental Conservation*, **21**, 49–56.

Auld, T.D. (1987) Population dynamics of the shrub *Acacia suaveolens* (Sm.) Willd.: survivorship throughout the life cycle, a synthesis. *Australian Journal of Ecology*, **12**, 139–51.

Barzetti, V. (ed.) (1993) *Parks and progress: protected areas and economic development in Latin America and the Caribbean*, Washington, DC: IUCN.

Bedward, M., R.L. Pressey and D.A. Keith (1992) A new approach to selecting fully representative reserve networks: addressing efficiency, reserve design and land suitability with an iterative analysis. *Biological Conservation*, **62**, 115–25.

Dyer, M.I. and M.M. Holland (1991) The biosphere reserve concept: needs for a network design. *BioScience*, **41**, 319–25.

Ehrlich, P.R. (1988) The strategy of conservation, 1980–2000. In M.E. Soulé and B.A. Wilcox (eds) Conservation biology: an evolutionary–ecological perspective. Sunderland, MA: Sinauer Associates, pp. 329–44.

Ehrlich, P.R. and A.H. Ehrlich (1992) The value of biodiversity. *Ambio*, **21**, 219–26.

Franklin, J.F. (1993) Preserving biodiversity: species, ecosystems, or landscapes? *Ecological Applications*, **3**, 202–5.

Freemuth, J.C. (1991) *Islands under siege: national parks and the politics of external threats*. Lawrence, KS: University Press of Kansas.

Grove, R.H. 1992. Origins of western environmentalism. *Scientific American*, **267**(1), 22–7.

IUCN (The World Conservation Union) (1994) *Guidelines for protected area management categories*. Gland, Switzerland: IUCN.

Janzen, D.H. (1983) No park is an island: increase in interference from outside as park size decreases. *Oikos*, **41**, 402–10.

Leader-Williams, N., J. Harrison and M.J.B. Green (1990) Designing protected areas to conserve natural resources. *Science Progress*, **74**, 189–204.

Morgan, G. and J. Terrey (1992) *Nature conservation in western New South Wales*. Sydney: National Parks Association of New South Wales.

Myers, N. (1988) Threatened biotas: 'hot spots' in tropical forests. *The Environmentalist*, **8**, 187–208.

Noss, R.F. and A.Y. Cooperrider (1994) *Saving nature's legacy: protecting and restoring biodiversity*. Washington, DC: Island Press.

Parr, J.W.K., N. Mahannop and V. Charoensiri (1993) Khao Sam Roi Yot – one of the world's most threatened parks. *Oryx*, **27**, 245–9.

Pressey, R.L. (1994) *Ad hoc* reservations: forward or backward steps in developing representative reserve systems? *Conservation Biology*, **8**, 662–8.

Pressey, R.L. (1995) Conservation reserves in New South Wales: crown jewels or leftovers? *Search*, **26**, 47–51.

Pressey, R.L. and S.L. Tully (1994) The cost of *ad hoc* reservation: a case study in western New South Wales. *Australian Journal of Ecology*, **19**, 375–84.

Pressey, R.L., C.J. Humphries, C.R. Margules, R.I. Vane-Wright and P.H. Williams (1993) Beyond opportunism: key principles for systematic reserve selection. *Trends in Ecology and Evolution*, **8**, 124–8.

Pressey, R.L., I.R. Johnson and P.D. Wilson (1994a) Shades of irreplaceability: towards a measure of the contribution of sites to a reservation goal. *Biodiversity and Conservation*, **3**, 242–62.

Pressey, R.L, M. Bedward and D.A. Keith (1994b) New procedures for reserve selection in New South Wales: maximizing the chances of achieving a representative network. In P.L. Forey, C.J. Humphries and R.I. Vane-Wright (eds) *Systematics and conservation evaluation*. Oxford: Clarendon Press, pp. 351–73.

Rebelo, A.G. and W.R. Siegfried (1992) Where should nature reserves be located in the Cape Floristic Region, South Africa? Models for the spatial configuration of a reserve network aimed at maximising the protection of floral diversity. *Conservation Biology*, **6**, 243–52.

Reid, W.V. and K.R. Miller (1989) *Keeping options alive: the scientific basis for conserving biodiversity*. Washington, DC: World Resources Institute.

Runte, A. (1979) *National parks: the American experience*. Lincoln, Nebraska: University of Nebraska Press.

Sætersdal, M., J.M. Line and H.J.B. Birks (1993) How to maximize biological diversity in nature reserve selection: vascular plants and breeding birds in deciduous woodlands, western Norway. *Biological Conservation*, **66**, 131–8.

Saunders, D.A., R.J. Hobbs and C.R. Margules (1991) Biological consequences of ecosystem fragmentation. *Conservation Biology*, **5**, 18–32.

Soulé, M.E. (1991) Conservation: tactics for a constant crisis. *Science*, **253**, 744–50.

Strom, A.A. (1979) Some events in nature conservation over the last forty years. *Parks and Wildlife*, **2**(3–4), 65–73.

Terborgh, J. (1992) Maintenance of diversity in tropical forests. *Biotropica*, **24**(2b), 283–92.

Theberge, J.B. (1989) Guidelines to drawing ecologically sound boundaries for national parks and nature reserves. *Environmental Management*, **13**, 695–702.

Ward, G.C. (1992) India's wildlife dilemma. *National Geographic*, **181**(5), 2–28.

CHAPTER 15 *Ex situ* conservation

DAVID WORLEY

Introduction

The protection of species within their natural habitat (*in situ*) is the main aim of conservation programmes. However, habitat protection alone cannot ensure the survival of all species or, as the 1980 *World Conservation Strategy* points out, the maintenance of biotic diversity (IUCN *et al.*, 1980). Consequently, *ex situ* (off-site) conservation is increasingly being used to provide an important support system for *in situ* programmes. *Ex situ* programmes for animals generally take place in zoos, aquaria, specialized breeding centres or semi-reserves. Plants are usually housed in botanic gardens, university research areas, specialized breeding centres (e.g. the International Rice Research Institute) and semi-reserves.

Species conserved in *ex situ* programmes are inevitably separated from other components of their natural habitat, such as the prevailing weather conditions, predation and interaction with other species. This might have serious evolutionary implications and certainly makes reintroduction of species more difficult (Box, 1991).

However, *ex situ* conservation can prevent the immediate extinction of species, e.g. the the black-footed ferret (*Mustela nigripes*) (Thorne and Oakleaf, 1991). *Ex situ* breeding and propagation programmes can rapidly increase the number of individuals in a population and help to protect their genetic diversity. They can also provide additional opportunities for interactive management programmes, e.g. reintroduction and restocking programmes. Finally, *ex situ* conservation can promote public awareness and provide opportunities for education and research beneficial to the species concerned. The World Zoo Conservation Strategy states, '*Ex situ* programmes in support of *in situ* conservation are indispensable for the continued existence of an ever increasing number of critically endangered species' (IUDZG/IUCN, 1993). The challenge for *ex situ* conservation centres (zoos, aquaria, botanic gardens, etc.) is to respond to the threats facing wildlife by conserving endangered species until they can be used to re-establish viable populations in the wild.

An example that illustrates the value of zoos and their captive breeding programmes concerns Przewalski's horse (*Equus przewalskii*) (Figure 15.1). In 1879 the Russian explorer Nicolai Przhevalsky was visiting the Kazakhstan border town of Zaysan where he was presented with the skull and hide of a horse. It had been shot by native hunters in the Dzungarian Desert and Przhevalsky immediately suspected it belonged to a wild horse, possibly a tarpan. The specimen was sent to the Academy of Sciences in St Petersburg where it was later described as a new species and given the name *Equus przewalskii* (Weeks, 1977). The wild horses were probably in decline when Przhevalsky made his discovery. Unable to compete with domestic livestock, they were forced into their last stronghold in the Takhin Shara Nuru or Yellow Horses Mountain in south-west Mongolia from where they eventually disappeared. The last wild-caught Przewalski's horse was a female, trapped in Mongolia in 1957. Since the mid-1960s, Przewalski's horses have been presumed extinct in the wild. However, there are currently over 1000 Przewalski's horses in captivity around the world and plans are now in motion to return some of them to their former range in Mongolia (Seal *et al.*, 1990; Sattaur, 1991).

IUCN threatened species categories

The World Conservation Union's (IUCN) Red Data Books contained lists of threatened species. In those

Fig. 15.1 Przewalski's horse (*Equus przewalskii*). Przewalski's horses are extinct in the wild. In zoos they can be distinguished from domestic horses by their colour and markings and the presence of an upright mane. Unlike domestic horses, which have 66 chromosomes, Przewalski's horse has only 64. (Photograph: D. Worley.)

Red Lists each species was assigned a category indicating the level of threat they faced. The categories included Extinct (EX) (species not seen in the wild for 50 years), Endangered (E) (species in danger of extinction if the causal factors were not removed), Vulnerable (V) (species likely to become endangered if the causal factors were nor removed) and Rare (R) (species with small world populations, not currently 'Endangered' or 'Vulnerable' but which may be at risk). Other categories were 'Indeterminate' (I), 'Insufficiently Known' (K), 'Threatened' (T), and 'Commercially Threatened' (CT) (IUCN, 1993). As the number of threatened species grows, so too does the need for conservationists to manage their resources, e.g. people, land and finances, more effectively. Consequently, the Red Lists are increasingly being used to set conservation priorities. This situation has highlighted a number of weaknesses in the current system and prompted a review of the categories (see Mace in IUCN, 1993). For example, the African hunting dog (*Lycaon pictus*) and the Siberian tiger (*Panthera tigris altaica*) are both currently listed as Endangered species but the present IUCN category 'Endangered' does not indicate which is more likely to become extinct first and consequently require priority action.

Mace and Lande (1991) suggested three new categories of threat (plus 'Extinct') based on estimated risks of extinction. In their proposed scheme a species defined as 'Critical' would have a 50% probability of extinction within five years or two generations, whichever is longer. An 'Endangered' species would have a 20% probability of extinction within 20 years or 10 generations, whichever is longer, and a 'Vulnerable' species would have a 10% probability of extinction within 100 years. Mace and Lande recognized the difficulty of assigning species to the correct category and suggested using quantitative measurements of actual and effective population size to help distinguish between levels of threat. Their system combines an assessment of risk with time and thus gives some indication as to when a species might become extinct.

However, it is not easy to predict the chances of extinction. In some cases the data are simply not available. Where information does exist, fluctuating parameters such as environmental stochasticity, catastrophic events and the effects of inbreeding make population predictions unreliable. Population Viability Analysis (PVA) is one approach that uses computer modelling techniques to predict a population's chances of survival within a given time frame (see Chapter 17). The models incorporate information on population dynamics, e.g. number of individuals, age structure and growth rates; population characteristics, e.g. genetic variation, behaviour and patterns of dispersal; and environmental effects, e.g. the state of the remaining habitat, rates of habitat loss, species interactions and human interference. PVAs allow researchers to analyse a large number of hypothetical case studies relatively quickly but the modelling is only as good as the data available.

Building upon the Mace and Lande categories, IUCN has produced a set of draft proposals for Red Lists using new categories that are currently being tested. Briefly, the new categories are Extinct (EX), Extinct in the Wild (EW), Critical (CR), Endangered (EN), Vulnerable (VU), Conservation Dependent (CD), Susceptible (SU), Low Risk (LR), Data Deficient (DD) and Not Evaluated (NE). The categories Critical, Endangered, Vulnerable and Susceptible are all described as threatened. To be placed in any of the threatened categories (excluding susceptible), one or more criteria must be met. There are five criteria for each category describing aspects of population decline, extent of occurrence, estimates of population size and quantitative analysis showing the probability of extinction in the wild. Armed with the new categories, it should be easier to make reasoned judgements about the status of a given species and its need for conservation action. At the time of writing, the new categories are only in draft form. When they do appear there will be a transition period when both old and new categories will be in use.

Botanic gardens

There are approximately 1600 botanic gardens around the world attracting 150 million people each year. Botanic Gardens Conservation International (BGCI) is the world's largest botanic gardens organization. Founded in 1987, its role is to gather and disseminate information on issues relating to plant conservation. The work being done in botanic gardens varies depending upon their location and the resources upon which they can draw, but they should have certain common objectives, namely to maintain essential ecological processes, to preserve genetic diversity and to promote the sustainable use of species and ecosystems (IUCN *et al.*, 1980). Implicit in these objectives is the importance of conserving species in the wild. Expansion of the role of botanic gardens is also important to ethnobotany: there are opportunities to strengthen collections of plants with traditional cultural importance and establish links with local communities (Given and Harris, 1994).

About 60 000 plant species could disappear over the next 50 years (IUCN, 1989). Some are known to be of economic value, e.g. *Prunus africanus* (Cunningham and Mbenkum, 1993); others, not yet studied, will simply vanish, taking their genetic potential with them (Koopowitz and Kaye, 1990).

The Botanic Gardens Conservation Strategy (IUCN, 1989) describes ways in which botanic gardens can help to conserve plants and suggests five categories of plants in need of protection. These are summarized as:

1 Rare and Endangered.
2 Economically important, e.g. medicinal, oils, intoxicants, fuelwood, forage and wild relatives of crops (Rhoades, 1991).
3 Species required for restoration of ecosystems.
4 Keystone species, i.e. those that are known to be important in the maintenance and stability of ecosystems.
5 Taxonomically isolated species with scientific value.

It is not possible to save every species from extinction; consequently care must be taken to ensure that limited resources are used effectively. Considerations for selecting individual species for *ex situ* conservation measures include:

1 An assessment of the extinction risk, e.g. using the new IUCN Red List categories and PVA analysis.
2 The suitability of the plant for *ex situ* conservation. Some species, e.g. the Brazil nut tree (*Bertholletia excelsa*), do not grow well if removed from their natural habitat (see page 31).
3 The value of the plant to the ecosystem or to humans, e.g. yams (*Dioscorea* spp.) have provided chemicals used in birth control pills, and the periwinkle (*Catharanthus roseus*) has given us vincristine, an alkaloid successfully used to treat children with acute leukaemia.
4 The ease with which the species can be collected in the wild. Difficult terrain, political unrest, etc., could make collection hazardous.
5 Available funds. There is little point in starting a project that cannot be completed due to lack of funds or other resources.
6 Chances of success. If the probability of success is slim then resources might be better used elsewhere.

The mechanism for identifying priority cases and initiating plant conservation programmes is currently under review. The Species Survival Commission (SSC) of the IUCN (World Conservation Union) has a number of taxon advisory groups, e.g. for orchids and carnivorous plants, whose task is to gather information on threatened species and, ultimately, to produce action plans for their conservation. However, at present, the initiative for plant conservation lies with individual organizations, government agencies and botanic gardens. These bodies form their own working groups and develop action plans for their chosen species.

Gene banks

If a species cannot be protected *in situ* then plant material can be collected and stored elsewhere. This may be as whole plants or as seeds, pollen, eggs, ovules and various tissues. The storage facility is referred to as a gene bank and will usually be associated with a botanic garden, university research centre or plant conservation centre. Each species will have specific collecting requirements which will vary depending upon the season, weather conditions and the material taken, e.g. seeds or leaf cuttings. In the case of orchids, seed capsules can be harvested before they split. This ensures that the seeds within are sterile, an important consideration if *in vitro* techniques are to be used. Seeds from split capsules may be contaminated by fungi or bacteria and will need to be properly treated and dried before sowing.

Fig. 15.2 Looking for a green future. A researcher examines seeds in a seed bank that may contain valuable genetic information. In crop plants this might include genes for disease or drought resistance, improved colour, taste and texture, or greater productivity. (Photograph: Royal Botanic Gardens, Kew.)

Seed banks

Most seeds are small, easy to store and contain a full complement of viable genetic material (Figure 15.2). A tiny container, holding perhaps a few thousand seeds, is able to store most of the genetic variability of a whole plant population. Compare this to the equivalent space required to maintain 200 endangered palm trees or even 50 giant redwoods! However, the seeds of different species may require different storage techniques, an important area of study as an increasing number of species require preservation (Gunn, 1991). Prior to storage, the seeds are cleaned and dehydrated to approximately 5% moisture content.

The removal of excess water helps prevent ice crystals forming inside the seed and rupturing vital cell membranes. They are then cooled to a temperature of about −20°C, depending upon the species. Cooling lowers the seeds' metabolic rate and allows them to be stored for longer periods of time. However, the seeds do not remain viable forever and must be periodically germinated and fresh seeds obtained. The frequency of germination depends upon the species and may be somewhere between 5 and 25 years. However, the potential for storage is illustrated by a collection of lotus flower seeds that were recovered from an ancient peat bog and successfully germinated after a dormancy of 1000 years! In some cases the soil around recently extirpated populations may contain seeds (a soil gene bank) and provide material for *ex situ* propagation.

Species that do not readily produce seeds or have oily or hard seeds, and those with recalcitrant seeds (which cannot be cooled or dried without injury), are poor candidates for seed banks. Though comparatively rare, these groups include some important plants such as tea, breadfruit and coconuts. Consequently, such plants should be given higher priority for *in situ* conservation (or storage in field gene banks) whilst researchers look for new strategies for long-term storage.

Field gene banks

Field gene banks are collections of growing plants rather like plantations. They provide refuge for those species that are unsuitable for seed banking and whose existence is still threatened in the wild, e.g. coffee, bananas and peach palm. Field gene banks take up a lot of space and usually retain far less genetic diversity than most seed collections can achieve. They are particularly vulnerable to the spread of disease (as monocultures are) and are constantly exposed to environmental disasters such as fire or hurricanes. Yet they do offer some insurance against genetic erosion and extinction and provide an important base for researchers to study plant development (Hoyt, 1988).

Pollen and spore banks

Pollen and spores (from non-flowering plants) can be treated in much the same way as seeds. Vast numbers, and hence enormous genetic variation, can be cryogenically stored in very small vessels. Botanists at the University of California's Irvine Arboretum have put this technique to good use. They found that their five specimens of the rare African amaryllid *Cyrtanthus obliquus* were self-sterile and would not flower simultaneously. Pollen from any one plant was simply wasted on itself. But by storing the pollen and using it when other individuals flowered they were able to produce thousands of seedlings (Koopowitz and Kaye, 1990). Freeze drying is a technique that combines the drying and freezing stages into one 'quick freeze' process. It has proved successful with some pollen grains and may eventually provide a cheaper and faster alternative to drying and freezing material in separate stages.

Micropropagation

Micropropagation represents a revolution in plant management. Using freshly collected material or plant material taken from a gene bank, scientists are developing increasingly ingenious methods of *in vitro* (in glass) propagation. Gone are the rows of potted seeds whose germination is subject to the capricious conditions found on a greenhouse bench. Instead, tiny dishes containing culture medium provide support and nutrition, not only for seeds, but for pollen, spores and a variety of vegetative plant material that might otherwise prove difficult to propagate. The cultures, also known as tissue cultures, are placed in 'special care units' where optimum light, temperature and humidity are provided (Fay, 1992). In this way, many thousands of plants can be produced from a minimum amount of original plant material, an important consideration when the species concerned is itself reduced to just a handful of individuals. Tissue culture is particularly suited to plants that do not set seed readily, that flower infrequently or have recalcitrant seeds. It also allows field botanists to collect living material from plants, *in situ*, that are not ready to set seed. Large numbers of young plants can be produced from a small sample of parent material causing minimum disturbance to the wild population.

If vegetative material is propagated, the resulting plants will be identical clones. Such plants can be of great value to commercial growers who may require large numbers of identical plants for sale to the general public. Cloning threatened species can relieve the pressure on wild populations by undercutting the prices asked, by unscrupulous dealers, for wild plants. Plant breeders can also use micropropagation techniques to 'cleanse' their stock by culturing uninfected tissue taken from diseased plants. Having produced identical, though disease-free, plants the infected parent stock can be discarded. Compared to traditional cultivation methods, micropropagation can reduce the risk of disease and protect against contamination and the associated problems of hybridization. It takes up little space and is well suited to mass cloning of rare species. *In vitro* cultures can also be used for long-term storage of plant material at low temperatures to reduce growth rates (i.e. a cryo-genebank). Theoretically, such storage could be indefinite, although it is expensive and labour intensive because of the need to regularly subculture the specimens (Groombridge, 1992).

Genetic fingerprinting

Plant breeders are increasingly directing their efforts towards careful breeding programmes that aim to minimize inbreeding and retain the maximum amount of genetic variability in their plant populations (see the section on captive breeding programmes later in this chapter). An important development in plant breeding is genetic fingerprinting (Newbury and Ford-Lloyd, 1993). Using DNA taken from plant cells, researchers are looking at ways of assessing population variability and developing studbooks along the lines of their animal breeding counterparts. Plants whose genes are under-represented in the gene pool can then be incorporated into a breeding programme. Those plants with over-represented genes can be 'rested' or discarded.

Plant reintroductions

Reintroduction can be defined as the 'intentional movement of an organism into part of its native range from which it has disappeared or become extirpated in historic times as a result of human activities or natural catastrophe' (IUCN, 1987). Where species are moved from one part of their range to another they are said to be 'translocated', whilst species that are released into areas outside their historic range are 'introduced' (also see Chapter 17). Islands often have a high level of endemic species that have proved particularly vulnerable to the depredations of introduced species (Bourne *et al.*, 1992; Seitre and Seitre, 1992). For example, *Memecylon floribundum* and *Clidemia hirta* have been introduced to some areas of the Seychelles where they form dense foliage that prevents other (local) species from growing (Gerlach, 1993). The aim of reintroduction programmes is to re-establish self-sustaining populations of plants and animals in the wild without adversely affecting the native species already present.

Ex situ propagation of rare plant species is a well established and rapidly developing science. In many cases this means that plants are available for reintroduction into their former range if local conditions are favourable. Since individual, rooted plants are immobile, it is relatively easy to survey prospective replanting sites and monitor the progress of plants after reintroduction. However, unlike animals, plants are restricted to the researcher's exact choice of site; if that proves unsuitable, they cannot relocate to a more suitable area. On the other hand, the unpredictable nature of animals and their dependence on learning are complications (especially during reintroduction) that botanists are spared.

Having prepared a site for reintroduction it is desirable to conduct field trials to test the respective merits of various propagules, e.g. seeds, seedlings or adult plants. In some cases seeds might seem a natural choice but may be more vulnerable to predation and competition. Alternatively, seedlings may have a better survival rate once planted but prove more difficult to transport and handle and must withstand the shock of transplanting. Also, seedlings may require a period of acclimatization prior to planting out. Maunder (1992) points out the very real danger of introducing pathogens from nursery stock into wild populations. He cites the case of *Tecomanthe speciosa*, from New Zealand. With just one known plant surviving in the wild, it would be ironic to kill it with a pathogen carried by a reintroduced specimen destined to support its survival. Legislation to protect wild species makes it increasingly difficult to move plants across international borders. By using *in vitro* micropropagated material, this problem is partly overcome since the plants are sterile.

The legume *Sophora toromiro*, from Easter Island, was first described in 1917 from a single remaining specimen. This plant died in the 1960s leaving just a few scattered survivors in various European and Chilean botanic gardens. Attempts are now being made to reintroduce the toromiro into its former range. Often, though, it is the habitat that has declined and restoration involving multiple reintroductions and restocking is more appropriate than individual species programmes. For example, the endemic Bermudan cedar (*Juniperus bermudiana*) suffered massive losses after an accidental introduction of two scale insect species. Attempts to re-establish the cedars on Nonsuch Island have included (where possible) the elimination of introduced flora and fauna, the importation of native plants from the main island and overseas botanic gardens, and the creation of additional habitats and ecological niches for wildlife. Scale-resistant variants and biological control methods have also been tried since the original ecotype is still vulnerable to the disease. Where a habitat has been radically altered, for example where marginal land has been turned to desert, there may be a case for a careful introduction programme outside the plant's normal range.

Zoos

Zoos are not a new phenomenon. In the twelfth century BC, the Chinese emperor Wen-Wang created a 600-hectare 'Garden of Intelligence' in which to house his animal collection. Menageries were also to be found in ancient Egypt, in the amphitheatres of Rome and at the court of Alexander the Great. Many of the royal families of Europe held private animal collections including the emperor Maximilian II of Austria and Louise XIV of France. But few of these collections opened their gates to the general public and fewer still embodied the aspirations of the Zoological Society of London. On 27 April 1828 this illustrious organization opened London's Regents Park Zoo to the public. The aim of its founders, including Sir Stamford Raffles and Sir Humphry Davy, was to develop a collection of animals dedicated to scientific research. This was an important step in the evolution of zoos (Box 15.1) and laid the foundation for future progress in animal welfare and conservation.

Zoos of the nineteenth century were in some ways, an extension of the Victorian passion for collecting. They reflected a 'stamp collection' mentality whereby enclosures often consisted of neat rows of small, bare cages. It was easy to see the occupants but little attention was paid to the animals' welfare. To be fair, the science of animal welfare was in its infancy and many years were to pass before the needs of the captive animals were to be fully appreciated. Carl Hagenbeck (1844–1913), an animal dealer by profession, did, however, have the vision to display his animals in naturalistic surroundings. His famous African Panorama at the Tierpark in Hamburg (1902) consisted of separate exhibits designed in layers to create the illusion of a single scene. In fact species such as zebra, antelope, ostrich and lion were separated by invisible moats and strategically placed rock formations (Ehrlinger, 1990). The effect on the zoo visitor must have been stunning but, more importantly, Hagenbeck established the concept of naturalistic enclosure design that was to re-emerge in the late twentieth century as part of the move towards better housing and presentation of captive species.

With a growing number of species finding themselves threatened, zoos are increasingly providing a sanctuary for many of them. Some animals, like Przewalski's horse (*Equus przewalskii*) and the scimitar-horned oryx (*Oryx dammah*), have no wild populations left. Many other species, e.g. the red wolf

Box 15.1 The evolution of zoos

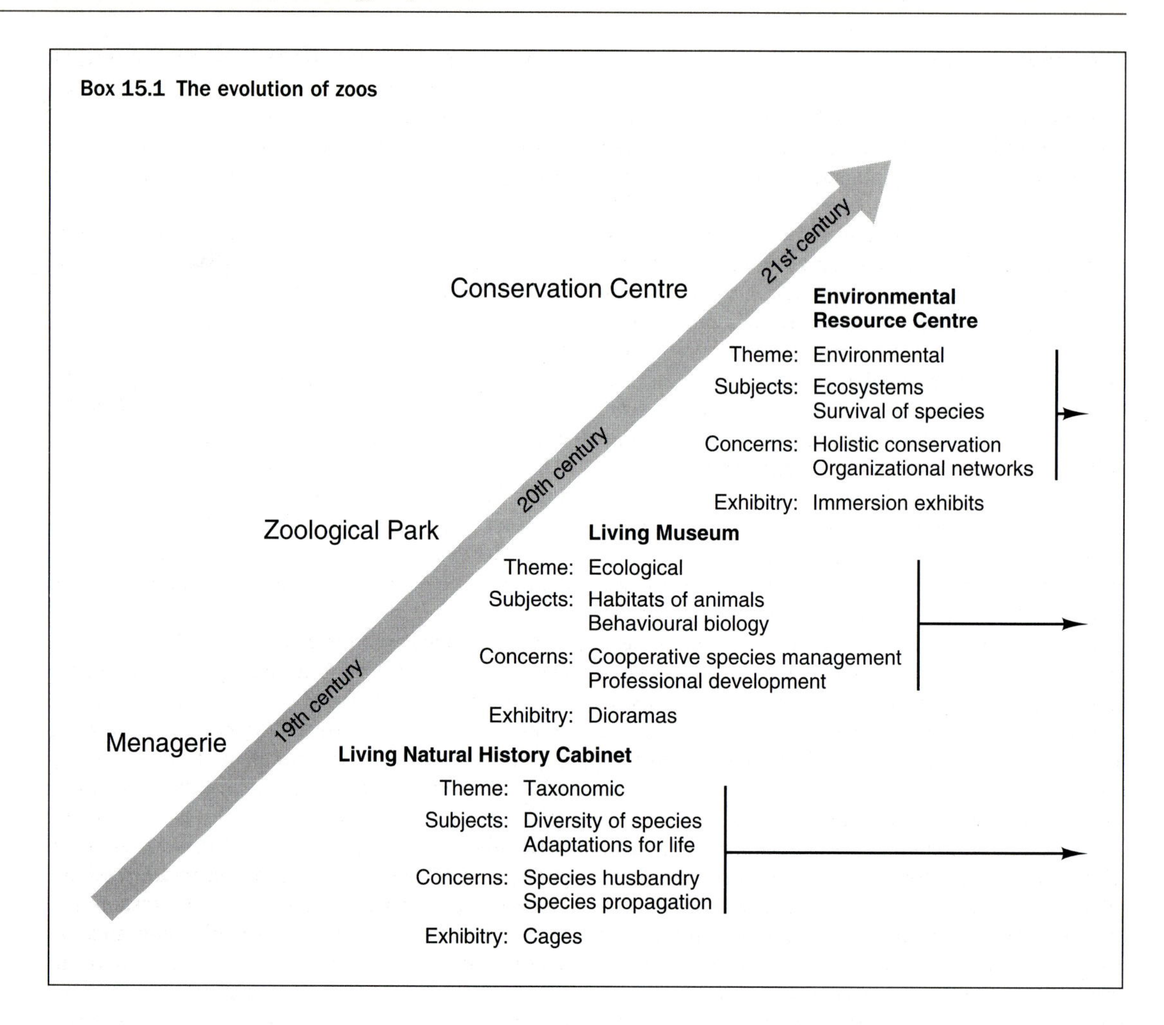

(*Canis rufus*), have been reduced to such small numbers in the wild that their only hope of survival is through intensive captive breeding programmes in zoos. At the same time, a better-informed public is asking important questions about the ethics of caging animals and the role of zoos in conservation. Zoo education programmes, once thought to be 'side shows', have consequently been drawn into the main arena of zoo conservation work to help provide some of the answers. The aims of the World Zoo Conservation Strategy (IUDZG/IUCN, 1993) herald a new thinking for zoos:

1 To identify the areas in which zoos and aquaria can make a contribution and determine how zoos can support and consolidate the processes leading to nature conservation and sustainable use of natural resources.
2 To develop understanding and support for the conservation potential of zoos and aquaria, from national, supranational and global authorities, as well as other social and political bodies and organizations.
3 To convince local zoo and aquarium authorities and conservation agencies that presently the greatest purpose to be served by the existence of these institutions is the contribution they can make to conservation, both directly and indirectly.
4 To assist zoos and aquaria in the formulation of policies wherein priorities relating to conservation are incorporated.
5 To indicate how contributions by the individual zoo and aquarium can be augmented by extending and intensifying contacts in the global zoo and aquarium network and other conservation networks.

Room in the ark

Of the 4000 species of mammal alive today, approximately 800 might need *ex situ* conservation in the near future if they are to survive (Tudge, 1992). Given that there are about 1000 zoos that are currently able and willing to participate in international breeding programmes, that makes approximately 500 000 available animal spaces, i.e. an average of 500 places per zoo multiplied by the number of zoos. If each breeding programme requires 250–500 individuals, then 500 000 spaces could house between 1000 and 2000 species (IUDZG/IUCN, 1993). However, these figures are little more than intelligent guesses, they only refer to mammals and they depend upon the full support of all the participating zoos in providing appropriate enclosure space. No allowance has been made here for subspecies or races, a point exemplified by tigers (*Panthera tigris*) that have five clearly defined subspecies (Indian, Siberian (Amur), South China, Indo-Chinese and Sumatran). But these figures do show that zoos can make a significant contribution to the protection of the world's threatened mammal species. Many zoos also play an important role in conserving birds, reptiles, amphibians and fish, as well as numerous invertebrate species. But they cannot save them all. Choosing species for conservation can prove a difficult task (IUDZG/IUCN, 1993). One particular problem relates to present classification systems which do not always reflect systematic differences relevant to the species' evolution. Therefore, in selecting the most appropriate phylogenetic level for conservation purposes, e.g. species, subspecies, race or population, the idea of an evolutionarily significant unit or ESU has been proposed (Ballou and Cooper, 1992). However, without additional funding from governments and Non Government Organizations (NGOs) many thousands of species will disappear forever. In the meantime zoos, in association with many other organizations, are implementing their own captive breeding programmes.

Education

According to the World Zoo Conservation Strategy, there are approximately 1000 zoos worldwide attracting in excess of 600 million visitors each year – in other words, 10% of the entire world population visit zoos annually (IUDZG/IUCN, 1993). Zoos, therefore, have the potential to be a major force in conservation education. This point is well illustrated in the strategy document which describes the present state of zoo education with numerous examples of working practices. For an account of a small zoo education programme see Woodhouse (1988). For a more general discussion on conservation education see Chapter 7 in this book.

The role of zoos has changed from simply exhibiting animals to breeding and reintroducing species to the wild. However, their image is still one of caged animals denied their freedom. Such a distorted view of zoos is damaging their credibility, yet many zoos have been slow to acknowledge this fact and to recognize the value of education in proclaiming the positive work that they are doing. The Federation of Zoological Gardens of Great Britain and Ireland (FZGGI) recently released a leaflet aimed at redressing the balance (FZGGI, 1994). Entitled *Education in Zoos*, it summarizes the role of education and provides a useful list of British zoos that are members of the Federation.

Education is, of course, far more than a public relations exercise. If species are to be saved, people need to know why animals are becoming endangered and how zoos can help them. They need to know what captive breeding and reintroduction programmes are, what research the zoo is undertaking and what successes it is having in saving species. To answer these questions it is necessary for education to pervade every aspect of zoo life, continually seeking to inform people and promote their understanding. But education is far more dynamic than this. It operates at the interface between the public and the work of zoos, drawing in information from around the world and relaying it to the zoo visitor. Animals are a unique living resource and zoo education programmes can really inspire individuals to take an active part in wildlife conservation. For example, 'The Ecosystem Survival Plan (ESP)', operating in some North American zoos, raises funds for specific conservation projects such as buying plots of tropical rainforest.

Zoo educationalists can be very progressive in their outlook and embrace conservation in its widest sense, recognizing the need for cooperation and *in situ* measures to balance their own zoo-based efforts. Robinson (1988), for example, has promoted the idea of 'bioparks' where zoos and botanic gardens show plants and animals in ecological systems. These biodisplays could provide a far greater understanding of the species involved. Robinson has also suggested that education could become a political force to bring about environmental change (Robinson, 1989). He argues that a better-informed public would bring greater pressure to bear on the governments of the richer nations to help those in poorer countries to protect their environment.

Conservation Assessment and Management Plans (CAMPs)

The Species Survival Commission (SSC) of the World Conservation Union (IUCN) oversees a number of Taxonomic Advisory Groups (TAGS) plus specialist groups such as a Reintroduction Specialist Group and a Captive Breeding Specialist Group (CBSG). These groups work together to identify animals in need of priority action and look at ways of protecting them. Together, they formulate Conservation and Assessment Management Plans (CAMPs) for threatened species. CAMPs apply to wild and captive populations and indicate where captive breeding would be desirable. The production of CAMPs, for a given group of animals, gives rise to Global Action Plans (GAPs) which seek to implement the recommendations made in the CAMPs. A coordinator is nominated to oversee individual GAPs and advise zoos on the implementation of their (*ex situ*) part in the plan (called Global Captive Action Plans or GCAPs). An important development in this field is the idea of integrating research proposals with the CAMPs process, thus directing research to priority areas.

Captive breeding programmes

Captive breeding programmes (Box 15.2) form part of the GCAPs strategy and are coordinated through the International Union of Directors of Zoological Gardens (IUDZG) and the IUCN. By implementing the GCAPs, IUDZG and IUCN aim to protect the genetic variability of captive populations in the hope that they can be used to restock the wild at some later date. In the case of the scimitar-horned oryx (*Oryx dammah*) individuals have already been returned to the wild (see box). In order to utilize the full gene pool, individuals of a given species, held in zoos around the world, are managed as a single population. Using information from various sources, including the International Species Information System (ISIS), the International Zoo Year Book (IZY) and individual zoos, a studbook is compiled for a selected species (Box 15.3). The studbook contains information on individual members of that species that have ever been held in captivity including their births, deaths, matings, sex, and parentage. Using the studbook, the species coordinator (who may also be the studbook keeper) can identify those animals that are genetically over-represented in the zoo population and remove them from the breeding programme. Conversely, animals whose genes are under-represented in the population are encouraged to breed. However, during breeding some genetic loss is inevitable and, in practice, zoos hope to conserve 90% of the available genetic variability of a species for 100–200 years. After this time scientists hope that advanced technologies will offer new solutions for captive management of species. For example, cryopreservation of eggs and sperm should allow the indefinite 'cold storage' of some species without loss of genetic material and without the cost of maintaining the live animals. The present state of captive breeding is reviewed in ZSL (1991).

Genetics of small populations

Small and declining populations of animals may suffer from a number of genetic disorders, resulting from a lack of breeding opportunities and leading to inbreeding depression (see Chapter 10). For example, the present wild population of African cheetahs (*Acinonyx jubatus*) are all genetically similar. This fact, in association with historical records, suggests a recent dramatic decline in numbers that has forced the cheetahs into a genetic bottleneck (Shafer, 1990). With loss of vigour and fertility problems, wild cheetahs may already be on the road to extinction (but see comments on cheetahs in Chapter 10). Not only are zoo populations vulnerable to bottlenecks, they can also experience genetic drift, for example where animals suited to captive conditions survive (and become increasingly domesticated) whilst those less suited to captivity die. The result might be a subtle change in the gene pool that could be detrimental if the animals were introduced into the wild.

To ensure a broad genetic base to the captive population, founder animals, i.e. those that begin the breeding programme, should ideally consist of as many unrelated individuals as possible. Whilst six individuals could in theory encompass 90% of the genetic variation in a population, this figure does not allow for the innumerable difficulties that could beset individual animals, e.g. incompatibility, injury and disease (Tudge, 1992). In the case of Przewalski's horses there were only 13 founding animals and one of those was a domestic mare that had interbred with the wild horses to produce cross-bred offspring.

Box 15.2 Establishing a captive breeding programme

The figure shows the chronology of establishing a conservation programme for a species such as the scimitar-horned oryx (*Oryx dammah*). The timing of various kinds of management is shown beneath the graph.

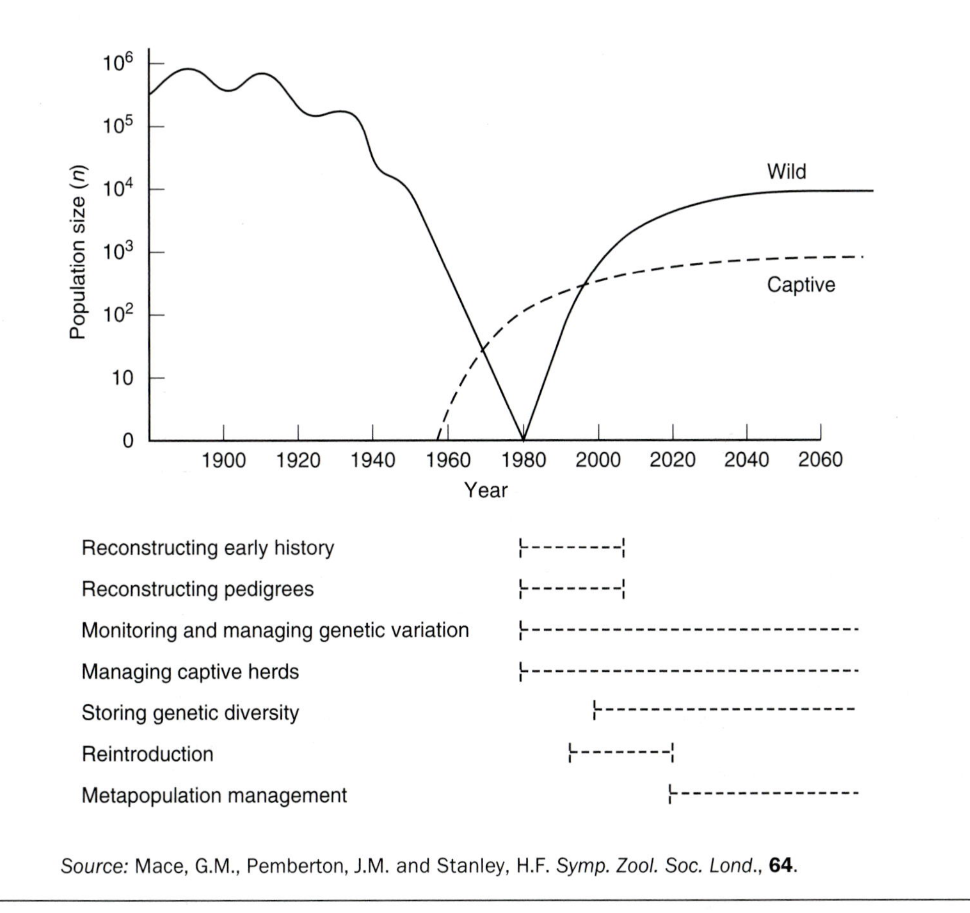

Source: Mace, G.M., Pemberton, J.M. and Stanley, H.F. *Symp. Zool. Soc. Lond.*, **64**.

Having established a founder group, the population must be quickly increased to its minimum viable (effective) size; for mammals this is usually between 250 and 500 animals (IUDZG/IUCN, 1993). The effective population size for Siberian tigers (*Panthera tigris altaica*), with a generation time of 7 years, is 136 animals. The Caribbean flamingo (*Phoenicopterus ruber ruber*), whose generation time is 26 years, requires only 37 animals to maintain 90% of the population's genetic diversity over a period of 200 years.

The sex ratio should be maintained at 1 : 1. Where male animals hold a harem of several females, e.g. in Chapman's zebra (*Equus burchelli chapmani*), this ratio may seem unnatural. However, to maintain the total genetic variability of a population, each individual animal must make an equal contribution. If a single male is allowed to breed more frequently, then the genetic contribution of the non-breeding males would be lost.

Managing numbers

Once the effective population size is reached, various techniques are employed to delay further reproduction since each successive mating inevitably loses a small amount of genetic material. This is because the chromosomes in unfertilized eggs and unused sperm, which may carry some unique genes, are lost to the gene pool. Many of the genes will occur on the

Box 15.3 Studbook for scimitar-horned oryx

A section from the studbook of the scimitar-horned oryx (*Oryx dammah*). Note that the animals shown are all captive-bred and are exchanged between zoos for breeding purposes. Those whose final location is Tunisia have been returned to the wild.

SCIMITAR-HORNED ORYX Studbook
(Oryx dammah)

Stud#	Sex	Birth Date	Sire	Dam	Location	Date	Local ID	Event	Birth-Origin
333	M	7 Jun 1985	197	112	MARWELL	7 Jun 1985	373	Birth	Captive Born
					MCALPINE	13 Nov 1986	UNK	Transfer	
					MARWELL	4 Oct 1988	373	Transfer	
						14 Jun 1990		Death	
334	F	7 Jun 1985	197	108	MARWELL	7 Jun 1985	372	Birth	Captive Born
					PERTH	15 Feb 1986	UNK	Transfer	
335	M	13 Jun 1985	197	183	MARWELL	13 Jun 1985	378	Birth	Captive Born
					TUNISIA	9 Dec 1985	1	Transfer	
336	F	14 Jun 1985	197	175	MARWELL	14 Jun 1985	375	Birth	Captive Born
					TUNISIA	9 Dec 1985	3	Transfer	
337	M	15 Jun 1985	228	248	EDINBURGH	15 Jun 1985	856B11	Birth	Captive Born
					MARWELL	11 Nov 1985	UNK	Transfer	
					TUNISIA	9 Dec 1985	8	Transfer	

Source: Compiled by Simon Wakefield at Marwell Zoological Park.

chromosomes of other members of the group, but this is not always the case, especially in small populations. Consequently the offspring have less genetic variation than their parents. To reduce matings, and therefore decrease the rate of genetic loss, the generation time can be extended by separating breeding pairs. If this procedure disrupts the group's behaviour, the use of birth control methods, such as a slow release contraceptive 'pill' inserted under the skin, might be considered. To achieve the right balance, in terms of population size, age and sex ratios, it is sometimes necessary to euthanase animals (humanely destroy them). No one wants to see their precious animals destroyed, but superfluous animals use up valuable resources, such as space, time and money, which may be better spent on other endangered species.

Many of the animals required for breeding programmes are already in zoos. Where species are taken from the wild, the practice is strictly enforced by international agreements. In fact the whole concept of breeding programmes is international. It is neither desirable nor likely that an individual zoo will house the total population of an endangered species. Instead, zoos maintain their own small populations of animals and exchange individuals with other zoos around the world in accordance with the wishes of the species coordinator (Tudge, 1988).

Biotechnology

Breeding programmes are not without their difficulties, e.g. stress induced in animals whilst in transit and the incompatibility of mating partners. Continuous advances in animal husbandry are helping to resolve some of these problems but recent developments in biotechnology promise far greater rewards for the breeder and species alike. For example, where males are incompatible with their female partners, artificial insemination can be used to impregnate the female. *In vitro* fertilization techniques, more commonly associated with human infertility problems, are now being tested on other mammals and may ultimately prove more reliable than insemination. Intraspecific embryo transplants, i.e. within the same species, have been successfully performed with species such as eland (*Taurotragus oryx*) and Przewalski's horse

(*Equus przewalskii*). This opens up the possibility of rescuing embryos from sick or injured animals in the wild and rearing them (*ex situ*) in surrogate females. This technique also allows ovules to be collected from females with a rare genotype and transplanted into less valuable surrogate mothers who then rear the offspring with the valuable genes. Interspecific embryo transplants, i.e. between species, have also been successfully demonstrated using, for example, a bongo (*Tragelathus eurycercus*), and an eland as recipient. Genome Resource Banking (GRB) is the term used to describe the collection, storage and application of biological material including gametes, embryos, tissues and blood. For a good review of the present situation see ZSL (1992).

Biotechnology has many other important implications for zoo animal breeders. For example, genetic fingerprinting enables breeders to identify individual animals and develop breeding programmes that avoid unnecessary inbreeding (Ashworth and Parkin, 1992). Reproductive technology enables breeders to rapidly increase the number of animals whilst giving greater control over the sex ratio. Generation times can be extended by storing fertilized eggs instead of continually having to rear offspring. In future, the routine movement of sperm, eggs and embryos across international borders will reduce the need to move animals. This will greatly decrease transport costs, stress to the animals and problems of quarantine (since the eggs and gametes are transported in sterile containers). A herd of zebra could pass through customs in a container carried in someone's pocket! Cryopreservation (freezing tissues) raises the intriguing prospect of a 'frozen zoo' where endangered species are maintained in test tubes as opposed to enclosures. Imagine rows of eggs, embryos and sperm, all neatly stored in liquid nitrogen at a temperature of −196°C where they would remain in suspended animation until the world is ready to receive them again. To some this is scientific scaremongering, to others it is the only way to ensure the survival of thousands of endangered species over extended periods of time, perhaps 500–1000 years.

Reintroduction programmes

There are many factors to take into account when reintroducing animals into the wild including:

1. The need for a viable captive population that can provide candidates for reintroduction.
2. The removal of threats in the wild and restoration of habitat where necessary.
3. Training and acclimatization of candidates.
4. Political goodwill and commitment.
5. Adequate resources including equipment, skills and finance to cover, for example, preliminary research, medication, transport, monitoring, education programmes and salaries.
6. The impact that the reintroduced species will have on local flora and fauna.

(See also the section on plant reintroductions earlier in this chapter.)

Even a well-planned reintroduction programme can have its problems. The Zoological Society of London's symposium *Beyond Captive Breeding* (ZSL, 1991) provides an excellent summary of the problems together with case studies for scimitar-horned oryx (*Oryx dammah*), which has been reintroduced into Tunisia, and the red wolf (*Canis rufus*), which has been reintroduced into North America. Lieberman *et al.* (1993) have discussed the reintroduction of Andean condors (*Vulture gryphus*) into Colombia, and Ounsted (1991) has considered some of the problems involved in reintroducing migratory whooping cranes (*Grus americana*) and bean geese (*Anser fabalis*) into the wild.

Golden lion tamarins (*Leontopithicus rosalia*) provide one of the classic examples of a species that has undergone a captive breeding and reintroduction programme. In the early 1970s, there were only a few hundred golden lion tamarins left in the wild and their habitat, the Atlantic coastal forests of Brazil, was under increasing pressures from an expanding human population (Tudge, 1992). In a final bid to save the species from extinction a collaborative conservation programme was established between the United States National Zoological Park/Smithsonian Institution, the Instituto Brasileiro Desenvolvimiento Florestal (IBDF) and the Centro de Primatologia do Rio de Janeiro (CPRJ). With about 80 tamarins in captivity at that time, the international studbook keeper, Devra Kleiman, set about establishing standardized husbandry procedures that all zoos could follow. The aim was to improve the survival rate of captive golden lion tamarins and increase the size of the zoo population. Specific zoos were invited to take part in the breeding programme and to manage their tamarins as directed by an international management committee.

As the number of captive animals increased, so did the possibility of reintroducing them into the wild. Attention focused on the golden lion tamarins still living wild in the Poço das Antes Biological Reserve, near Rio de Janeiro. This 5000-hectare (12 000-acre) site was later chosen as the centre for an ambitious reintroduction programme. At that time, tamarins were still being taken from the wild as pets and their habitat was declining due to human encroachment and deforestation (Mallinson, 1989). The reserve also suffered heavy loss of vegetation from uncontrolled forest fires. So, whilst zoos were concentrating on their breeding programmes, other workers were studying the wild tamarins and looking at ways of protecting and restoring their natural habitat.

During May and July 1984 fourteen captive-bred tamarins were released into the wild. By June the following year eleven were dead or had been rescued. Most of the animals succumbed to disease, though exposure, snakebite and hunger contributed to some of the deaths (Tudge, 1992). This salutary lesson highlighted the need to provide the tamarins with some kind of training before release and adequate support after release. Zoo bred tamarins, therefore, had to be taught how to forage effectively, recognize and deal with unprocessed foods (whole fruits, leaves, live insects, etc.) and climb on flimsy branches similar to those encountered in the wild. Water was provided in bromeliads rather than pools on the ground where the tamarins are more vulnerable to predators.

Prior to release, each tamarin was given a full medical examination. This procedure helped to eliminate weaker animals and identify those that might otherwise carry infectious diseases to the wild population. For a limited time after their release, the tamarins were provided with food to give them time to acclimatize to their new surroundings (this is called a 'soft' release). The post-release period also involved intensive monitoring of all the tamarins, some of whom wore radio collars to help researchers locate them in the dense vegetation. An education programme was established that took the conservation message to local schools and businesses. As the programme developed, the tamarins became celebrities, acquiring the status of a 'flagship' species with the power to promote conservation in the national and international arenas of the world.

Between 1984 and 1991, a total of 91 golden lion tamarins were released into the wild, of which 33 survived. They produced 57 offspring, of which 38 survived. The losses of adults and juveniles were attributed to a number of factors, including theft by humans, predators, starvation, disease and exposure. By 1991 the wild tamarin population had increased by 71 animals as a direct result of the reintroduction programme. The estimated cost was $22 000 per surviving tamarin (Kleiman *et al.*, 1991).

Since the captive-born tamarins were released into areas where wild tamarins still existed, the project could more accurately be described as a restocking programme. But reintroduction programmes have implications far beyond the re-establishment of single species in the wild. By conserving the habitat for tamarins, the monkeys are acting as an 'umbrella' species, conferring protection on many thousands of other plants and animals.

Behaviour enrichment

The natural behaviour of species in the wild generally ensures their physical and mental well-being, exactly the situation one would like to see in captive animals. Until recently zoos argued that animals were better off in captivity because they were protected from bad weather, disease and the threat of attack. However, for those captive animals destined for reintroduction an appreciation of the wild is essential, since fear and hunger promote behaviour responses that are essential for the animals' survival. Zoos must therefore try to strike a balance between the behavioural needs of captive animals and realistic management regimes. Sadly, for many animals, the scales are tipped in favour of the latter. Animals in captivity tend to lead predictable lives with routine feeding and cleaning times whilst their enclosures are often less complex (therefore less challenging) than the animal's natural habitat (Shepherdson, 1988). Poor enclosure design, lack of appropriate stimulation and social deprivation are some of the factors that can lead to boredom and frustration. In extreme cases animals can exhibit stereotypic behaviour which is characterized by the animal performing the same activity repeatedly, often leading to physical and mental stress (Zoo Check, 1994). Examples include neck stretching in giraffes, pacing in big cats and head swaying in elephants. Parrots are prone to feather plucking and chimpanzees and gorillas have been recorded eating their own vomit and faeces. Large carnivores, such as polar bears (*Ursus maritimus*), seem particularly prone to stereotypic behaviour (Ames, 1993).

Behavioural enrichment is one way in which zoos are seeking to counteract abnormal behaviour patterns and improve the lives of their animals. For example many studies on wild mountain gorillas (*Gorilla gorilla beringei*) have shown that their diet consists of over 50 different plant species. Jersey Zoo

(UK) is using this information to provide a more natural diet for their gorillas, one that promotes interest in the food and extends the search and reward time. A number of methods have been employed at different zoos to enrich the lives of captive tigers (*Panthera tigris*) including enlarging their enclosure, providing more cover and resting platforms, keeping several tigers together to promote social contact and providing tree trunks and tyres for the animals to play on. Tigers also enjoy playing in water and, in Copenhagen Zoo, they were given live fish in their pool to encourage hunting behaviour. Increasingly, the public want to see animals in naturalistic surroundings. This concept has been fully developed by San Diego Zoo where their penguins are housed in a replica of an Antarctic ice field, with real ice and the water at a cool 7°C. The gorilla house at Port Lympne Zoo (UK) takes a different view of enclosure design (Figure 15.3). The Universities Federation for Animal Welfare (UFAW) has released an interesting video (UFAW, 1990) that discusses behaviour enrichment and shows a variety of behaviour enrichment projects.

Fig. 15.3 The gorilla house at Port Lympne Zoo, Kent, UK. This spacious 'tree house' design will please enthusiasts of modern zoo architecture but will it satisfy the behavioural needs of the animals which occupy it? (Photograph: D. Worley.)

The 'mega-zoo'

Habitats, and therefore some animal populations, are becoming increasingly small and fragmented. For example, in 1990 there were less than 1000 Sumatran rhinos (*Dicerorhinus sumatrensis*) living in scattered and isolated groups in the wild (Neesham, 1990). These mini-populations are extremely vulnerable to inbreeding, environmental catastrophe and human disruption (see Chapter 13). To promote conservation their gene pools can be linked by 'green corridors' (Spellerberg and Gaywood, 1993) or the translocation of animals from one area to another. But for many species, these approaches are either impractical, too costly or too late. One alternative is to protect the species in captivity. However, this approach is not without its critics (Travers, 1993).

The Javan rhino (*Rhinoceros sondaicus*) has been reduced to a single population of about 70 animals (1990 figure). They live in the Ujong Kulong National Park and appear to be perfect candidates for *ex situ* conservation (Neesham, 1990). Yet some scientists question the wisdom of removing animals from an already stressed population, in the hope that they *may* be able to establish a captive breeding group elsewhere. Others fear that once a species is established in captivity, the habitat loses its significance and can be further exploited. Reintroducing animals also has its dangers since captive animals may transmit potentially lethal diseases to the wild population.

Many of the major international organizations including the World Conservation Union (IUCN), the Wildfowl and Wetlands Trust, the World Wide Fund for Nature (WWF) and the United Nations Conference on Environment and Development (UNCED) have expressed their support for *ex situ* conservation measures. However, they all stress that the two approaches to conservation (*in situ* and *ex situ*) are not independent, they are opposite ends of the same spectrum. For many species, a combination of both strategies will be the most effective way of ensuring their survival. In the case of the Javan rhino, their habitat is already a managed system dependent upon government intervention to protect them from poachers, farmers, tourists and catastrophe. Further proposals put forward to protect the rhinos have included translocation of some animals to a second reserve and the establishment of two captive breeding populations. Because of the small number of animals involved, it has been suggested that the gene pools of the wild and captive populations should be managed as a single unit. This type of animal management is referred to as a 'mega-zoo' and incorporates a total species management concept (Olney *et al.*, 1994). In the future, wildlife reserves may become 'mega-zoos' whose species will require management procedures similar to their captive counterparts.

The future for *ex situ* conservation

Plants in botanic gardens are, in effect, captive populations and they face many of the problems experienced by captive animal populations including genetic drift, bottlenecks and inbreeding. In addressing these problems the importance of studbooks, proper management procedures and cooperation between breeding centres cannot be overstated. Neither can the role of education and research in supporting the main conservation aims of zoos and botanic gardens. Considering the present rate of habitat loss, it is unlikely that additional wild areas will become available for reintroduction programmes, on a large scale, for 300–500 years. Therefore captive breeding programmes (for plants and animals) must be designed to accommodate such enormous time scales.

Where *ex situ* conservation is necessary, it should be seen as supporting *in situ* efforts. In part, this is because it is very difficult to protect the whole of a population's gene pool in *ex situ* facilities. Parasites, predators and climate may change and present insurmountable difficulties to species that have bred for many generations in captivity prior to their release into the wild. Young animals with complex learning behaviour, e.g. social primates, big cats and elephants, require experienced adults to teach them and captive parents may not have that experience. Consequently, protecting species in the wild must remain the primary aim of conservationists.

The role of zoos and botanic gardens in the twenty-first century will become increasingly important as the number of threatened species rises. The idea of conserving isolated species in captivity will gradually yield to a more integrated approach that recognizes the relationships that exist between species and their place within the ecosystem. Enclosure design, research and education programmes will reflect this development and promote a far greater understanding of conservation biology; species will be treated as interdependent rather than independent as they often are today. Gene pools will increasingly be managed as 'lakes' that incorporate wild and captive populations and the concept of a mega-zoo will broaden to include botanic gardens. Cooperation, partnership and, in some cases, fusion will occur where plant and animal conservation is promoted under one roof. These 'Bio-Parks', 'Eco-Gardens' and 'Life Centres' will combine the needs of conservation and captive breeding with 'state of the art' educational technology. Going to Zoo 2050 will be more like visiting animals in the wild except that for many species there will be no wild.

But there is a real danger in simulated experiences and, as the 'frozen zoo' and the 'cryogenic gene bank' move ever closer, their very existence may pose an additional threat to wildlife. After all, if species can be saved *ex situ* and in their entirety, why worry about saving wild populations? Colin Tudge, in his book *Last Animals at the Zoo* (1992), discusses this point and finally concludes that species should be saved, not because of their value to humans or even their role in maintaining a healthy ecosystem, but because it is the right thing for human beings to do.

'There is still time to save species and their ecosystems. It is an indispensable prerequisite for sustainable development. Our failure to do so will not be forgiven by future generations' (WCED, 1987).

Conclusion

Zoos and botanic gardens spearhead measures to conserve endangered species in *ex situ* facilities by developing strategies for captive breeding programmes, education, research and fundraising. However, their success in sustaining populations of threatened species depends, to a large extent, upon maintaining the genetic diversity of the species in their care. This in turn is determined by the number of 'founder' members in the breeding population and the amount of genetic variability they possess.

Advances in biotechnology have provided a variety of important techniques which facilitate the collection, transport, storage and utilization of germ plasm (e.g. sperm, eggs, pollen and embryos) which can dramatically improve a species' chances of survival. As their numbers grow, captive-bred plants and animals can be used to restock dwindling wild populations and in some cases may be used to re-establish extirpated wild populations. Important though these projects are, the number of species in need of *ex situ* conservation measures far exceeds those that might benefit from reintroduction programmes. For many species it may be hundreds of years before suitable habitat is available for their release into the wild.

Consequently, zoos and botanic gardens will become increasingly important repositories of the world's genetic resources. This responsibility will be combined with a more active role in *in situ* conservation measures and a far greater degree of cooperation between establishments and organizations. However, in their struggle to protect species, zoos and botanic gardens must be sensitive to the criticisms levelled against them. Conversely, those that condemn *ex situ* conservation must face the reality of dealing with an increasing number of species that simply cannot survive in the wild without *ex situ* intervention.

References

Ames, A. (1993) Environmental enrichment for captive polar bears. *Biologist*, **40**(3), 130–1.

Ashworth, D. and D.T. Parkin (1992) Captive breeding: can genetic fingerprinting help? In H.D.M. Moore, W.V. Holt and G.M. Mace, (eds) *Biotechnology and the conservation of genetic diversity*. Symposia of the Zoological Society of London, No. 64. Oxford: Clarendon Press, pp. 135–47.

Ballou, J.D. and K.A. Cooper (1992) Genetic management strategies for endangered captive populations: the role of genetic and reproductive technology. In H.D.M. Moore, W.V. Holt and G.M. Mace, *Biotechnology and the conservation of genetic diversity*. Symposia of the Zoological Society of London, No. 64. Oxford: Clarendon Press, pp. 183–206.

Bourne, W.R.P., M. de L. Brooke, G.S. Clark and T. Stone (1992) Wildlife conservation problems in the Juan Fernández Archipelago, Chile. *Oryx*, **26**(1), 43–51.

Box, H.O. (1991) Training for life after release: simian primates as examples. In J.H.W. Gipps (ed.) *Beyond captive breeding*. Symposia of the Zoological Society of London, No. 62. Oxford: Clarendon Press, pp. 111–21.

Cunningham, A.B. and F.T. Mbenkum (1993) *Sustainability of harvesting* Prunus africana *bark in Camaroon*. Paris: UNESCO.

Ehrlinger, D. (1990) The Hagenbeck legacy. *International Zoo Yearbook*, **29**, 6–10.

Fay, M. (1992) Conservation of rare and endangered plants using *in vitro* methods. *In vitro Cell Dev. Biol.*, **28**, 1–4.

FZGGI (1994) *Education in zoos*. London: The Federation of Zoological Gardens of Great Britain and Ireland.

Gerlach, J. (1993) Invasive Melastomataceae in Seychelles. *Oryx*, **27**(1), 22–26.

Given, D.R. and W. Harris (1994) *Techniques and methods of ethnobotany as an aid to the study, evaluation, conservation and sustainable use of biodiversity*. London: The Commonwealth Secretariat.

Groombridge, B. (ed.) (1992) *Global biodiversity: Status of the Earth's living resources*. WCMC (World Conservation Monitoring Centre). London: Chapman & Hall.

Gunn, S. (1991) Banking for the future. *Kew*, spring 1991/1, 16–21.

Hoyt, E. (1988) *Conserving the wild relatives of crops*. Rome: IBPGR.

IUCN (1987) *The IUCN position statement on translocation of living organisms*. Gland: IUCN.

IUCN (1989) *The Botanic Garden Conservation Strategy*. Gland: IUCN.

IUCN (1993) *1994 IUCN Red List of Threatened Animals*. Gland: IUCN.

IUCN, UNEP, WWF (1980) *The World Conservation Strategy. Living resource conservation for sustainable development*. Gland: IUCN.

IUDZG/IUCN (1993) *The World Zoo Conservation Strategy*. Illinois: Chicago Zoological Society.

Kleiman, D., B. Beck, J. Dietz and L. Dietz (1991) Costs of a reintroduction and criteria for success: accounting and accountability in the golden lion tamarin conservation programme. In J.H.W. Gripps (ed.) *Beyond captive breeding*. Symposia of the Zoological Society of London, No. 62. Oxford: Clarendon Press, pp. 125–42.

Koopowitz, H. and H. Kaye (1990) *Plant extinction*. 2nd edn. Bromley: Christopher Helm.

Lieberman, A., J.V. Rodriguez, J.M. Paez and J. Wiley (1993) The reintroduction of the Andean condor into Colombia, South America: 1989–1991. *Oryx*, **27**(2), 83–90.

Mace, G.M. and R. Lande (1991) Assessing extinction threats: towards a reevaluation of IUCN threatened species categories. *Conservation Biology*, **5**(2), 148–57.

Mallinson, J. (1989) *Travels in search of endangered species*. Newton Abbot: David and Charles.

Maunder, M. (1992) Plant reintroduction: an overview. *Biodiversity and Conservation*, **1**, 51–61.

Neesham, C. (1990) All the world's a zoo. *New Scientist*, **127**(1730), 31–5.

Newbury, H.J. and B.V. Ford-Lloyd (1993) The use of RAPD for assessing variation in plants. *Plant Growth Regulation*, **12**, 43–51.

Olney, P., G. Mace and A. Feistner (1994) *Creative conservation*. London: Chapman & Hall.

Ounsted, M.L. (1991) Re-introducing birds: lessons to be learned from mammals. In J.H.W. Gipps (ed.) *Beyond captive breeding*. Symposia of the Zoological Society of London, No. 62. Oxford: Clarendon Press, pp. 75–84.

Rhoades, R.E. (1991) The world's food supply at risk. *National Geographic*, **179**(4), 74–105.

Robinson, M. (1988) Bioscience education through bioparks. *Bioscience*, **38**(9), 630–4.

Robinson. M. (1989) The zoo that is not: education for conservation. *Conservation Biology*, **3**(3), 213–15.

Sattaur, O. (1991) Rare horses ready to return to Mongolian home. *New Scientist*, **129**(1751), 26.

Seal, U.S., T. Foose, R.C. Lacy, W. Zimmermann, O. Ryder and F. Princee (1990) *Przewalski's horse global conservation plan.* Gland: IUCN.

Seitre, R. and J. Seitre (1992) Causes of land bird extinctions in French Polynesia. *Oryx*, **26**(10), 215–22.

Shafer, C.L. (1990) *Nature reserves: Island theory and conservation practice.* Washington and London: Smithsonian Institution Press.

Shepherdson, D. (1988) Environmental enrichment in the zoo. In Universities Federation for Animal Welfare, *Why zoos?* London: Ennisfield, pp. 45–54.

Spellerberg, I. and M.J. Gaywood (1993) *Linear features: linear habitats and wildlife corridors.* English Nature Research Report No. 60. Peterborough: English Nature.

Thorne, T. and B. Oakleaf (1991) Species rescue for captive breeding: black-footed ferret as an example. In J.H.W. Gipps (ed.) *Beyond captive breeding.* Symposia of the Zoological Society of London, No. 62. Oxford: Clarendon Press, pp 241–58.

Travers, W. (1993) *Prisoners of the mind.* Coldharbour: The Born Free Foundation.

Tudge, C. (1988) Breeding by numbers. *New Scientist*, **119**(1628), 68–71.

Tudge, C. (1992) *Last animals at the zoo.* Oxford: Oxford University Press.

UFAW (1990) *Environmental enrichment* (Video). Potters Bar: Universities Federation for Animal Welfare.

WCED (1987) *Our common future.* Oxford and New York: Oxford University Press.

Weeks, M. (1977) *The last wild horse.* Boston: Houghton Mifflin.

Woodhouse, H. (1988) Educational use of zoos. In Universities Federation for Animal Welfare, *Why zoos?* London: Ennisfield, pp. 37–44.

Zoo Check (1994) *Animal welfare.* Fact Sheet No. 2. Coldharbour: The Born Free Foundation.

ZSL (1991) *Beyond captive breeding.* Symposia of the Zoological Society of London, No. 62. Oxford: Clarendon Press.

ZSL (1992) *Biotechnology and the conservation of genetic diversity.* Symposia of the Zoological Society of London, No. 64. Oxford: Clarendon Press.

CHAPTER 16

Restoration and conservation gain

RICHARD PYWELL and PHILLIP PUTWAIN

Introduction

Opportunities to restore damaged parts of ecosystems to their original state continually arise. For example, the recent shift in European agricultural policies towards less intensive management and taking farmland out of production could mean that potentially large areas of countryside may become available for habitat restoration. Similarly, the move away from traditional heavy industries and the associated extraction of raw materials will continue to provide derelict and degraded sites which require the reconstruction of many components of the original ecosystem. Habitat restoration can be broadly defined as the reassembly or reconstruction of a degraded ecosystem to its original state and function. The terminology of habitat restoration strategies is both interchangeable and confusing. These can be divided into three types depending upon the degree of disturbance or degradation the site has previously undergone and the mitigation measures required to restore the fundamental ecosystem processes (Box 16.1).

If we are to successfully restore these communities we require a fundamental understanding of the factors (including edaphic and climatic factors) controlling the assembly and succession of these species into populations and communities. Only then can we hope to manipulate and accelerate these processes in order to achieve the desired end-point. Much can be learnt from the study of natural plant and animal colonization and the detailed study of species autecology. However, in the highly altered and fragmented landscapes of developed countries, restoration by the processes of dispersal and natural colonization has proved to be slow and unreliable. Furthermore, the extent of the changes in the chemical and physical characteristics of the substrata caused by human activities often imposes severe limitations on the establishment of these species. In such cases, it is often necessary to intervene to alleviate the limiting factors and speed the restoration process.

In this chapter we discuss the typical constraints on habitat restoration following a variety of different types of disturbance. These vary from the severe physical limitations on species assembly which result from quarrying and construction activities, the chemical constraints imposed by industrial wastes and the biological constraints encountered on former arable farmland. The relative cost-effectiveness of a range of techniques to overcome these constraints will be considered using case studies of grassland, heathland, wetland and woodland restoration schemes.

Design of habitat restoration schemes

The decision process in the design and implementation of habitat restoration schemes for conservation gain is summarized in Box 16.2. Consideration of these factors should ensure that the appropriate plant community is selected, established and maintained. In most cases the scheme must strike a balance between what is desirable to fulfil the objectives of habitat reconstruction and what is practical within the limitations presented by the site, substrate and resources available.

Selecting the appropriate species to restore on a site

If habitat restoration is to benefit conservation then it is important that species and communities

Box 16.1 Habitat restoration: definitions

1 *Habitat enhancement:* management or introductions which attempt to enhance the ecological potential and biodiversity of existing, but degraded or impoverished habitats where some of the desired species already occur.
2 *Habitat creation:* the establishment of plant communities *de novo*, usually on bare substrata where none or few of the desired species are present, or the reinstatement of a community that was formerly present (*re-creation*).
3 *Habitat translocation:* the transfer of the entire plant and animal communities from the donor site to a receptor site.

appropriate to a given site are reinstated. Several criteria can be used to select these, namely:

1 *Species biogeography.* The pattern of species distribution at the local and national scale can be used to determine which species should be restored at a given location. Similarly, information on the co-occurrence of species in phytosociological groupings or communities (Ellenberg, 1988; Rodwell, 1991, 1992) can be invaluable both in defining targets for restoration and in tracking its subsequent success.
2 *Species' ecological requirements and attributes.* These should be closely matched to the abiotic characteristics of the site, e.g. geology, soils, hydrology and climate (see Chapter 8). For example, soil maps have been used to identify areas of hydric soils which once supported wetland vegetation in Iowa, USA (Thompson, 1992). Such information is summarized in restoration texts (e.g. Schramm, 1978; Harker *et al.*, 1993) which contain lists of species suitable for the restoration of given habitats.
3 *Past and present land-use pattern.* Semi-natural communities adjacent to the disturbed area can be used to define a 'template' for restoration. Such sites may also act as potential sources of propagules for natural colonization. Similarly, details of former land use and management (e.g. Sheail, 1974) can also be invaluable in defining a suitable restoration 'goal' and can also be a good predictor of the likelihood of successful restoration.

Defining other objectives for habitat restoration

The choice of communities and habitat type should also take account of other important end-use objectives which may include:

- *protection*: of remaining semi-natural habitats by creating buffer zones between them and an adjacent, potentially damaging land use (e.g. intensive agriculture);
- *amenity and education*: to provide sites for the study of communities and ecosystems, and for public access, recreation and tourism;
- *aesthetic*: to improve and complement the visual quality of a landscape (colour, texture and form) whilst retaining any existing attractive features;
- *bioengineering*: rapid revegetation of unstable, riparian substrata (e.g. road verges and spoil heaps);
- *genetic resource*: maintaining and increasing the reservoir of genetic material which is essential for the long-term survival of species and populations.

Site evaluation

Site evaluation should aim to characterize the topographic, hydrological and ecological features of the site and how they relate to the proposed restoration scheme. It is important that the ecological survey should identify the type and range of plant communities which are still present at a site. Assessment should be made of the conservation value of the existing habitats in terms of rarity and biodiversity using established criteria (see Chapter 3). It is important to integrate any remnant semi-natural habitats into a restoration scheme, so that they are retained to provide foci for natural colonization of the newly created habitats. The survey should also attempt to characterize the potential physical, chemical and biological constraints on habitat regeneration or restoration at the site. Basic methodologies for the survey and evaluation of damaged and derelict sites are described in texts such as Bradshaw and Chadwick (1980) and Harker *et al.* (1993).

Every effort should be made to integrate the newly restored habitat in the cultural and socio-economic framework of local societies. Where possible, provision should be made for the use of the area for sustainable fishing, hunting and grazing, thus creating employment for local people. Finally, the scheme should conform to any legal and planning requirements. For example, in the US any modification of the hydrological regime of a site for wetland restoration must obtain the agreement of the US Army

Box 16.2 The steps involved in the design and implementation of ecological restoration

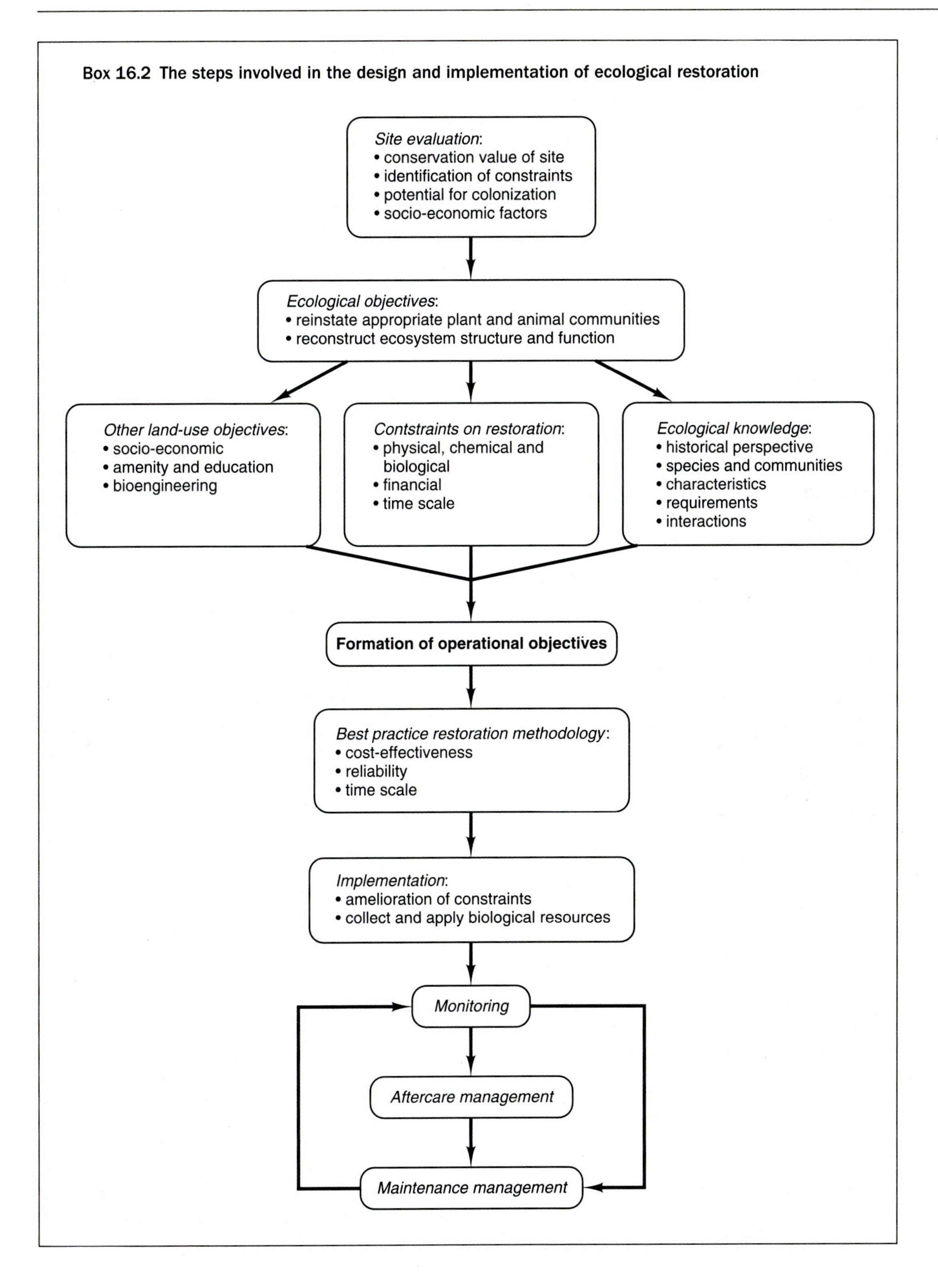

Corps of Engineers (ACE), and in the UK of the National Rivers Authority (NRA).

Targeting sites with potential for habitat restoration

Where there are many potential sites for habitat restoration and resources are limited, it is important to carefully target sites where there is a high probability of successful reinstatement of the desired community and the benefits to conservation are maximized. Spatially referenced data on soils, geology and land use can be successfully combined into a computerized Geographic Information System (GIS) to aid the location of sites with a high potential for habitat reconstruction.

Maps showing the historical distribution of lowland heath in Dorset, UK in 1811 and 1960 (see also Box 2.2) have been combined with current land-use data derived from satellite imagery in a GIS (Veitch *et al*., 1995). Interpretation of these data has allowed the ranking of different land-use categories for the relative ease of restoring heathland vegetation (Plate 6). For example, conifer plantations and improved grasslands which were heathland up to 1960 are considered to have a relatively high potential for successful heathland restoration because the physical structure and chemical characteristics of the original heathland soil are likely to be relatively unchanged, and they are likely to contain a relict heather seed bank. In contrast, heathland which is now arable land (both pre- and post-1960) and pre-1960 grasslands are considered to have a lower potential for restoration because the longer period of more intensive agricultural use is likely to have increased the soil fertility and destroyed any heathland seed bank.

Similarly, data on the spatial distribution and interconnectivity of the remnant heathland patches in Dorset together with their surrounding land-use types allows further site selection criteria to be used: for example, priority could be given to the restoration of heathland in areas adjacent to existing heathland with the aim of increasing the area of these patches and creating a protective buffer around them. Proximity to the existing heathland should enhance the rate of colonization of the newly restored areas by the heathland fauna. In addition, heathland should be reinstated in areas which link existing fragments. This may maximize the area of existing habitat and may reduce the chances of local extinctions of some invertebrates (Webb and Thomas, 1994).

Design and implementation of restoration strategy

Once the site has been located and the potential constraints on habitat restoration identified, it is possible to devise a strategy to achieve the desired end-point. The choice of restoration strategy should consider the following:

- the speed of attainment (how long to restore the desired habitat);
- reliability of attainment between different sites;
- economic cost (£/ha);
- resources of manpower and machinery;
- stability of restored plant communities and the long-term management requirements.

In order to successfully implement a habitat restoration strategy it is important to have good communication between the client and the contractors carrying out the work. Interested parties should be provided with clear, detailed instructions and sensitive operations should be supervised by a trained ecologist in order to achieve the desired end-point.

In the past, many habitat restoration schemes have been poorly monitored and managed. It is important that ecological end-points are defined together with finite time scales within which to achieve these objectives. Only by regular monitoring can we determine if the end-points have been achieved and if further remedial action is required.

Natural colonization

The study of plant communities which develop naturally on bare and derelict land and their colonization by animals can be important because of what it tells us about the processes controlling species assembly and succession. The critical step is the ability of species to arrive at a site, which depends on the proximity of a source habitat and adaptation of the species for dispersal. Propagules can also disperse through time in persistent seed banks which enable the species to avoid considerable disturbance. For example it has been shown that seed banks of some wetland species in North America can remain viable for up to 20 years following drainage (Weinhold and van der Valk, 1989). Colonists must also be adapted to the abiotic environment of a site in order to become established. The characteristics of a site may have changed considerably as a result of disturbance and dereliction. Where extreme environmental stress occurs, only species which have been able to evolve suitable adaptations to these stresses may colonize. Such multi-tiered selection processes can result in very distinctive plant communities containing rare and specially adapted species which are of considerable conservation interest in their own right. Examples of this are provided by the metal-tolerant plant communities found on mine wastes (Johnson *et al.*, 1978) and the calcicole flora of alkaline Leblanc wastes (Greenwood and Gemmell, 1978).

Following colonization, ecosystem development on bare and skeletal substrata can be a very slow process, for example taking between 30 and 70 years for woodland to develop on glacial moraines in Alaska, USA (Crocker and Major, 1955) and up to 120 years for the revegetation of china clay wastes in Cornwall, UK (Roberts *et al.*, 1981). Soil formation and the accumulation of nutrients are the most important processes. These involve allogenic processes, such as the weathering of soil minerals and inputs of nutrients from rainfall, which are unrelated to the developing ecosystem. Other changes in the substrate are autogenic and are due mainly to the activities of plants and soil fauna. These include changes in soil structure and texture, nutrient cycling, the fixation of atmospheric nitrogen and the accumulation of organic matter. For example, nitrogen has been shown to be a limiting factor on ecosystem development on china clay wastes (Roberts *et al.*, 1981).

Potential constraints on habitat restoration and their amelioration

The success of habitat restoration depends largely on the identification and amelioration of the physical, chemical and biological constraints on the germination and establishment of the desired plant species (Bradshaw and Chadwick, 1980; Table 16.1, Boxes 16.3–16.6). What is a severe constraint on the reconstruction of a given habitat on a site may only be a slight constraint for another type of habitat. For example, high residual soil fertility can be a severe constraint on restoring heathland vegetation which is associated with infertile substrata, but may have little effect on the restoration of woodland and scrub communities which occur on more fertile soils.

Biological resources for habitat restoration

In many cases the damage to semi-natural habitats has been so severe that it has resulted in the depletion or complete loss of the original flora and fauna. The availability of biological resources largely determines what will be practically feasible for restoration. There are many different sources of plant propagules for habitat restoration, each of which varies considerably in its cost and effectiveness (Table 16.2). It is also important to consider the genetic provenance of the material (see Chapter 10). The physical and chemical characteristics of both the propagule donor and recipient sites should be carefully matched so that the introduced species are well adapted and have adequate competitive ability (Hodgson, 1989). Where possible, the species composition and viability of any source of plant propagules should be tested. This can easily be achieved by simple germination trials.

Enhancing the ecological potential of existing habitats

Often semi-natural habitats are not totally destroyed, but are damaged and degraded by changing or neglected management practices. For example, the intensive use of herbicides and fertilizers has caused the floristic impoverishment of many arable fields and grasslands in the UK (e.g. Hopkins and Wainwright, 1989). Such sites often contain some of the desired species which are worthy of conservation. In these cases it is often more appropriate and cost-effective to apply techniques which will enhance the ecological potential and biodiversity of the existing communities rather than create new ones. Furthermore, the retention of the established flora and fauna has numerous 'bioengineering' and aesthetic advantages over techniques which seek to create habitats from bare ground.

Natural colonization by the desired plant species from seed bank or through dispersal can be facilitated or accelerated by restoration management which aims to create suitable niches for germination and recruitment within the existing vegetation. These can be achieved by different types of cultivation, carefully controlled livestock grazing and cutting or the application of herbicides. The timing of microsite creation should coincide with the optimum time for germination and establishment of the desired species.

Table 16.1 The physical, chemical and biological constraints on the restoration and enhancement of semi-natural habitats

Constraints:

Severe deficiency	Moderate deficiency	Slight deficiency	Adequate	Slight excess	Moderate excess	Severe excess
○○○	○○	○	◦	●	●●	●●●

Treatments: □ beneficial; ± in some cases; ■ harmful

Damaged land category or treatment	Physical				Chemical					Biological			
	Texture and structure	Stability	Water supply	Surface tempera-ture	pH	Nutrients	Ion exchange capacity	Toxic materials	Salinity	Propagules of desired species	Invasive species	Establish-ment	Soil flora and fauna
Colliery spoil	○○○	○○○/◦	○/◦	◦/●●●	○○○/◦	○○○/◦	○○○/◦	◦	◦/●●	○○○/○	◦/●	○○○/◦	○○○/◦
Strip mining	○○○/◦	○○○/◦	○○/◦	◦/●●●	○○○/◦	○○○/◦	○○○/◦	◦	◦/●●	○○○/◦	◦/●	○○○/◦	○○○/◦
Fly ash	○○/◦	◦	◦	◦/●●●	●/●●●	○○○	○○○	●●	◦/●●	○○○/○	◦	○○○/○	○○○/○
Oil shale	○○	○○○/◦	○○	◦	○○/◦	○○○	○○○/◦	◦	◦/●●	○○○/○	◦/●	○○○/◦	○○○/○
Iron ore mining	○○○/◦	○○/◦	○/◦	◦/●●	◦	○○	○○○/◦	◦	◦	○○○/◦	◦/●	○○○/◦	○○○/◦
Bauxite mining	○○/◦	◦	◦	◦	◦	○○	○○	◦	◦	○○○/○	◦/●	○○○/◦	○○○/○
Metalliferous mine spoil	○○○	○○○/◦	○○/◦	◦	○○○/●	○○○	○○○	●/●●●	◦/●●●	○○○/○	◦	○○○/○	○○○/○
Gold wastes	○○○	○○○	○	◦	○○○	○○○	○○○	◦	◦	○○○/○	◦/●	○○○/○	○○○/○
China clay wastes	○○○	○○	○○	◦	○	○○○	○○○	◦	◦	○○○/○	◦/●	○○○/○	○○○/◦
Acid rocks	○○○	◦	○○	◦	○	○○	○○○	◦	◦	○○○/◦	◦/●	○○○/◦	○○○/◦
Calcareous rocks	○○○	◦	○○	◦	●	○○○	○○○	◦	◦	○○○/◦	◦/●	○○○/◦	○○○/◦
Sand and gravel	○/◦	◦	◦	◦	○/◦	○/◦	○○○/◦	◦	◦	○○○/◦	◦/●	○○○/◦	○○○/◦
Coastal sands	○○/◦	○○○/◦	○/◦	◦	◦	○○○	○○○	◦	◦/●●	○○○/◦	◦/●	○○○/◦	○○○/◦
Land from sea	○○	◦	◦	◦	◦/●	○○	◦	◦	●●●	○○○/◦	◦/●●	○○○/◦	○/◦
Urban sites	○○○/◦	◦	◦	◦	◦	○○	○○/◦	○/●●	◦	○○○/◦	◦/●●	○○○/◦	○○/◦
Industrial sites	○○○/◦	◦	○○/◦	◦/●	○○○/●●●	○○	○○/◦	○/●●●	◦/●	○○○/◦	◦/●●	○○○/◦	○○○/◦
Roadsides	○○○/◦	○○○	○○/◦	◦	○○/◦	◦/●●	◦/●●	◦	◦/●●	○○/◦	◦/●●●	○○/◦	○/◦
Arable land	◦	◦	○/◦	◦	○○○/◦	◦/●●●	◦/●●●	◦	◦/●●	○○○/◦	●/●●●	○○/◦	○/◦
Improved grassland	◦	◦	◦	◦	○○/◦	◦/●●	◦/●●	◦	◦	○○/◦	●/●●●	○○/◦	○/◦
Plantation forestry	◦	◦	◦	◦	○/◦	◦/●	◦/●	◦	◦	○/◦	◦/●●	○/◦	○/◦

Treatments:													
Weathering	□±	■			□±	□±	□±	□±	□±				
Burial with subsoil/topsoil	□±	□±■	□±	□±	□±	□±	□±	□±			□±■		
Grading and demolition	□±	□±■	□±										
Cultivation (rip, plough, scarify)	□±	±■	□±■	□±							□	□±	□±
Compaction	□±■	□±	□±■										
Soil improver (organic matter, marl)	□		□±	□±	□±	□±	□±	□±					
Companion species	□	□	±■	□±		□±	□±				□±	□	
Stabilizer/mulch	□±	□	□±	□±							□±	□	
Polymer	□±		□										
Irrigation	±■		□	□		□±			□±■				
Drainage	±■	□±■	±■	±■									
pH increase (liming)					□±	□±■		□±					
pH reduction (sulphur, pyritic wastes, etc.)					□±	□±■							
Fertilizers						□±							
Add propagules (seed, topsoil, turf, plant)										□			□±
Selective herbicide											□	□±	
Cutting/grazing/burning											□±	□±	

Source: Modified after Bradshaw and Chadwick (1980).

Box 16.3 Potential constraints on habitat restoration

Physical constraints

- **topography**
  Steep slopes are often unstable and dangerous, and make access difficult.
- **soil structure**
  Compaction causes poor infiltration of water and air leading to seasonal waterlogging and drought, increased run-off and erosion, reduced root growth and microbial activity.
- **soil texture**
  Coarse-grained particles (>2 mm diameter) are often excessively drained and lack organic matter which provides nutrient exchange sites and water-holding capacity. Fine-grained particles (<0.2 mm diameter) are prone to compaction and surface capping.
- **stability**
  Unconsolidated substrata are liable to erosion by wind and water, and slippage. This is made worse by the lack of vegetation and steep topography of many mineral waste heaps.
- **temperature**
  Lack of vegetation cover means that dark, bare substrata can reach high temperatures in the summer.

Chemical constraints

- **extremes of pH**
  Alkaline substrata (pH > 8) have an excess of bases, such as calcium. Under these conditions phosphates, manganese and iron become unavailable to plants. Acid substrata (pH < 3) have a deficiency of bases, but aluminium and manganese become more soluble and are toxic. Aluminium and iron also combine with phosphate to form insoluble compounds. Microbial activity such as the mineralization of nitrogen is reduced. The cation exchange capacities of clay minerals are reduced which diminishes the ability of the soil to retain nutrients.
- **extremes of substrata fertility**
  There is usually a shortage of nutrients essential for plant growth in mineral wastes and subsoil. Lack of nitrogen is often critical in preventing the establishment of vegetation on such materials. Nutrients are often in excess in agricultural soils because of prolonged use of fertilizers. Transfer of these nutrients, as eroded soil particles, drainage water and atmospheric deposition, can pollute adjacent habitats. The progressive accumulation of nutrients (eutrophication) can lead to the dominance of more competitive plant species and an increased susceptibility to succession to scrub and woodland.
- **toxicity**
  Low concentrations of a few μg/g of heavy metals (e.g. lead and zinc) are toxic to plants. These toxins are produced by the weathering of metalliferous mine spoil, or atmospheric deposition and effluent from industrial processes. Such sites can remain toxic for many years.
- **salinity**
  High concentrations of salts (e.g. chlorides) are toxic to many plant species. Irrigation in arid regions can lead to an accumulation of salt in the soil through the process of evaporation. Also, the oxidation of pyrite found in some colliery spoil and peat produces sulphuric acid. Subsequent neutralization by carbonates may lead to a build-up of sulphate salts.

Biological contraints

- **propagules**
  The lack of the propagules of desired species is typical of virtually all degraded and damaged habitats, whereas an excess of propagules of ruderal and competitive species can be an acute problem of agricultural land, and urban and industrial sites.
- **invasion and competition**
  Problems of invasion and competition by undesirable plant species are likely to be greater on substrata with high fertility.
- **soil biota**
  Soil flora and fauna are depleted or absent from some degraded sites. These are responsible for key nutrient cycling processes: decomposition, mineralization, nitrogen fixation. Plant associations with mycorrhizal fungi increase the efficiency of nutrient capture and are therefore especially beneficial to species growing on impoverished soils. They may also present a physical barrier to toxic metals entering the plant.
- **excessive grazing, trampling, pests and disease**
  These can reduce plant fitness, inhibit vegetation establishment and increase erosion.

Box 16.4 Amelioration of physical constraints

- **burial with topsoil and subsoil**
  Burial of degraded substrata with topsoil, subsoil, or spoil with a texture and structure suitable for plant growth.
- **grading and shelter belts**
  Surface contouring and grading to reduce erosion and create variation in relief which is commensurate with the proposed habitat. Constructing physical shelter belts or planting biological ones to alleviate the effects of wind erosion and desiccation.
- **cultivation**
  Cultivation (e.g. ripping, ploughing, harrowing, rolling) to relieve compaction, improve drainage and aeration, and create a suitable seedbed. The effects of cultivation are usually only temporary, so it is important to establish vegetation soon after treatment of the substrate.
- **soil improvers**
  Incorporation of materials with a high organic matter content (e.g. composed domestic refuse and farmyard manure) to improve the structure of the soil and increase the nutrient and moisture retentive capacity of the substrata.
- **polymers**
  Polyacrylamide gels can absorb 100 to 400 times their own weight of water or nutrient solution. This is gradually released as the substrata dry, buffering plants against drought and lack of nutrients. Polymers can remain effective for 1 to 4 years in temperate regions.
- **chemical stabilizers**
  Stabilizers, such as Verdyol and Curasol AH, can be used for the short-term stabilization of bare and eroding mineral substrata. They may be used in hydraulic seeding mixtures in combination with rapidly growing companion species for stabilization of inaccessible slopes, but their effectiveness has been questioned.
- **mulches, meshes and mats**
  Surface mulches of coarsely chipped woody materials can reduce erosion and moisture loss, and suppress the growth of weed species. Meshes and mats of biodegradable materials, such as jute and reclaimed textile fibres, can be secured to slopes to stabilize the substrate.
- **companion species**
  Rapidly establishing companion species can stabilize eroding substrata and provide protected microsites for recruitment of the desired species. They can also improve the substrate characteristics by increasing infiltration and aeration, adding organic matter and, if leguminous species are sown, increasing the nitrogen content. Carefully chosen companion species can improve the visual impact of the site in the early stages of restoration.

Case study 1. The diversification of lowland, acidic grassland and heathland in Dorset, UK by turf removal

The loss and fragmentation of the lowland heaths of Dorset, UK have been largely due to the conversion to agriculture which continued up to the 1970s. Recent changes in agricultural policy offer opportunities to take land out of agricultural use and to restore heathland vegetation. In 1983 a series of replicated treatments were carried out to increase the frequency of heathland plant species present in abandoned pasture which had been heathland 20 years previously in Dorset (Smith *et al.*, 1991). Treatments included rotary cultivation, and cultivation with the addition of heathland topsoil and litter, and the removal of the top 30–50 mm of soil. After six years, turf removal was found to have significantly increased the rooted frequency of heathland species within the grassland (Figure 16.1). Subsequent investigation revealed a diminished seed bank of heathland species which had survived disturbances caused by agricultural use for 20 years.

Other management techniques, such as cutting, grazing and burning, can be used to direct succession and enhance the ecological value of the habitat (Bakker, 1989; Luken, 1990). For example, regular haymaking has been successfully used to restore fen meadows in The Netherlands (Bakker and Olff, 1995). and in the North American Midwest controlled burning of degraded prairie has been used to encourage the regeneration of prairie species from the seed bank (e.g. Holtz and Howell, 1983).

Case study 2. Ecological enhancement of existing, mesotrophic grasslands in Cambridgeshire, UK by slot-seeding and container-grown transplants

Where seed banks of the desired species are absent and there is unlikely to be significant wind dispersal of seed, it is possible to deliberately introduce seed into

Box 16.5 Amelioration of chemical constraints

- **adjustment of pH**
  pH can be raised to that required by the desired species by the incorporation of ground limestone or alkaline wastes and reduced using additions of sulphur, bracken litter or pyritic spoil.
- **lack of nutrients**
  This can be readily overcome by regular, carefully timed additions of the deficient nutrients in the form of inorganic or organic fertilizers. Organic fertilizers, such as sewage sludge, release nutrients more slowly and act as soil improvers, but their nutrient content is variable and they may contain toxins. Nitrogen applied as inorganic fertilizer is rapidly leached from skeletal substrata. Problems of long-term supply of nitrogen can be solved by sowing leguminous species, but these species require adequate phosphorus and moisture, and are not generally tolerant of very acid substrata.
- **excess nutrients**
  (a) maximizing nutrient losses from the soil by repeated cropping without addition of nitrogen fertilizer, or by grazing or burning;
  (b) the direct removal of nutrient pools by soil stripping and turf cutting;
  (c) manipulation of the stores and fluxes of nutrients within the soil by altering the pH or hydrology to reduce the availability of some nutrients; dilution of the nutrient pools by mixing subsoil or other inert materials with the topsoil; fallowing together with the regular destruction of the vegetation may accelerate nutrient losses by leaching (N), soil erosion (P, K) and the fixing of nutrients into insoluble and/or unavailable forms.
- **toxicity**
  The toxicity of some metalliferous spoils can be reduced by the incorporation of organic matter to form complexes with the metal, rendering it temporarily unavailable to the plant. However, the toxicity is likely to return following the decomposition of the organic material. A more permanent solution is burial with topsoil or subsoil, although this may cause the loss of species of conservation interest. Alternatively, naturally evolved metal-tolerant plant populations can be propagated and used to revegetate the site. Colonization may be accelerated by the amelioration of other physical and chemical constraints.

the existing vegetation after the deliberate creation of suitable microsites. Several techniques are available to achieve this, including slot-seeding (Boatman *et al.*, 1980). Slot-seeding uses only a fraction of the wild flower seed required for the complete reseeding of a site and represents a significant saving in cost and time. Cutting of the sward following seeding is essential to reduce competition from the grass species, and to maintain and create microsites for the germination and spread of the introduced species.

Transplanting container-grown forbs into existing vegetation can also be used to enhance the species diversity of relatively small areas (<0.5 ha) and in so doing create an instant visual improvement. This

Box 16.6 Amelioration of biological constraints

- **application of plant propagules**
  Propagules of the appropriate plant species can be applied to a suitably prepared seedbed in the form of seeds, cut vegetation bearing mature seeds, topsoil containing both dormant buds and seeds, and chopped or entire turves.
- **competition**
  Disturbance and cultivation associated with habitat restoration will provide ideal conditions for the invasion and growth of ruderal and competitive species. This problem is worse on sites with high residual soil fertility (e.g. farmland and road verges). Ruderal species may need to be controlled by further cutting, grazing, hand pulling, or the use of selective herbicides.
- **inoculation with soil biota**
  The introduction of the appropriate soil fauna and mycorrhiza can be achieved by the incorporation of small amounts of soil from the nearest similar semi-natural habitat, either at the site or in the nursery beds of container-grown species.
- **pests and disease**
  Slugs, snails and other pests and diseases can be controlled with chemical pesticides or preferably biological control agents.
- **aftercare management**
  Invasive and competitive species can be controlled by cutting, pulling or careful application of herbicides (e.g. with a weed wipe); excessive grazing and trampling by large herbivores can be prevented by fencing.

Table 16.2 Relative effectiveness of different biological resources for habitat restoration

Biological resource	Advantages	Disadvantages
Natural regeneration: seed banks/seed rain	No donor site required; correct ecotype	Slow and uncertain, may require treatment to facilitate or accelerate regeneration
Seed collection	Renewable; control over species composition and abundance. Native seed of many wild flowers and grasses is now commercially available	Time consuming to collect and clean seed; several visits to site required to get representative mixture. Should avoid seed of non-native cultivars. Need careful site preparation as seed susceptible to erosion
Mowings, hay and brashings	Renewable; rapid, mechanized collection; woody material and stems act as mulch	Little control over composition of seed mixture; some species missing or present in low numbers
Topsoil	Can contain seed and vegetative propagules of many species; good growing medium; soil conditioner, nutrients, inoculum of soil fungi (mycorrhizae and bacteria) and invertebrates. The area restored can be up to five times that destroyed by topsoil collection	Often damages the donor site; species absent or in low numbers, composition of seed bank rarely similar to that of vegetation. Often contains large quantities of undesirable species, such as *Rumex* spp. and *Cirsium* spp.
Turf translocation	Entire plant community–soil system is immediately re-established	Donor site destroyed, changes in species composition of the turves, costly to carry out
Container-grown plants	Renewable resource providing potentially reliable establishment of desired species with control over community composition	Expensive, time-consuming propagation and planting, aftercare irrigation often necessary, best suited to enrichment of existing habitats

technique is also appropriate for species with scarce supplies of seed, or which have seed dormancy mechanisms that prevent them from germinating rapidly, or for those which propagate vegetatively. Such transplanted seedlings appear to have low relative mortality (Table 16.3) and are readily and cheaply propagated from seed and cuttings in pots. These plants can be transplanted into established swards using a bulb planter. Planting densities will vary from 0.25 to $4/m^2$ according to the species, cost and type of habitat required. This technique has been used for introducing species which will not tolerate competition in the latter stages of prairie restoration and for enriching degraded fragment of prairie in the US (Egan, 1994).

Case study 3. Wetland restoration in North America

Wetlands are generally regarded as land where the soil is saturated or covered with shallow water for at least part of the year. They are of considerable ecological importance due to the diverse and rare communities of plants and animals which inhabit them. Furthermore, they can be of importance in erosion protection and preserving groundwater quality. Wetlands are amongst the most vulnerable habitats to experience damage through drainage and pollution. It has been estimated that overall, the United States has lost somewhere between 30 and 50% of its wetlands, excluding Alaska (Mitsch and Gosselink, 1986). Similarly, over 40% by area of UK wetlands have been lost in the last 60 years (RSPB, 1994). Wetland ecosystems are complex and difficult to restore. Current methodologies are summarized in Harker *et al.* (1993). Successful restoration depends upon a fundamental understanding of the hydrological regime of the site (e.g. water levels and chemistry, periodicity of inundation, and sedimentation). This enables the manipulation of the existing drainage structures or, by the construction of new systems, the re-establishment of the correct hydrological function. Such management usually involves penning ditches, plugging tile drains and excavating ponds for standing water. Wetland restoration studies in Iowa, USA, have shown that seeds of some wetland species have survived several decades of drainage (Thompson, 1992). These wetland species usually become established within a few years of raising the water table. Other species may arrive from adjacent wetlands during winter flooding, by wind dispersal and as birds and other animals utilize areas of open water. The restoration of wetlands which have been drained and in agricultural use for many years is more difficult and

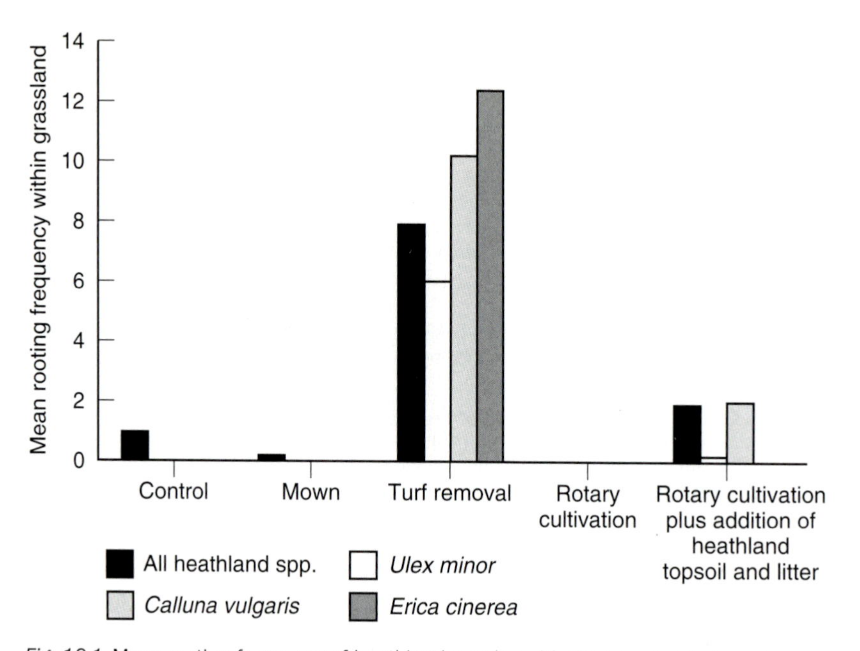

Fig. 16.1 Mean rooting frequency of heathland species with the grassland of an abandoned pasture in Dorset, UK. (Data from Smith *et al.*, 1991.)

requires the addition of propagules to the site. These may have to be collected from existing wetlands as relatively few wetland species will be available from commercial seed suppliers. It is important that species are selected carefully according to a defined water table regime.

Habitat (re-)creation

On severely damaged or disturbed sites it is often necessary to reconstruct habitats from bare substrata (e.g. mineral wastes and arable farmland). On such sites, the constraints on establishing vegetation are often severe and require more sophisticated and costly restoration techniques. Lack or scarcity of propagules of the desired species and the physical and chemical characteristics of the substrate often preclude rapid colonization and establishment of vegetation and associated fauna.

Case study 4. Prairie restoration in North America

Prairie grasslands once extended over large areas of the North American Midwest. Following the Euro-American settlement, much of this habitat was damaged and destroyed by agricultural use. For example, it is estimated that some 1.2 million hectares (3 million acres) of prairie originally occurred in Iowa of which only 12 100 hectares (30 000 acres) remain today (Pearson, 1990 pers. comm. in Thompson, 1992). Since the 1930s there has been a long history of prairie restoration. The first of these was the pioneering work of Leopold at the University of Wisconsin who attempted to restore prairie using minimal site preparation techniques, such as sod turning, based on the assumption that the introduced prairie species would successfully displace the undesirable species (Jordan *et al.*, 1987). Later work identified the key role which fire plays in the restoration and maintenance of this ecosystem (Curtis and Patch, 1948). More recent studies have emphasized the importance of rigorous site preparation by means of cultivation and the use of herbicides to create a firm, weed-free seedbed prior to seed drilling. This was based on the idea that this would give the germinating prairie species a competitive advantage over the later colonizing weeds (Schramm, 1978). Both approaches have been successful, but careful seedbed preparation does seem to speed the restoration of prairie vegetation. Prairie-cut hay has been used successfully as an alternative to seed

Table 16.3 Percentage survival of pot-grown plants transplanted into existing grassland in June 1984 at Monks Wood, Cambridgeshire, UK

Species	No. pots inserted	% survival August 1984	% survival June 1985	% survival July 1985
Anthyllis vulneraria	42	95.2	4.6	0
Campanula glomerata	36	91.7	16.6	2.8
Centaurea nigra	51	100.0	82.7	72.5
Centaurea scabiosa	51	100.0	65.3	33.3
Leucanthemum vulgare	45	100.0	80.4	66.7
Knautia arvensis	43	97.7	84.1	84.1
Leontodon hispidus	50	100.0	52.0	36.0
Lychnis flos-cuculi	52	96.2	84.6	84.6
Ononis spinosa	50	98.0	42.0	16.0
Ranunculus acris	51	94.1	58.8	17.6
Ranunculus bulbosus	45	53.3	35.6	13.3
Plantago media	55	98.2	74.1	52.7
Primula veris	52	100.0	96.1	86.5
Stachys officinalis	41	97.6	92.5	87.8

Source: Wells *et al.* (1989).

drilling since the 1940s. Prairie hay is an excellent, low-cost source of seed of species, some of which are not available commercially, and the hay has a mulching effect which provides some protection from erosion (Ries *et al.*, 1980). Annual prairie species have been sown as companion species to accelerate the rate of natural secondary succession on overburden from surface coal mining in North Dakota. The following year, prairie hay was mulched over the stubble of the annual species and anchored by disking to create a community which closely resembled that of the original prairie.

Like the restoration of species-rich grassland in Europe, some prairie species are more readily established from seed than others, with some, e.g. wild indigo (*Baptisia leucantha*), appearing in abundance in the early years of restoration and later disappearing. Most prairie restoration takes 3–5 years to achieve a vegetation resembling a prairie. Non-native weed species, e.g. curled dock (*Rumex crispus*) and white melilot (*Melilotus alba*), can be a major problem of prairie restoration, especially following soil disturbance.

Case study 5. Heather moorland creation on china clay wastes in Devon, UK

The china clay extraction industry in Devon and Cornwall, UK began in the late 1700s and thus there is a long history of disturbance and dereliction of the land on a large scale. These deposits are continuously worked to ever greater depths, hence the waste products cannot be disposed of in disused pits. Following the kaolin extraction process there remains a sandy waste which lacks plant nutrients, especially nitrogen, is liable to erosion and has a deficiency of clay and silt-sized particles which give it low water-holding and ion exchange capacities. The natural colonization of such material takes many years to develop a climax community of heathland or acid oak woodland.

In 1976 an experiment was established to examine different methods of restoring heathland vegetation to china clay waste in Devon (Putwain and Rae, 1988). Topsoil was removed to a depth of 50 mm from an area of existing heath which was to be buried under the sand waste. Germination tests on the topsoil showed it to be a very rich source of seeds of almost all the heathland plant species present at the donor site (Putwain and Gillham, 1990). Fertilizer (17% N: 17% P_2O_5 : 17% K_2O) was applied and the following companion species were sown: *Agrostis castellana* and *Lolium perenne*. In addition, the same experiment was replicated on sand waste within a stock-proof fence to prevent grazing by ponies and cattle.

After seven years the highest percentage cover of *Calluna*, *Erica tetralix* and *E. cinerea* was found on the plots sown with the *A. castellana* companion species plus the high rate of fertilizer input (Figure 16.2(a)). The untreated topsoil controls had the lowest. This suggests that companion species and addition of nutrients were beneficial in the establishment of heathland vegetation on such skeletal substrata. Furthermore, the application of subsoil

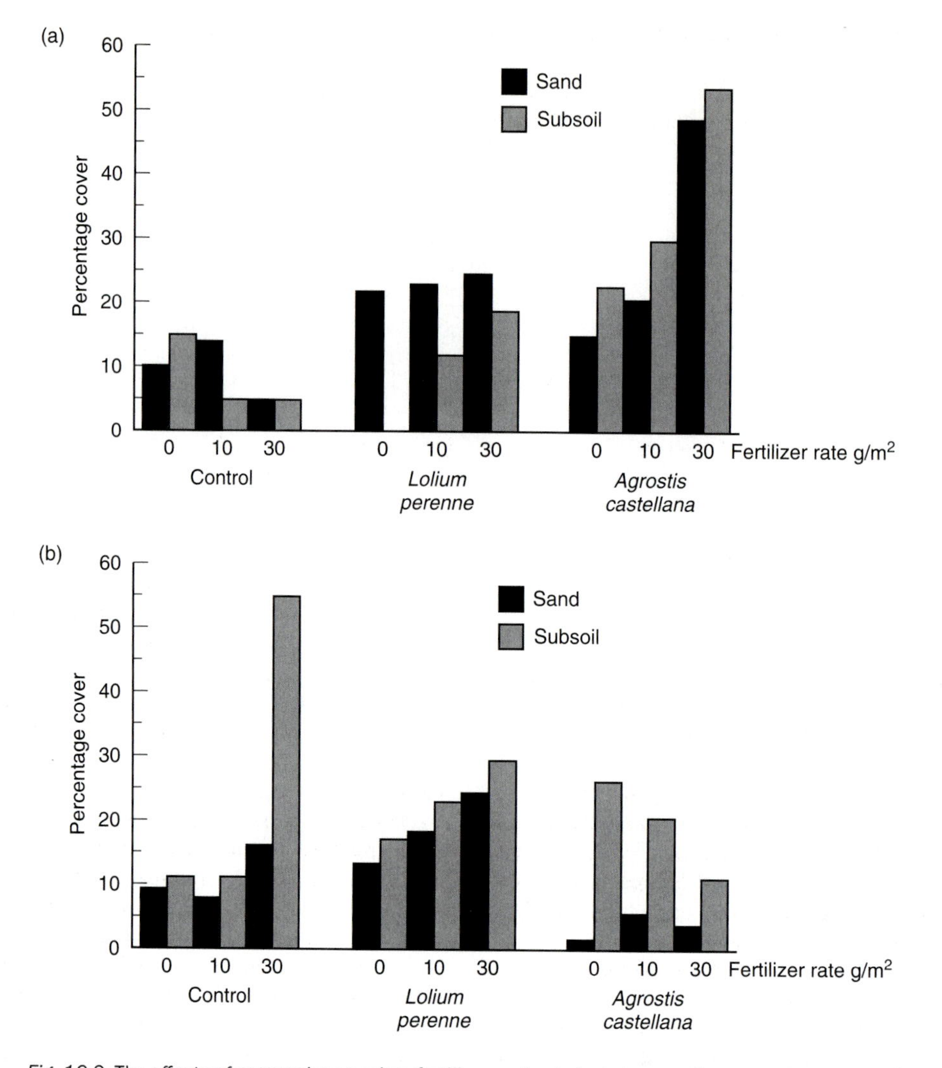

Fig. 16.2 The effects of companion species, fertilizer and substrate type on the percentage cover of heathland species on china clay waste in Devon, UK: (a) *Calluna vulgaris* and *Erica* spp.; (b) *Agrostis curtisii*.

was found to be beneficial to the recruitment of the native heathland grass *Agrostis curtisii* (Figure 16.2(b)). Finally, protection from grazing animals for the first five years was found to be essential to the restoration of heathland vegetation.

Habitat translocation

The translocation of entire communities is a way of conserving habitats of high conservation value which are threatened by activities such as road construction and mineral extraction. The transfer of large (up to 8 m^2) turves can rapidly recreate the entire plant–soil system. This approach has an immediately favourable visual impact and also provides effective, long-term stabilization of slopes. Theoretically, the turves may act as refugia for the invertebrate fauna which may colonize adjacent areas which are being established more slowly from seed. However, turf transfer results in the destruction of the donor site and, unless the turves are spaced out, this technique does not increase the area of a habitat. Translocation is relatively expensive (Pywell, 1991), and usually is only an option for large-scale construction and mineral extraction projects where earth-moving machinery is already available on the site and distances between the donor and receptor sites are relatively short. A potential drawback of this technique is that failure to match the hydrological, physical and chemical characteristics of the donor site with those of the receptor site can lead to wholesale change in the species composition of the turves.

Case study 6. Calcareous grassland translocation in County Durham, UK

The grassland community which occurs on the magnesium limestone in County Durham, UK contains up to 40 species of plant and is unique in Western Europe (Rodwell, 1992). An extension to an existing quarry threatened to destroy one of the remaining fragments of this habitat at Thrislington Plantation. Between 1982 and 1989 some 5.5 ha of this grassland was progressively translocated a distance of 800 m to a receptor site where the topsoil had been removed to expose the subsoil (Park, 1989). Turf relocation was carried out using earth-moving machinery which was equipped with a specially designed transplant bucket measuring 4.75 m × 1.75 m. The turves were placed on the subsoil and the 200–300 mm interstitial gaps between them were filled with topsoil and sown with a grass/forb mixture or allowed to colonize naturally. The plants and invertebrates were carefully monitored in the grassland both before and after translocation. Only two species of plant – autumn gentian (*Gentianella armarella* and field scabious (*Knautia arvensis*) – have disappeared from the turves since relocation, although there were changes in the relative abundances of different species, including the appearance of four orchid species.

Aftercare and management

Many restoration schemes have deteriorated because of the inadequate provision of aftercare management. Virtually all semi-natural habitats require some form of intervention to prevent the normal successional processes from taking place. This can be by controlling the growth of the introduced species and either preventing or limiting the invasion of the site by other species (Bradshaw, 1989). Some newly restored habitats will be more susceptible to succession than others. For example, former farmland is likely to have a high residual soil fertility and a large seed bank of ruderal and competitive species which can exploit these conditions. In contrast, it may take tens or hundreds of years for sufficient nutrient capital to accumulate in quarry substrata and succession to become a problem.

Traditional management practices such as cutting, grazing and burning can all be applied in order to maintain newly restored habitats (Bakker, 1989). Cutting and grazing can be used to restrict the growth of the more competitive species. Annual cutting has been used to successfully maintain a diverse chalk grassland sown onto ex-arable land in Hertfordshire, UK over 20 years ago (T.C.E. Wells pers. comm.). Cutting has the additional benefit of removing nutrients from the system and is often the only management option in urban areas. However, cutting requires vehicle access and can be costly. Unlike grazing, cutting is non-selective and can lead to an accumulation of litter which can be detrimental to low-growing species. Burning can also be effective in arresting successional processes on heathland and tall grass prairie, but its use is also restricted near urban areas. If management by these traditional techniques is unsuccessful then it may be necessary to control problem species by hand-pulling or by spot treatment with herbicide.

Timing and monitoring management are critical. Newly restored habitats are especially sensitive to disturbance or damage. If management is implemented too soon or the practices are too severe there may be detrimental effects on the establishing vegetation. This is especially true of bare and eroding substrata. Regular monitoring of the vegetation and substrata characteristics is essential for the rapid identification of shortcomings in the aftercare programme which can then be remedied.

The cost of restoration

The cost of habitat restoration varies enormously and depends on the following factors:

- the type of habitat desired;
- the site and seedbed preparation necessary to overcome the constraints on restoration;
- the cost of seed collection, transport and application;
- aftercare and maintenance costs;
- the cost of monitoring and supervision.

Table 16.4 A comparison of the relative cost-effectiveness of different habitat restoration techniques

Restoration technique	Site preparation (£/ha)	Collection of propagules (£/ha)	Application of propagules (£/ha)	Total costs (£/ha)	Time to establish	Reliability of attainment
Habitat enhancement:						
Creation of niches	Scarification: £30; soil disturbance: £40–£80; turf stripping: £250	n/a	n/a	£30–£250	2 years to decades	Low to moderate
Container-grown plants at planting density of 5/m^2	n/a	Plants: £0.14–£0.70 each = £7000–£35 000	Labour £0.10 per plant = £5000	£13 000–£40 000	Immediate	High
Slot-seeding	Mowing site: £20–£50	Seed mixture (18 forbs): 1-2 kg/ha, £200–£400/ha	Slot-seeding + cost of paraquat: *c.* £35–£45	£260–£500	1–2 years	Moderate to high
Habitat translocation:						
Turves	Grading: soil pit excavation: £250–£1000	Turf lifting: £3500–£5000	Turf setdown: £3500–£5000	£7250–£11 000	Immediate	High
Habitat creation:						
Seed	Seedbed preparation: £60–£70; plough/harrow/roll	Seed mixture: 50 kg/ha, £1150-£1750	Drilling: £40–£60	£1250–£1880	1–2 years	Moderate
Hay and brashings	Seedbed preparation: £60–£70; plough/harrow/roll	Forest harvester: £75–£200; baler: £30–£50	Pitchfork: £200–£300	£290–£570	1–2 years	Moderate
Topsoil	Seedbed preparation: £60–£70; plough/harrow/rolll	Soil stripping: £1300	Excavator: £450	£1820	1 year	Moderate to high

Table 16.4 is a summary of the relative cost-effectiveness of the different restoration strategies. In general terms, techniques which rapidly and reliably reconstruct the desired habitat, such as turf translocation and container-grown plants, are more expensive than those slower and less certain techniques which aim to facilitate natural regeneration (Pywell, 1991).

Severely damaged habitats, such as quarries, may require large-scale engineering both to produce varied but stable topography, and to relive compaction. The costs of these treatments will be very high, but on such sites the machinery required is often freely available. Other treatments to produce suitable seedbeds using standard agricultural techniques (e.g. ploughing and rolling) are relatively cheap to carry out. Treatments to alleviate the chemical constraints on vegetation establishment are often more costly because of the need to purchase materials such as topsoil, fertilizers or sulphur. However, the major cost of most habitat reconstruction schemes is that associated with the acquisition of propagules and their application to the site. Commercial wild flower seed mixtures can cost as much as £1200/ha (£500/acre). The collection of propagules from existing habitats, such as hay and topsoil, can also be expensive. Similarly, only a small area of turf can be transported in a load and this is reflected in the higher cost of transport. Some or all of the costs of habitat restoration may be met by agricultural subsidy payments. Schemes such as Countryside Stewardship and the Environmentally Sensitive Areas offer farmers payments of £250–£450/ha (£100–£180/acre) to restore habitats of conservation value on land taken out of agriculture in the UK. Similarly, derelict land grants are available for the restoration of mineral workings and industrial sites in continental Europe and the USA.

Reinstatement of the invertebrate and vertebrate fauna

If habitat restoration is to be successful all of the component flora and fauna must be reinstated to a given site. This chapter has shown that it is feasible to restore plant species and communities to many different types of derelict and damaged land. However, very little is known about the recolonization of newly restored habitats by their associated invertebrate and vertebrate species. In order to achieve this we require an understanding of the dispersal and habitat requirements of animal species (Webb and Thomas, 1994), together with the detailed monitoring of the progress of the restored fauna as well as flora.

Some groups of animals are highly mobile (e.g. birds) and are able to rapidly locate and colonize suitably restored and managed habitats (e.g. Hill, 1989). However, other groups are relatively immobile (e.g. some phytophagous invertebrates) and may require introduction to a restored area using transplanted material from existing sites to act as foci for dispersal. Of the habitat reconstruction methodologies described, the disturbance and wholesale changes in plant species composition and architecture associated with habitat *creation* make it likely that this technique will be the least likely to reinstate the associated fauna in the short term. However, this is not to say that the transitional habitat will not contain an interesting fauna component. The more gradual changes in composition and structure associated with habitat *enhancement* are likely to provide a better chance of natural colonization by the desired fauna.

Conclusions

It should be stressed that the restoration of semi-natural habitats, however successful, is not a substitute for their conservation and management in the first instance. However, changes in land-use policies now offer opportunities to restore semi-natural habitats on potentially large areas which have been damaged by activities such as intensive agriculture and mineral working. Given the limited biological and financial resources for restoration, it is important to target sites with the greatest likelihood of success and those whose restoration would be of maximum benefit to conservation through linking isolated fragments of existing habitat and providing buffer zones to protect existing habitats. The collation and manipulation of spatial data on habitats within GIS can provide a useful tool in the site selection process at a landscape and local scale.

If habitat restoration is to benefit the conservation of species and communities, clear ecological objectives must be defined which include the reinstatement of the appropriate plant and animal communities to a given site over known time trajectories. The case studies have shown that the key to successful restoration is a fundamental understanding of the ecology and environmental requirements of species together with the identification and amelioration of the constraints on their establishment. Most of the amelioration techniques described are based on our knowledge of the natural processes of establishment, colonization and succession. In order to improve the reliability and speed of attainment of the desired habitat, there is still a need for further, detailed information on species autecology and community ecology including attributes such as seed viability, dormancy, germination, seed dispersal and longevity, together with how individuals interact with one another.

Given the potentially large areas of derelict and damaged land, the high costs of site amelioration, and the limited supplies of plant propagules, it is important to examine the cost-effectiveness of each restoration technique. This can only be achieved by carefully designed and monitored field trials of the type described in this chapter. Also, it should not be forgotten that restoration can be a slow process which requires a long-term commitment to aftercare and management. Finally, the establishment of vegetation is only the first step in the reconstruction of an ecosystem. More work is required on ways of facilitating and accelerating the colonization of such restored areas by their native invertebrate and vertebrate fauna. Only then will a habitat or ecosystem be truly restored to something resembling its original state.

References

Bakker, J.P. (1989) *Nature management by grazing and cutting*. Dordrecht: Kluwer.

Bakker, J. P. and H. Olff (1995) Nutrient dynamics during the restoration of fen meadows by hay-making without fertilizer application. In B.D. Wheeler, S. Shaw and W. Fojt (eds) *Restoration of temperate wetlands*. Chichester: Wiley.

Boatman, N.D., R.J. Haggar and N.R.W. Squires (1980) Effects of band-spray-width and seed coating on the establishment of slot-seeded grass and clover. *Proceedings of 1980 British Crop Protection Conference*.

Bradshaw, A.D. (1989) Successional processes. In G.P. Buckley (ed.) *Biological habitat reconstruction*. London: Belhaven Press, pp. 68–78.

Bradshaw, A.D. and M.J. Chadwick (1980) *The restoration of land*. Oxford: Blackwell Scientific.

Crocker, R.L. and J. Major (1955) Soil development in relation to vegetation and surface age at Glacier Bay, Alaska. *Journal of Ecology*, **43**, 427–48.

Curtis, J.T. and M.L. Patch (1948) The effects of fire on the competition between blue grass and certain prairie plants. *The American Midland Naturalist*, **39**(2), 437–43.

Egan, D. (1994) Growing green: Prairie nurseries raise native plants and consciousness in the Upper Midwest. *Restoration and Management Notes*, **12**(1), 26–31.

Ellenberg, H. (1988) *Vegetation ecology of central Europe*, 4th edn. Cambridge: Cambridge University Press.

Greenwood, E.G. and R.G. Gemmell (1978) Derelict industrial land as a habitat for rare plants in S. Lancs (V.C.59) and W. Lancs (V.C.60). *Watsonia*, **12**, 33–44.

Harker, D., S. Evans, M. Evans and K. Harker, (1993) *Landscape restoration handbook*. Florida.

Hill, D.A. (1989) Manipulation of water habitats to optimise wader and wildfowl populations. In G.P. Buckley (ed.) *Biological habitat reconstruction*. London: Belhaven Press, pp. 328–46.

Hodgson, J.G. (1989) Selecting and managing materials used in habitat construction. In G.P. Buckley (ed.) *Biological habitat reconstruction*. London: Belhaven Press, pp. 45–67.

Holtz, S.L. and E.A. Howell (1983) Restoration of grassland in degraded woods using the management techniques of cutting and burning. In R. Brewer (ed.) *Proceedings of the Eighth North American Prairie Conference*. Kalamazoo: Western Michigan University, pp. 124–9.

Hopkins, A. and J. Wainwright (1989) Changes in the botanical composition and agricultural management of enclosed grassland in upland areas of England and Wales, 1970–86, and some conservation implications. *Biological Conservation*, **47**, 219–35.

Johnson, M.J., P.D. Putwain and R.J. Holliday (1978) Wildlife conservation value of derelict metalliferous mine workings in Wales. *Biological Conservation*, **14**, 131–48.

Jordan, W.R., M.E. Gilpin and J.D. Aber (1987) Restoration ecology: ecological restoration as a technique for basic research. In W.R. Jordan III, M.E. Gilpin and J.D. Aber (eds) *Restoration ecology: a synthetic approach to ecological research*. Cambridge: Cambridge University Press, pp. 1–21.

Luken, J.O. (1990) *Directing ecological succession*. London: Chapman & Hall.

Mitsch, W.J. and J.G. Gosselink (1986) *Wetlands*. New York: Van Nostrand Reinhold.

Park, D.G. (1989) Relocating magnesium limestone grassland. In G.P. Buckley (ed.) *Biological habitat reconstruction*. London: Belhaven Press, pp. 264–80.

Putwain, P.D. and D.A Gillham (1990) The significance of the dormant viable seed bank in the restoration of heathlands. *Biological Conservation*, **51**, 1–16.

Putwain, P.D. and P.A.A. Rae (1988) *Heathland restoration: a handbook of techniques*. Southampton: British Gas

Pywell, R F. (1991) Heathland translocation and restoration. In M.H.D. Auld, B.P. Pickess and N.D. Burgess (eds) *History and management of southern lowland heaths*. Sandy: Royal Society for the Protection of Birds, pp. 18–26.

Ries, R.L., L. Hofmann and W.C. Whitman (1980) Potential control and value of seeds in prairie hay for revegetation. *Reclamation Review*, **3**, 149–60.

Roberts, R.D., R.H. Marrs, R.A. Skeffington and A.D. Bradshaw (1981) Ecosystem development on naturally colonized china clay wastes. I. Vegetation changes and overall accumulation of organic matter and nutrients. *Journal of Ecology*, **69**, 153–61.

Rodwell, J.S. (ed.) (1991) *British plant communities. Volume 2. Mires and heaths*. Cambridge: Cambridge University Press.

Rodwell, J.S. (ed.) (1992) *British plant communities. Volume 3. Grasslands and montane communities*. Cambridge: Cambridge University Press.

RSPB (1994) *Wet grasslands – what future? An account of wet grassland loss in the U.K.* Sandy: Royal Society for the Protection of Birds.

Schramm, P. (1978) The do's and don'ts of prairie restoration. In D.C. Glen-Lewin and R.Q. Landers (eds) *Fifth Midwest Prairie Conference Proceedings*. Ames: Iowa State University.

Smith, R.E.N., N.R. Webb and R.T. Clarke (1991) The establishment of heathland on old fields in Dorset, England. *Biological Conservation*, **57**, 221–34.

Sheail, J. (1974) The legacy of historical times. In A. Warren and F.B. Goldsmith (eds) *Conservation in practice*. London: Wiley, pp. 291–306.

Thompson, J.R. (1992) *Prairies, forests, and wetlands: the restoration of natural landscape communities in Iowa*. University of Iowa Press.

Veitch, N., N.R. Webb and B.K. Wyatt (1995) The application of Geographic Information Systems and remotely sensed data to the conservation of heathland fragments. *Biological Conservation*, **72**, 91–8.

Webb, N.R. and J.A. Thomas (1994) Conserving insect habitats in heathland biotopes: a question of scale. In R.M. May, P.J. Edwards and N.R. Webb (eds) *Large scale ecology and conservation biology*. British Ecological Society Symposium, **33**. Oxford: Blackwell, pp. 129–52.

Weinhold, C.E. and A.G. van der Valk (1989) The impact of duration of drainage on seed banks of northern prairie wetlands. *Canadian Journal of Botany*, **67**, 1878–84.

Wells, T.C.E., R. Cox and A. Frost (1989) Diversifying grasslands by introducing seed and transplants into existing vegetation. In G.P. Buckley (ed.) *Large scale ecology and conservation biology*. London: Belhaven, pp. 283–98.

CHAPTER 17

Risk and uncertainty: mathematical models and decision making in conservation biology

HUGH POSSINGHAM

Introduction

MacArthur and Wilson's (1967) theories of island biogeography represented the first time that mathematical models had a significant impact on conservation biology. Since then an increasing range of models have been used to address problems in conservation biology. In many cases, the early models provide results that are difficult to apply to real problems in conservation biology. Unfortunately most attempts to transfer traditional ecological theory and models to applied nature conservation have failed because they do not answer the kinds of questions that nature conservation managers ask. Fortunately a growing number of scientists are constructing models and theories that target real problems in applied nature conservation.

Those responsible for managing biodiversity often ask the question: why do we need mathematical models? What is wrong with experience and intuition in decision making? Some of the answers to these questions will become apparent as this chapter develops. However, here we briefly note three reasons:

1. *Conservation needs to address problems over large time scales and large spatial scales.* Experimental or observational methods over the scales of time and space that are typically addressed in nature conservation are either impossible or very expensive. For example, consider the issue of how best to conserve populations of the northern spotted owl (*Strix occidentalis caurina*) which occurs in the north-west USA (Wilcove, 1994). The northern spotted owl occupies territories of about 800 ha and is a long-lived animal; its persistence over the next few hundred years is not something amenable to experimental study because the spatial and temporal scales are too large. Models enable us to extrapolate from existing information and help us to understand the response of populations and ecosystems to different management strategies. Some of the population modelling examples we discuss below consider extinction probabilities within time frames of hundreds of years for species that may occupy thousands of square kilometres.
2. *Dealing with risks and managing complex systems.* Nature conservation is often concerned with rare events, like catastrophes, and extinction. We know that people are notoriously unreliable at placing the impact of rare events in the context of decision making. Typically we allow extremely rare, but devastating, events to play too big a role in our decision making. For example people allow the possibility of being struck by lightning to influence their lives disproportionately to being killed in a car accident. Decision theory and mathematical models provide us with a formal framework within which to make decisions in situations where the final outcome is unpredictable (Maguire, 1991).
3. *The modelling process: integrating existing information and pointing the way to further work.* A frequently ignored advantage of using a model in the process of conserving biodiversity is that the task of defining processes in a model and parametrizing those processes is beneficial in itself. The process of building a model forces us to explicitly state our objectives, define and describe factors affecting a situation, and focus on the collection of data that are relevant to decisions that we need to make. Not only is the modelling process beneficial to the model builder, but it helps them to unambiguously explain to others their thoughts about how an ecological system operates and the reason why they have made certain management decisions. It is wrong to think of modelling as a task

that is carried out after all the data have been collected. Ideally model construction is carried out simultaneously with data collection and experimentation. In this way modelling can not only organize our thoughts but also help us to guide further data collection and experimentation.

In this chapter we explore the role of mathematical modelling in conservation biology. In mathematical models we describe our ideas about an ecological system with mathematical equations. Verbal models, where these ideas are described without mathematics, are invariably precursors to mathematical models. The advantage of a mathematical model is that it is less ambiguous than a verbal model, although to some the implications of a system of equations are not transparent unless fully explained.

In managing natural systems there are no certainties; the outcome of any decision is unpredictable. Most of this chapter is concerned with decision making under these risky and uncertain circumstances. First we need to clarify the difference between the three concepts of risk, uncertainty and ignorance.

Risk, uncertainty and ignorance

In this section a hypothetical example is used to help us understand the concepts of risk, uncertainty and ignorance. Suppose we are concerned with the management of a population of twenty orchids in a patch of remnant vegetation. Fire has a major impact on the population dynamics of the species through its direct effect on the species and its effect on the rest of the vegetation. In the absence of fire, competition from the surrounding plants will slowly reduce the population; however, with every fire there is a risk that the entire population becomes extinct. Discussions with a local expert suggest that this information can be summarized by the following probabilities (Table 17.1). If there is no fire next year we are told that there is a 10% chance the population will be extinct and if it is not extinct its population is expected to be about 15 plants. If there is a fire there is a 20% chance that the population becomes extinct but if it survives it is expected to have a population size of about 40 individuals. The chance of a fire in any year is 25%.

Table 17.1 Probability of different outcomes for a given population of orchids if we do or do not burn

Management option	Expected population size		
	0	15	40
No burn	20%	0%	80%
Burn	10%	90%	0%
Perceived 'utility'	0	1	2

In this example the 'risk' is the probability that an unusual and usually unfavourable event will occur. Even if we know exactly the risk of an event, whether it will occur or not is unpredictable. In this example we are told that there is a 25% risk of fire; even if this estimate is exact we cannot say that a fire will or will not occur – there is an inherent element of chance in the system and our challenge will be to manage the population with this unavoidable risk.

If we are uncertain about a parameter (which may be a probability) then we do not know its value precisely. Although we have been told there is a 25% chance of a fire, this is likely to be an estimate based on fires in previous years. In this case we may know the fire history of the site over the past 20 years and have recorded five fires. This is a small set of data on which to base our estimate and the true probability of fire could reasonably be as high as 40% or as low as 10%. We are uncertain about the true fire probability. There will also be uncertainty about the local orchid expert's estimates of the expected orchid population size next year with or without fire. As this chapter progresses we will see how to manage populations, even with this uncertainty.

Finally, we may be entirely ignorant of other important processes affecting the population. In this case, for example, further information may lead us to believe that the season in which a fire occurs determines its effect on the orchid population – for example fires early in the season may cause population extinction but fires later in the season may be largely beneficial. There is little we can do about our ignorance aside from accepting that any models we construct, and any decisions we make, are contingent on the availability of existing information and will need to be continually updated as new information comes to light. Again the processes of modelling and data collection need to proceed together.

Utility – costs and benefits

For any decision there are costs and benefits. For ecological problems these costs and benefits will be measured in some ecological or environmental currency, like probability of population extinction or impact on human health. In all cases we also need to be mindful of other costs and benefits of a decision, like economic and social costs and benefits.

Where there is uncertainty about the outcome of a decision, which is normal in conservation biology because many events are inherently uncertain, then we need to assess the utility of all the possible outcomes of a decision. Utility is a term often used by economists to express the value of an outcome and it is equivalent to the notion of costs and benefits. Because costs are only negative benefits, the term utility is often more convenient although less transparent.

Let us return to our example of the orchid population and assume now that we decide whether or not there is a fire next year, consequently our management decision becomes: should we burn or not burn? The answer to the decision depends on the time frame over which we are managing the system and the utility of the three possible outcomes: extinction, an expected population of 15, and an expected population of 40. First assume we are only concerned with what happens next year and not about future years. The best decision depends on the value we assign to the three possible outcomes. If our objective is to minimize the chance of extinction next year, and we are unconcerned with its population size next year, then not burning gives the lowest probability of extinction – 10%. Alternatively we may wish to maximize the expected population size. If we do not burn, the expected population size is $15 \times 0.9 = 13.5$; if we do burn, the expected population size is a lot higher – $40 \times 0.8 = 32$. The apparent conflict between these results arises because we have assigned different utilities to different outcomes. Before we can make a decision we need to decide if a population of 40 is more valuable than a population of 15, and if so, how much more. Larger populations are more valuable because they will retain more genetic diversity and be less vulnerable to extinction from other causes (as we shall see later). For example we may arbitrarily assign a utility of 0 to a population of 0, a utility of 1 to a population of 15, and a utility of 2 to a population of 40. The expected utility of burning is now $2 \times 0.8 = 1.6$ and the expected utility of not burning is $1 \times 0.9 = 0.9$, and we would burn. In practice we would probably wish to look further into the future and design a fire management plan that maximizes long-term survival. This will only be possible with more knowledge about the population dynamics of the species.

Our example shows the importance of how we assign values to different outcomes. In conservation biology these values are often subjective; however, if the values are clearly stated and justified at least they can be discussed and revised. The example also illustrates the importance of time in decision making. We need to know over what time frame we are making decisions – are we managing for the next few years, the next few decades, or the next few centuries? Although we would like to think that the best decisions are long-term decisions, our uncertainty about processes and events in the future mean that it is often more pragmatic to concern ourselves with maximizing the short-term utility of our decisions.

The example we have explored here is extremely simple and the advantages of any model are not clear. In real conservation problems the processes are more complex, and the decisions available to the manager are usually more numerous. The advantages of an explicit mathematical model are greater as the problem becomes more complex. In the next section we introduce the idea of population viability analysis, and then return to ideas of risk, utility and decision making in the context of specific examples where maximizing population viability is the prime management objective.

Risk assessment for threatened populations

Before discussing how to assess the likelihood of populations becoming extinct, it is important to review the processes that contribute to species extinction (see Chapters 10 and 11).

Chance, often referred to by mathematicians as stochasticity, plays an important role in the extinction process. Before we consider the consequences of stochasticity, there are deterministic factors that lead to extinction. If all the available habitat for a species is being destroyed, or the size of a population declines every year because death rates are too high, the species is doomed to extinction. In these circumstances the appropriate management is to stop the forces leading to the decline, e.g. stop habitat destruction or reduce predation, and we do not need a model to enable us to make management decisions. For many species that have suitable habitat, extinction is still possible through a series of unfortunate events. The stochastic processes that can lead to extinction in this way can be classified into the following three categories (Shaffer, 1981).

Demographic stochasticity

The discrete nature of individuals means that predicting the properties of very small populations using average values will often be highly inaccurate. For example, if we are told that a park contains ten adult animals of a certain species, and the normal sex ratio is one male to four females, we would be foolish to assume that there are two males and eight females in the population. Indeed there is a 6% chance that there are no males in the park; in the absence of male immigrants this population would be doomed to extinction. Demographic stochasticity includes all those processes that operate independently on individuals and can be thought of as the chance nature of birth and death (see Chapter 11). For any reproducing female, there is a stochastic element to how many young that female produces, the health of those young, and their sex. These stochastic events mean that all populations must show some random fluctuations in abundance, even if environmental conditions are completely constant. Typically, in populations with fewer than 20 animals, demographic stochasticity may lead to extinction (Figure 17.1). For large populations demographic stochasticity causes relatively minor fluctuations. Modelling demographic stochasticity normally requires a simulation model that follows every individual separately, including its sex and age, although analytical approximations can be made (Goodman, 1987).

Environmental stochasticity and catastrophes

As we have just seen, populations will fluctuate even when conditions from year to year are identical. For large populations these fluctuations are relatively insignificant. Another form of stochasticity, usually called environmental stochasticity, causes more significant fluctuations in populations because all individuals are simultaneously affected (see Chapter 11). For most species every year is not the same – there may be good years and bad years for both reproduction and survival. These good and bad years usually affect the entire population and cause much more significant fluctuations (see Figure 17.2). For example the numbers of most species in Australia's arid and semi-arid environments fluctuate enormously with changing rainfall. Environmental variability can be modelled by assuming a mean and variance for birth and death rates, and picking a value for birth and death rates for each year by randomly sampling the appropriate distribution. Catastrophes are an extreme form of environmental variability that refers to the death of a significant fraction of the population in a single event like fire, flood or hurricane.

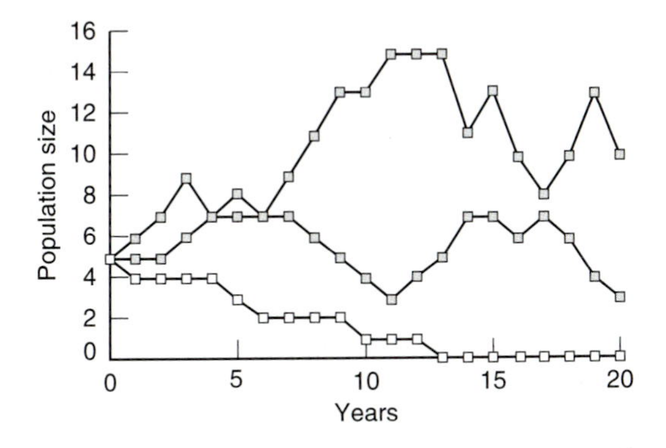

Fig. 17.1 Examples of the impact of demographic stochasticity on small populations. All three populations started with a population size of five and an annual rate of increase of 1.1 (that is on average they should increase by 10%). The three lines represent independent examples of the same random process.

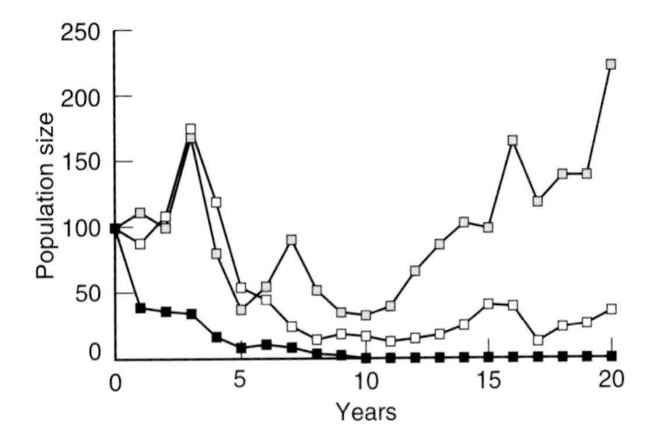

Fig. 17.2 Environmental stochasticity has a significant impact in large populations. The small population fluctuations caused by demographic stochasticity are masked by the environmental variability. The three lines represent independent examples of the same random process.

Genetic stochasticity

Genetics has important implications for the persistence of a population. In small populations individuals will often mate with related individuals, inbreeding, and this is known to decrease survivability and/or depress fecundity in most species. Populations that suffer decreases in population sizes may lose genetic diversity rapidly (see Chapter 10 and Franklin, 1980). Gilpin and Soulé (1986) think of the interaction of demographic stochasticity and inbreeding depression as an extinction vortex. If a

Table 17.2 Examples of population viability analysis in the literature

Species	Geographic range	Primary threats	Reference(s)
Grizzly bear	North America	Hunting	Shaffer (1983)
Florida panther	Florida, USA	Lack of habitat, road kills, inbreeding	Seal and Lacy (1989)
Bay checkerspot butterfly	California, USA	Habitat loss	Murphy *et al.* (1990)
Furbish's lousewort	USA, Canada	Habitat loss	Menges (1990)
Leadbeater's possum	Victoria, Australia	Logging, fire	Lindenmayer and Possingham (1995)
Monk seal	Mediterranean Sea	Disease, fish line entangling	Durant and Harwood (1992)
African elephant	Kenya	Poaching, habitat loss	Armbruster and Lande (1993)
Sooty shearwater	New Zealand	Predation, poaching	Hamilton and Moller (1993)
Samano monkey	South Africa	Natural disasters, allee effects	Swart *et al.* (1993)

Source: Adapted from Lindenmayer *et al.* (1993).

population's size fluctuates to a sufficiently low level, inbreeding depression may reduce birth and death rates which in turn reduces population size and so forth. (Notably some species show little or no apparent inbreeding depression – see Chapter 10, Table 10.1.) Although this picture is somewhat simplistic, and the relationship between the genetic structure of a population and its population dynamics is very complex, it serves as a timely warning about how we manage populations.

Population viability analysis

Population viability analysis (PVA) is a process in which we assess the viability of a population (see Chapter 15). That assessment usually comes in the form of predicting the probability that a population will become extinct within a certain time frame, given a set of assumptions about the factors that affect a species and a specific management regime. For example Lacy and Clark (1990) predicted that there is a 25% chance that eastern barred bandicoot will be extinct within the next 30 years, given that existing conditions continue into the future. Some examples of population viability analysis are shown in Table 17.2.

Shaffer (1981, 1983) was the first author to crystallize and promote the notion of population viability analysis (sometimes called population vulnerability analysis). He was concerned with the likelihood of extinction of the grizzly bear population in Yellowstone National Park, USA. His model and approach differed from previous work on population viability by playing down the role of genetics and focusing on the importance of demographic and environmental variability.

Population viability analysis is not just a tool for threatened species. We are often concerned with the management of populations that are threatened with extinction, but are of a species that is not necessarily threatened. As more countries move towards legislation that resolves to ensure species not only persist but also occupy most of their normal range, we become more concerned with remnant populations of species that are common elsewhere. Shaffer's (1983) work on the grizzly bear is a classical example of this – as the grizzly bear is still very common, as a species, over much of its range. The Yellowstone population is, however, relatively isolated, and one remnant of a once larger population in the United States of America. One of the most popular questions asked with population viability analysis is how much suitable habitat needs to be set aside to give a particular species a reasonable chance of persisting.

Area requirements for threatened species

For managing threatened species, often one of the first questions asked is how big a reserve has to be to adequately protect a species, that is to retain a viable population. This question immediately demands a definition of adequate protection. Most authors use one of the several definitions of 'viable' like a 99% chance of persisting for the next 1000 years (Shaffer, 1981). It is important that we make a clear definition even though it is necessarily subjective. Armbruster and Lande (1993) address the question of how big a reserve should be in semi-arid Africa to attain a 99% probability of persistence of a population of African elephants (*Loxodonta africana*) over a 1000-year period. They conclude that 1000 square miles is an adequate reserve size using a density-dependent demographic model that includes environmental stochasticity. Although genetic factors are not included in the model, they note that a reserve of this size will have an effective population size of about 500 animals, which is believed to be adequate for the retention of genetic diversity (Franklin, 1980).

Probably the best known example of the application of mathematical modelling to decide the size of conservation areas is that of the northern spotted owl (*Strix caurina occidentalis*). As part of the recovery plan for the subspecies several workers constructed models of the dynamics of the northern spotted owl. In these models each potential territory area is represented by a cell which may be occupied, unoccupied, or uninhabitable (as a consequence of logging, for example). For the purposes of the recovery process it was concluded that a habitat conservation area should contain about 20 suitable territories, and these territories must be sufficiently close (that is they must be embedded in a landscape where *at least* 30% of the landscape is potentially suitable habitat). This kind of research represents a major advance on previous work.

Although the question of how big a reserve should be is important in situations where a country must allocate its limited land resource between economic and conservation concerns, the amount of land and/or other resource that can be allocated to nature conservation is controlled by political, social and economic factors. Under these circumstances PVA can be used to answer a question of more specific practical concern: given that we can allocate a certain amount of land to the conservation of a particular species, should it all be in one place? And if not how should it be spatially arranged? Or more generally, given finite resources, e.g. money, land allocation, or time, how can we maximize the likelihood that a species persists?

Conservation strategies, reserve design and climate change

In a thought-provoking paper Peters and Darling (1985) raised the issue of climate change in the context of nature reserve design. Peters and Darling (1985) argued that many species inhabit particular climate zones, that is specific combinations of rainfall, temperature and other abiotic factors. If we reserve an area for the conservation of a suite of species, and then climate change causes the abiotic niche of those species to move across the landscape (away from the equator in most cases), how will the species deal with the change? Before broad-scale habitat destruction many species would have been able to shift geographic range naturally; however, in a landscape where reserves are islands in an inhospitable landscape, this natural population movement may not be possible.

To deal with these anticipated problems Peters and Darling (1985) suggested that north–south linkages in the reserve system would be essential for the successful movement of species ranges. They also noted that reserves with significant altitudinal gradients contained a high diversity of climate zones and species may find appropriate climates nearby more easily by moving up mountains. Of course the latter possibility is inadequate for those species that are already restricted to the tops of mountains. In the absence of natural movement to suitable habitat, another possibility is direct human intervention through transplantation of species, indeed entire assemblages. Hughes and Westoby (1994) argue that we should begin transplanting long-lived species into the expected position of their particular climate zone as a matter of urgency. Our ability to decide which species to move, and where, will rely on improved predictions about changes in rainfall and temperature, as well as a better picture of the climate niche of many species. We must remember that the arguments above are primarily relevant to species that respond directly to the abiotic environment – many vertebrates that are buffered against the abiotic environment, and which respond more to biotic attributes, may be relatively resilient to climate change.

Despite these concerns many species, particularly those that are short-lived and small, may be able to accommodate the rapid change (Possingham, 1993). Small species that are attuned to a particular microclimate may be able to find suitable microclimates without changing geographic range – for example some insects and small plants may move from north-facing to south-facing slopes (in the southern hemisphere) where conditions are cooler. Many of these smaller species have short generation times that facilitate rapid evolution in response to the changing climate.

Because of the uncertainty of climate change itself, and the responses of different species, whatever new conservation strategies we choose must be robust in the face of such uncertainty (Hughes and Westoby, 1994). Again, a risk-spreading and flexible approach to conservation is essential.

Using PVA to rank management options

Lindenmayer and Possingham (1995) modified a population viability analysis package (ALEX;

Fig. 17.3 Leadbeater's possum (*Gymnobelideus leadbeateri*). (Photograph courtesy of D. Lindenmayer and M. Tanton.)

Fig. 17.4 The square (70 km × 70 km) contains almost the entire world distribution of Leadbeater's possum.

Possingham *et al.*, 1992) to consider different forest management options for Leadbeater's possum (*Gymobelideus leadbeateri*), an endangered Australian marsupial (Figure 17.3).

Leadbeater's possum is virtually confined to the mountain ash (*Eucalyptus regnans*) forests of central Victoria, Australia, which are intensively harvested for timber (Figure 17.4). Aside from timber harvesting the major impact on the forest is fire. The objective of the recovery plan for the species is to find the strategy that maximizes the chance that it will persist within the constraints of economically feasible options. In particular they explore several different forest management options for one 3000-ha forest block.

The current management strategy is to reserve all old-growth forest, forest near streams, and forest on steep and/or rocky ground. With this current strategy and their baseline parameter set, the probability of extinction of this population is 33% in the next 150-year period, and 58% in a typical 150-year period beyond the immediate 150 years. (The two extinction probabilities differ because the area of old growth is likely to be reduced as a result of fires over the next 150-year period.) Figure 17.5 shows the extinction probabilities for a subset of the different forest management strategies they explored: *no salvage logging* means that if a fire does occur in old-growth forest the remaining merchantable timber is not harvested (the current strategy recommends that it is harvested), *reserves* are areas that are completely excluded from harvesting over and above the existing exclusions, and extending the *rotation time* to 200 years means that stands are typically harvested when around 200 years old whereas at present management plans to harvest forest that is 80–120 years old.

Figure 17.5 gives us a great deal of information about what can be done for this species:

1 In the short term the only management option that improves viability significantly (note figures in parentheses) is not to salvage-log old-growth patches that are burnt. This causes a reduction in extinction probabilities because we would expect some, if not many, of the existing old-growth patches to burn, and not salvage-logging them will enable them to retain some habitat quality. Most of the forest matrix is 55 years old, and because suitable habitat will only start to form in trees that are over 150 years old, increasing the rotation time and setting aside reserves realizes only limited short-term benefits.
2 In the long term, both reserves and an increased rotation time do significantly enhance viability, though reserves are clearly a better option. Setting aside 300 ha, only 10% of the forest block is better than a 200-year rotation time. The latter would require a cessation of logging for at least 100 years!

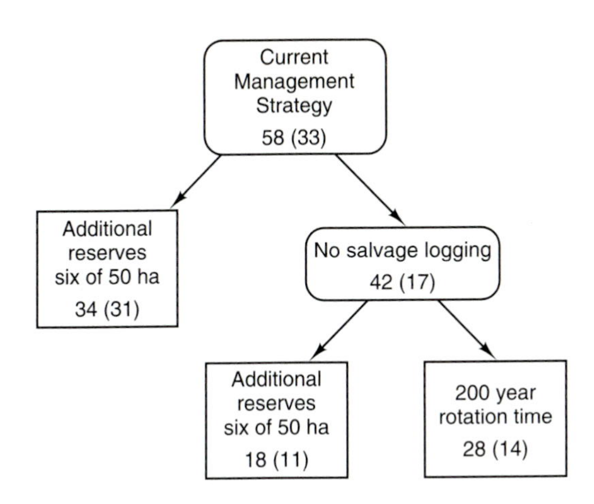

Fig. 17.5 A flow chart showing the impact of different management options – alone or in synchrony – on the extinction probability of Leadbeater's possum in the Steavenson forest block, Victoria. The first number is long-term extinction probability in a typical 150-year period, while the number in parentheses is the extinction probability in the next 150 years. The results were generated using ALEX, one of the stochastic computer simulation models that are publicly available.

3 If we decide that we can afford to set aside 300 ha we need to decide how that should be subdivided: should it be one block or several? Lindenmayer and Possingham (1995) considered a variety of ways of distributing the 300 ha: see Table 17.3. Given that there is an error of up to 2% on these figures the best option, assuming the baseline parameter set is accurate, is to set aside the 300 ha in 25 to 50 ha blocks. A small reserve is vulnerable to extinction through demographic and environmental stochasticity (25 ha is expected to contain about five colonies) but many reserves are unlikely to be burnt in the same fire. As we shall see later this 'risk-spreading' strategy relies on the ability of the possum to recolonize empty patches after they have become vacant. We would expect the best management strategy to depend on both the frequency and extent of fires, and also the movement capabilities of the species concerned.

Sensitivity analysis

The results and conclusions from a model will generally vary as we change the value of parameters in the model, or indeed alter the structure of the model. Consequently sensitivity analysis is an important part of all modelling.

Table 17.3 Probability of extinction of Leadbeater's possum in the Steavenson forest block, Victoria, Australia, over a typical 150-year period

	Arrangement of 300 ha of reserves			
	12 reserves of 25 ha	6 reserves of 50 ha	3 reserves of 100 ha	1 reserve of 300 ha
Baseline movement	17 (11)	18 (11)	22 (13)	27 (17)
Limited movement	58 (46)	52 (45)	55 (46)	63 (47)

Consider the example of Leadbeater's possum discussed in the previous section. Because Lindenmayer and Possingham (1995) know that there is uncertainty about how the possum recolonizes new habitat and also the future fire regime, they rank the management options with different values for these processes. Table 17.3 shows the extinction probabilities with baseline and limited movement (when movement was enhanced the results remain virtually unaffected). As expected, reducing the movement capabilities increases the extinction probabilities for any given strategy so we can say that the extinction probability is sensitive to movement. If we assume that 300 ha is to be set aside, the sensitivity of extinction probabilities is unimportant; it is the ranking of the management options that determines what we will do. In this case reduced movement suggests that the best size for reserves is closer to 100 ha, not 50 ha. Interestingly, the ranking of management options was relatively insensitive to the frequency and extent of fires.

Sensitivity analysis also helps to guide our data collection. If a conclusion of our model is found to be very sensitive to a particular parameter or process then we know that we need to estimate that parameter more accurately, or find out more information about that process. Conversely, if our results seem insensitive to a model parameter we may place a low priority on estimating it more accurately. This is another advantage of doing the modelling, natural history, and experimentation concurrently. Unfortunately a comprehensive sensitivity analysis is not as easy as it would appear.

Consider an ecological model with eight parameters. To determine the sensitivity of a result for each parameter in turn, assuming we vary each parameter up and down 10%, will require 17 repeats of the simulation (including the first experiment where no parameter is varied). This does not seem very many, but our protocol is not entirely adequate. Assume that we are asked what the result of the model is if two of the parameters are both 20% bigger than our best estimate. Let us assume that when we increase parameter A by 20% the model result increases by 10%, and when we increase parameter B by 20% the model result increases by 30%. What is the impact on the result if both A and B are increased by 20%? The answer is unlikely to be 40%. There is no way in which we can answer the question using sensitivity analysis on just single parameters. Indeed a comprehensive sensitivity analysis where we consider all situations in which each of the eight parameters may take one of three values (baseline, ±10%) involves 3^8 repetitions of the model! To fully explore the sensitivity of a model to parameters is a daunting process for which some special methods exist. For practical purposes we may be forced to make subjective decisions about which parameters to vary.

Spatial heterogeneity and the metapopulation concept

Early population models assumed that a population exists in a homogeneous landscape and the population is 'well-mixed'. Some of the most interesting advances in population modelling have been the incorporation of space – these are particularly important for applied conservation modelling. One class of spatial population model, called a metapopulation model, has led to some interesting implications for conservation biology (see Chapter 11). More recently the idea of source and sink populations (Pulliam, 1988) has also made people think about how best to manage a species that exists in a spatially complex landscape.

Most applied metapopulation models allow for differences between patches, in their position, size, and sometimes quality. Source patches are those patches where there is generally a positive population growth rate and an export of excessive individuals. Sink patches contain populations that have lower or negative growth rates, and may only persist as a result of continual immigration from source patches. Some important general messages emerge from the study of metapopulations (Doak and Mills, 1994; see also Chapter 11).

1. Although many suitable territories or patches may be unoccupied, the long-term viability of a species may depend as much on the unoccupied patches as on the occupied patches. The warning is that just because a patch is not currently occupied by a species does not mean that it is of no conservation value to that species.
2. Some patches may be sources and some may be sinks. The conservation of source patches may be essential to the conservation of the species, so much so that the removal of just a few key patches may cause the collapse of an entire species. From data on the density of animals in a patch it is not always possible to identify the source patches. In general we need to monitor life-history parameters for different patches and different habitats to be able to assess their utility. Places where birth rates are high and/or death rates are low may be particularly critical to the survival of a species.
3. For a metapopulation to persist, empty patches must be recolonized as fast as occupied patches suffer local extinction. This means that there are two general methods for enhancing the viability of a metapopulation – reduce the local extinction rate and increase the recolonization rate. The latter may be achieved by the provision of so-called 'wildlife corridors' or deliberate translocations.
4. Where catastrophic processes have the capacity to eliminate entire populations, or groups of populations, a conservation strategy must include an element of 'risk-spreading' – the risk of extinction must be spread across many populations so that the simultaneous local extinction of all populations is unlikely. Whether or not catastrophic events are correlated between patches is critical. Metapopulations where there is no correlation, even better negative correlation, between the catastrophic events and environmental fluctuations in different local populations will be more stable than metapopulations where catastrophic events tend to occur simultaneously across the entire range.

Factoring in conflicting demands

Decision making in conservation biology is rarely so simple that we need only to consider ecological impacts of a decision. Usually we have to take in to account sociological and economic issues. For the Leadbeater's possum situation wood production and jobs need to be considered alongside the viability of this threatened species. In this section we consider an example where sociological considerations interact with the desire to enhance the viability of an endangered population.

Maguire and Servheen (1992) take the process of making decisions about wildlife management a step further by integrating sociological concerns with biological concerns. The remnant population of grizzly bears (*Ursus arctos horribilis*) in the Cabinet–Yaak ecosystem of north-west USA requires the introduction of new individuals to give it a reasonable chance of long-term viability. Although such augmentation is physically and financially possible, there is a significant risk of conflict between the introduced bears and the human population surrounding the ecosystem.

Maguire and Servheen (1992) show how decision analysis is used to choose the sex and age of individuals to be translocated that attempts to reconcile biological concerns with threats to livestock and possibly even the human population. The difficulty in making the decision is exacerbated by uncertainties about the chance that a bear will conflict with humans. They address the problem by going through a sequence of steps:

1. Characterize the decision – should we translocate bears into the region and if so, what age, what sex and at what time of year?
2. Identify the objectives and criteria for evaluating the alternatives – the benefit to the existing bear population and the risk of conflict with people.
3. Specify the management options – the age, timing and sex of a translocated animal, or indeed whether any animals should be translocated at all.
4. Identify possible outcomes.
5. Estimate probabilities of these outcomes and uncertainties associated with these probabilities – what is the risk of conflict for a certain bear and how accurate is this estimate?
6. Assign utilities to the different outcomes – how do we weigh the advantages of the reproductive contribution of a bear against the probability that it will conflict with humans?

Using a panel of experts Maguire and Servheen (1992) were able to assign two values to each possible translocation that depended on the age, sex and timing of the bear. These two values were the expected reproductive contribution of the bear to the population (a high number is better) and the risk of conflict with humans (the lower the number the better). They reflect a conservation benefit and a sociological cost. Some typical values for four types of bear are shown in Table 17.4. If our task is to maximize reproductive value then we would choose the 4-year-old female translocated in the autumn, but if our objective is to minimize conflict we would choose the 8-year-old female translocated in autumn. Maguire and Servheen (1992) reconciled the conflicting demands by asking the experts to define equivalence relationships between reproductive value and the risk of conflict – that is to try to work out what value of one equals a certain value of the other. The experts considered that the following pairs are equivalent: a reproductive value of 0.33 and a conflict risk of 0.25, and a reproductive value of 0.18 and a conflict risk of 0.12. This enabled Maguire and Servheen (1992) to reconcile the conflicting demands and rank the four options.

Table 17.4 The reproductive value and conflict risk for different classes of translocated bear

Age	Sex	Timing	Reproductive value	Conflict	Rank
8	F	Autumn	0.22	0.21	1
4	F	Autumn	0.32	0.30	2
8	F	Spring	0.14	0.22	3
2	F	Autumn	0.27	0.32	4

The important message from the previous two sections is that PVA and other modelling tools are not ends in their own right. For applied management problems the first question must be to consider what management options exist. Often we may find that few, if any, realistic management options exist so the entire PVA process is largely academic (Possingham *et al.*, 1993). And finally it is worth re-emphasizing the advantages of the modelling process in terms of organizing thoughts, explicitly stating assumptions and communicating ideas to others.

Other examples of risk assessment in conservation biology

Risk assessment is a well-developed field, particularly in economic contexts. As countries grapple with rapidly changing environmental legislation ideas of risk and chance are playing an increasingly important role in environmental impact assessment. Often, say from the perspective of a dam overflowing, or a toxic chemical escaping into human water supplies, we are

Table 17.5 Impact of different management options on three forest values

	Probability of extinction of Leadbeater's possum	Water quality	Profit to timber industry
Current management	High	Moderate	Very high
No salvage logging	Moderate	Moderate	High
300 ha of reserves	Low	Moderate	Moderate
200-year rotation time	Moderate	High	Low

forced to first estimate levels of risk and then decide what is an acceptable level of risk. Notions of risk and uncertainty will play an increasingly important role in the way we manage ecosystems, and the perception of these ideas by both the scientific community and the public is becoming increasingly sophisticated. In this section we will briefly explore two case studies where risk assessment is being used in an ecological context.

The impact of introduced organisms to the endemic fauna of continents, particularly islands, has had a significant negative impact on global biodiversity. Although most countries try to control the introduction of exotic biota we continue to deliberately and accidentally transfer species around the globe. Townsend and Winterbourn (1992) were responsible for assessing the environmental risk imposed by introducing the channel catfish (*Ictalurus punctatus*) to New Zealand for fish farming. They explored aspects of its biology using information from its natural range and other places where it has been introduced, focusing in particular on attributes that may be detrimental to the endemic fish fauna. Aside from an unknown possibility of introducing new diseases they predicted that there would be a high likelihood that the channel catfish would cause local and possibly global species extinctions through predation and/or competition if it escaped from farms. Based on existing data they also argued that there would be a high risk that the species would escape. The high likelihood of both escape and detrimental impact led them to conclude that the introduction is unwise. They argue that the risk assessment process should include publicity and independent peer review if the final implementation process was likely to be acceptable.

In the examples so far we have mainly concentrated on the ecological costs of a management action. Quantitative Ecological Risk Assessment is a framework in which we can make decisions in risky circumstances while considering a range of 'values'. The technique is borrowed from disciplines where risk assessment is commonplace. It can be used in any circumstances where management options can be quantitatively assessed with respect to a variety of specified values. Let us consider this decision-making framework by extending the Leadbeater's possum example.

In Figure 17.5 we considered extinction probabilities for a local population of Leadbeater's possum subject to three management options – current management strategy, no salvage logging, 200-year rotation time (with no salvage logging) and 300 ha of reserves (with no salvage logging). For each management option we have a risk of local extinction (assuming the local population is closed). However, we have no assessment of the impact of these options on other forest values. In Table 17.5 we consider two other forest values, local water quality and timber output. For each option we give a hypothetical qualitative outcome (these could be quantitative if there was an appropriate model).

Each column of Table 17.5 is generated by a different model – verbal or mathematical. Associated with each outcome is a level of uncertainty, for example we may be particularly uncertain about the impact of changing rotation time on water quality. Decision making must take into account not only the expected outcomes, but also this uncertainty. Table 17.5 shows three issues of concern – Leadbeater's possum extinction risk, water quality and profit to the timber industry – only one of which is a risk. An alternative to water quality is risk of unacceptably low water quality, expressed as a probability, while an alternative to profit to the timber industry is probability of industry collapse. Framing the issues of concern as risks can completely change the complexion of a problem. For example we may agree that a management option that allows a high risk in any of the three categories is unacceptable.

The final decision as to the best management is now left to the public as no amount of mathematics can enable us to place the three criteria, whether they be risks or average outcomes, into a common currency.

Such a decision will involve an implicit trade-off between the possum, water and timber, a trade-off that cannot be made by a mathematical model.

Conclusions

Although extinction theory, island biogeography, population genetics and metapopulation theory have all made positive contributions to applied nature conservation, they suffer to differing extents from being existing theories that have been transferred to conservation issues. There are many questions in conservation biology that are not addressed by traditional theory and there are many opportunities for new models and new theories that specifically address applied conservation problems. For example, relatively straightforward questions about metapopulation management have not even been posed – such as, for an endangered metapopulation is the construction of new habitat, or the translocation of animals to an empty but suitable patch, more effective in decreasing the extinction probability? For an introduced pest, is the control of peripheral, recently colonized populations the best method of stopping its spread, or the control of larger central populations?

The methods for solving many of these problems will come from techniques in optimization that are more widely used in economics and engineering, and in the biological arena in fisheries modelling. For example, Markov decision theory is a tool for making optimal state-dependent decisions in a stochastic environment, yet it has not been applied to any problem in nature conservation.

Adaptive management is another concept that is considered important in the management of fisheries but, as yet, has not been utilized to help us manage nature conservation (Walters, 1986). In the examples we have considered so far, our objective was to find the best management option and then enact that option. For example, in the Leadbeater's possum example explored in this chapter the best size for reserves was found to be between 50 ha and 100 ha. With this result it is tempting to make all reserves for this species this size. Think now fifty years into the future when scientists wish to assess the effectiveness of such a reservation strategy. Because all the reserves are roughly the same size there is no way that we can test whether or not this was the optimal strategy. An increasingly popular alternative is to think of management as an experiment, where the objective is twofold – maximize the chance of the species persisting and maximize the gain of information. The latter can only be achieved by creating a variable-sized reserve system, with some very big reserves and maybe some very small reserves as well as the 50–100 ha reserves. Such variation, especially if the 'treatments' are carefully designed, will allow us to test the model, evaluate the management strategy and make modifications as new information is gathered. If new information is to be useful we must be in a position to adapt our strategies accordingly. For example, if fewer big reserves are found to be the most effective strategy, we need to be able to implement that new strategy. Problems like this are not straightforward, yet the possibility for applying the principles of adaptive management to problems in conservation biology is unexplored and exciting.

Conservation biology is a new and rapidly changing field that is about making decisions to achieve nature conservation objectives. Given that it is only recently that these objectives have become clear, it is perhaps not surprising that the level of sophistication with which we make conservation decisions is low. However, as in all the applied sciences, we can be sure that mathematical modelling will have a central role in successful applied conservation biology.

References

Armbruster, P.A. and R. Lande (1993) A population viability analysis for the African elephant (*Loxodonta africana*): how big should reserves be? *Conservation Biology*, **7**, 602–10.

Doak, D.F. and L.S. Mills (1994) A useful role for theory in conservation? *Ecology*, **75**, 615–26.

Durant, S.M. and J. Harwood (1992) Assessment of monitoring and management strategies for local populations of the Mediterranean Monk Seal *Monachus monachus*. *Biological Conservation*, **61**, 81–92.

Franklin, I.A. (1980) Evolutionary change in small populations. In M.E. Soulé and B.A. Wilcox (eds) *Conservation biology: an evolutionary–ecological perspective*. Sunderland, MA: Sinauer Associates, pp. 135–50.

Gilpin, M.E. and M.E. Soulé (1986) Minimum viable populations: processes of species extinction. In M.E. Soulé (ed.) *Conservation biology: the science of scarcity and diversity*. Sunderland, MA: Sinauer Associates, pp. 19–34.

Goodman, D. (1987) Consideration of stochastic demography in the design and management of nature reserve. *Natural Resource Modelling*, **1**, 205–34.

Hamilton, S.A. and H. Moller (1993) Population viability analysis of Sooty Shearwaters (*Puffinus griseus*) for efficient management of predator control, harvesting and long-term population monitoring. In A.J. Jakeman and M. McAleer (eds) *International Congress on Modelling and Simulation*. Perth, Australia: Uniprint, pp. 135–50.

Hughes, L. and M. Westoby (1994) Climate change and conservation policies in Australia: coping with change that is far away and not yet certain. *Pacific Conservation Biology*, **1**, 308–18.

Lacy, R.C. and T.W. Clark (1990) Population viability assessment of the eastern barred bandicoot in Victoria. In T.W. Clark and J.H. Seebeck (eds) *Management and conservation of small populations*. Chicago: Chicago Zoological Society, pp. 131–45.

Lindenmayer, D.B. and H.P. Possingham (1995) *The risk of extinction: Ranking management options for Leadbeater's possum using Population Viability Analysis*, Centre for Resource and Environmental Studies, The Australian National University, Canberra, Australia.

Lindenmayer, D.B., T.W. Clark, R.C. Lacy and V.C. Thomas (1993) Population viability analysis as a tool in wildlife management: a review with reference to Australia, *Environmental Management*, **17**, 745–58.

MacArthur, R.H. and E.O. Wilson (1967) *The theory of island biogeography*. Princeton, NJ: Princeton University Press.

Maguire, L.A. (1991) Risk analysis for conservation biologists. *Conservation Biology*, **5**, 123–5.

Maguire, L.A. and C. Servheen (1992) Integrating biological and sociological concerns in endangered species management: augmentation of grizzly bear populations. *Conservation Biology*, **6**, 426–34.

Menges, E. (1990) Population viability analysis for an endangered plant. *Conservation Biology*, **4**, 52–62.

Murphy, D.M., K.E. Freas and S.T. Weiss (1990) An environment–metapopulation approach to population viability for a threatened invertebrate. *Conservation Biology*, **4**, 41–51.

Peters R.L. and J.D.S. Darling (1985) The greenhouse effect and nature reserves. *Bioscience*, **35**, 707–17.

Possingham, H.P. (1993) Impact of elevated CO_2 on biodiversity: a mechanistic population-dynamic perspective. *Australian Journal of Botany*, **41**, 11–21.

Possingham, H.P., I. Davies, I.R. Noble and T.W. Norton (1992) A metapopulation simulation model for assessing the likelihood of plant and animal extinctions. *Mathematics and Computers in Simulation*, **33**, 367–72.

Possingham, H.P., D.B. Lindenmayer and T.W. Norton (1993) A framework for the improved management of threatened species based on Population Viability Analysis (PVA). *Pacific Conservation Biology*, **1**, 39–45.

Pulliam, H.R. (1988) Sources, sinks and population regulation. *The American Naturalist*, **132**, 652–61.

Pulliam, H.R., J.B. Dunning Jr and J. Liu (1992) Population dynamics in complex landscapes: a case study. *Ecological Applications*, **2**, 165–77.

Seal, U.S. and R.C. Lacy (1989) *Florida Panther population viability analysis*. Report to the US Fish and Wildlife Service. Captive Breeding Specialist Group, Species Survival Commission, IUCN, Apple Valley, Minnesota.

Shaffer, M.L. (1981) Minimum population size for species conservation. *Bioscience*, **31**, 131–4.

Shaffer, M.L. (1983) Determining minimum viable population sizes for the grizzly bear. *International Conference on Bear Management*, **5**, 133–9.

Swart, J, M.J. Lawes and M.R. Perrin (1993) A mathematical model to investigate the demographic viability of low-density Samango Monkey (*Cercopithecus mitis*) populations in Natal, South Africa. *Ecological Modelling*, **70**, 289–303.

Townsend, C.R. and M.J. Winterbourn (1992) Assessment of the environmental risk posed by an exotic fish: the proposed introduction of Channel Catfish (*Ictalurus punctatus*) to New Zealand. *Conservation Biology*, **6**, 273–82.

Walters, C. (1986) *Adaptive management of renewable resources*. New York: Macmillan.

Wilcove, D.S. (1994) Turning conservation goals into tangible results: the case of the spotted owl and old-growth forests. In P.J. Edwards, R.M. May and N.R. Webb (eds) *Large-scale ecology and conservation biology*. Oxford: Blackwell Scientific, pp. 313–29.

GLOSSARY

These definitions have purposely been kept as brief as possible. In most cases, more extensive explanations are found within the main text.

Abiotic. Not biotic, not of life. Part of the environment which is not biological, that is water, soil, climate, geology.

Adaptation. The way in which an organism has evolved to become fitted for its way of life in terms of its behaviour, ecology, physiology, etc.

Allele. Different forms of a gene occupying the same position on a chromosome.

Alloenzyme (Allozyme). One of the number of forms of the same enzyme.

Allogenic process. A process caused by external factors (opp. Autogenic process).

Amphibian. A group of vertebrates that includes frogs, toads, newts and salamanders.

Angiosperms (Angiospermae). The flowering plants; seed-bearing plants with the seeds enclosed in a fruit; the major group of plants or flowering plants (cf. Gymnosperms).

Autogenic process. A self-produced, independent process (opp. Allogenic process).

Autotroph (Autotrophic). An organism capable of synthesis of organic compounds from inorganic molecules using energy derived from chemical energy (green plants).

Biogeography. The study of the geographical distribution of organisms and their habitats.

Biological community (Biotic community). Populations of different species living in the same geographical area within which there are interactions between the populations.

Biological conservation. An activity which aims to ensure the continued existence of all levels of biological diversity.

Biological diversity (Biodiversity). The variety of life at different levels of biological organization.

Biomass. An estimate of the total mass of organisms within a given area.

Biome. A geographical region which is classified on the basis of the dominant or major type of vegetation and the main climate. For example the temperate biome is that geographical area with a temperate climate with forests composed of mixed deciduous tree species.

Biota. A general term for all living organisms.

Biotope. The smallest geographical unit of the biosphere, or of a habitat, that can be delineated by its characteristic biota.

Bryophytes. The group of simple plants that includes mosses and liverworts. Together with the algae and the fungi, these comprise the non-vascular plants.

Calcicolous. Tolerance of high levels of chalk or lime.

Chloroplast. Organelles found in some cells and which contain chlorophyll; this is the place where photosynthesis takes place.

Chromosome. Found in the cell nucleus, the chromosome is a small rod-shaped body (single long molecule of DNA associated with proteins) which carries the genes in a linear order.

Climax community. The stable biological community at the end of successional changes.

Coevolution. The independent evolution of two species (or more) in which there is a beneficial interaction or relationship between the species. Examples are found amongst the flowers of some plants and specialist bird or insect pollinators.

Conservation (in a biological sense). The continued existence of species, habitats and biological communities and the interactions between species and their environment.

Conservation biology. The integrated use of several social and science disciplines to achieve conservation of biological diversity.

Coppice. Area of small trees. Coppicing is traditional management where trees are periodically cut at the base. This promotes new growth of shoots from the stool. Widely practised in Europe between 1500 and 1800. There has been renewed interest in coppicing as a method of sustainable management of woodlands.

Cultivar. A plant variety maintained by cultivation.

Cytoplasm. All the material in a cell except the nucleus.

Demography. The study of populations including analysis of birth and death rates, age structure, growth rates and population movements (immigration and emigration).

Dicotyledon. Member of the group of flowering plants with an embryo which has two leaves (cf. Monocotyledon).

Diploid (organism). An organism with cells which have a double set of homologous (similar in position and structure) chromosomes; two copies of the genetic material (cf. Haploid).

DNA. Deoxyribonucleic acid, the main constituent of chromosomes of all organisms, which contains genetic 'codes' unique to each individual organism.

DNA fingerprinting. A unique 'fingerprint' can be obtained by extracting DNA from an individual and separating the parts of the DNA by electrophoresis and detecting the pattern.

Ecological succession. A process where over time there is a sequence of changes in the biological communities leading eventually to a climax community.

Ecology. The science of the interrelationships between living organisms and their environment (other organisms and the physical environment including soil, air, climate).

Ecosystem. A term encompassing the living communities (biological communities) and their physical environment and the processes therein, including the flow of energy through the system.

Ecotope. A habitat type within a larger geographical area.

Edaphic. Pertaining to the soil; edaphic factors are those caused by the ground, the soil.

Electrophoresis. A method which uses an electric field across a prepared substance to produce different rates of movements of protein molecules and other materials.

Endemic. A species which is restricted in its distribution to a particular region (cf. Indigenous).

Environment. All the surroundings, but could be differentiated between the natural environment and the human-made environment.

Enzyme. A protein which acts as a catalyst in all biochemical reactions. There are many kinds and they are essential in all cells; different kinds are specific to different biochemical reactions.

Exotic species. A species introduced to one region from another geographical region. Alien species.

Fauna. A collective term for all kinds of animals.

Flora. A collective term for all kinds of plants.

Formation. The highest category in the classification of vegetation, the major regional climax vegetation type or grouping of plant communities with similar climatic and environmental characteristics.

Fungi (singular Fungus). A group of organisms which absorb nutrients from rotting and from living organisms. There are three main groups of which toadstools and mushrooms are one.

Gene. The basic unit of inheritance. At the molecular level a gene consists of a length of DNA. Different forms of a gene are alleles.

Gene pool. The total amount of all the genetic material in a breeding population.

Genetic diversity (variation). The heritable variation in a population as a result of different variants (the alleles) of any gene.

Genetic drift. Progressive loss of alleles; the occurrence of random changes in the genetic frequencies of small isolated populations which are not due to selection and mutation.

Genetic fingerprinting. See DNA fingerprinting.

Genetics. The science of variation and heredity in organisms.

Genome. The genetic complement of a living organism or a single cell.

Genotype. The genetic constitution of an organism (cf. Phenotype).

Grassland. A natural biological community composed mainly of species of grasses. The many kinds of grassland communities may be classified on the basis of the climate, the geology of the area and the dominant grass species.

Gymnosperms (Gymnospermae). Seed-bearing plants with seeds exposed; includes cycads (palm-like in appearance) and conifers (cf. Angiosperms).

Habitat. The typical locality or area in which a population of a species lives.

Haploid (organism). An organism with cells with only one set of chromosomes (cf. Diploid).

Heathland. A lowland biological community dominated by shrubs, mainly heathers.

Heterozygosity. A measure of the genetic diversity in a population. The proportion of heterozygotes for a given locus in a particular population.

Heterozygous. A diploid organism that has inherited different alleles of a particular gene from each parent. The different alleles are at corresponding sites on homologous chromosomes.

Homozygous. A diploid organism that has inherited the same allele (of a particular gene) from both parents.

Indigenous (species). A species which is native to a particular region (cf. Endemic).

Isozyme (Isoenzyme). One of many different forms in which some enzymes may be found, each having similar enzyme characteristics but differing in certain properties such as the optimum pH.

Keystone species (Key species). A species in a community which interacts with other species and upon which many other species depend.

Life form. The characteristic appearance of a species when adult or at maturity.

Lithology. The study of rock-forming processes.

Locus (plural Loci). Position of a gene on a chromosome.

Meiosis. A type of division of the nucleus in a cell that then produces two nuclei each with half the number of chromosomes.

Metapopulation. A group of sub-populations linked in a functional way and sustained by individuals dispersing between the sub-populations.

Mitochondria. Elongated structures found in cells; these are organelles which are known as the 'power houses' of the cell; the site of major energy conversions in cells.

Monocotyledon. Member of the group of flowering plants which have only one first leaf; includes cereals, grasses, rice and many flowering bulbs (cf. Dicotyledon).

Monoculture. The dominance of one species in an area as in agriculture and forestry.

Morphometric. As used in morphometry, the measurement of external form.

Nature. The whole natural physical and living world.

Niche. The 'space' or 'ecological role' occupied by a species and the resources used by a species. Conceptually the niche is multidimensional and each resource (food, time of feeding, etc.) and each abiotic factor (salinity, temperature, etc.) can be considered a dimension of the niche.

Ombrogenous. Needing high rainfall for development.

Parthenogenesis. Reproduction without fertilization by a male gamete.

pH. The chemical unit of level of acidity or alkalinity. Neutral is 7, acidity increases below 7 and alkalinity increases above 7.

Phenotype (phenotypic). The physical or measurable characteristics of an organism (cf. Genotype).

Photosynthesis. The process (biochemical) in green plants that uses light energy (absorbed by chlorophyll) to synthesize carbohydrates from carbon dioxide and water.

Phylogenetic relationships. The evolutionary links between species.

Phylogeny. The evolutionary history of a species or other taxonomic group.

Phytosociology. 'Plant sociology', the study of and classification of plant communities based on their species associations and geographical distribution.

Plagioclimax. A vegetation community which is formed as a result of land-use practices or traditional management. Some grasslands and heathlands are grazed and periodically burnt to prevent further ecological succession.

Polymorphism. The co-occurence of several different forms.

Population. A collection of individuals (plants or animals) of all the same species in a defined geographical area.

Productivity. The rate at which biomass or organic matter is fixed (produced) over time.

Prokaryotes. Organisms with cells which lack a membrane-enclosed nucleus, as in bacteria and some algae groups.

Propagule. A plant reproductive unit (spore, seed, fruit or bud) which is capable of producing a new plant and of dispersing through space and time.

Protists. Organisms in the kingdom Protista. These include single-celled organisms, diatoms, primitive fungi and slime moulds.

Protozoa. The group of unicellular organisms (including amoebas) some of which are parasitic and cause disease.

Rendzina. Fertile lime-rich soil found typically under grass or open woodland.

Species. A group of organisms of the same kind which reproduce amongst themselves but are usually reproductively isolated from other groups of organisms.

Species composition. The species or assemblage of species in an area or a sample.

Species diversity. A measure of the relative abundance of individuals of different species in an area or a sample (a measure of the evenness). High species diversity occurs when all species are represented by the same number of individuals. Low species diversity occurs when one species or a few species are represented by large numbers of individuals and the other species by few individuals.

Species richness. The number of species in an area or a sample.

Stochastic. Randomly determined or with random probability.

Succession. See Ecological succession.

Sustainability. Used here in the sense of sustainable use, that is a use which can be continued through time without significantly changing the populations, species and habitats being used.

Taxonomic group (Taxon, plural Taxa). Any defined unit or group of organisms used in classification of organisms, e.g. species, genus, tribe, family, order, class, phylum.

Taxonomy. The scientific study and description of biological diversity and how that diversity arose.

Vascular plants. These are all the plants excluding mosses, liverworts, fungi, etc., and are plants with conducting tissue.

Wetland. A biological community in an area of wet ground; areas of marsh, peatlands or water whether permanently or temporary with water which is static or flowing, fresh or brackish. The classification of wetlands is based partly on the types of plant species found there and on the physical characteristics.

Wildlife. A term commonly used to refer to non-domesticated animals. In a biological sense wildlife means all kinds of living organisms which are not domesticated.

Woodland. A biological community dominated by trees. Each type of woodland is characterized by the species composition of the trees and other plants.

INDEX